MATHEMATIQUES
&
APPLICATIONS

Directeurs de la collection:
X. Guyon et J.-M. Thomas

43

Springer
Paris
Berlin
Heidelberg
New York
Hong Kong
Londres
Milan
Tokyo

Instructions aux auteurs:

Les textes ou projets peuvent être soumis directement à l'un des membres du comité de lecture avec copie à X. GUYON ou J.-M. THOMAS. Les manuscrits devront être remis à l'Éditeur *in fine* prêts à être reproduits par procédé photographique.

Ugo Boscain
Benedetto Piccoli

Optimal Syntheses for Control Systems on 2-D Manifolds

Springer

Ugo Boscain
Scuola Internazionale Superiore di Studi
Avanzati (SISSA)
Via Beirut 2–4
34014 Trieste, Italia
e-mail: boscain@sissa.it

Benedetto Piccoli
Istituto per le Applicazioni del Calcolo „M. Picone"
Viale del Policlinico 137
00161 Roma, Italia
e-mail: b.piccoli@iac.cnr.it

Mathematics Subject Classification 2000: 49J15, 49N35

ISBN 3-540-20306-0 Springer-Verlag Berlin Heidelberg New York

Springer-Verlag est membre du Springer Science+Business Media

springeronline.com
Imprimé en Allemagne

Imprimé sur papier non acide 41/3142/LK - 5 4 3 2 1 0 -

To Maria Silvana and Alessia

Preface

This book is devoted to synthesis theory of optimal control. The subject was taught by the authors at the International School for Advanced Studies of Trieste, in the period 1994-2003, as part of the Activity on Applied Math of the Sector of Functional Analysis. Both authors warmly thank the School and the Sector.

The book is suitable for graduate students in mathematics and in engineering, in the latter case with solid mathematical background, interested in optimal control and in particular in geometric methods. An introductory chapter to geometric control is contained, while the rest of the book focuses on synthesis theory for two dimensional manifolds. The text can be used both for a short half semester course based on the classification of optimal synthesis, and for a long one semester course. In the latter case one can focus either on properties of value function or on projection singularities. The book contains exercises (to Chapters 1, 2, 3) and bibliographical notes (to Introduction and Chapters 1, 2).

The authors are greatly indebted to their families for the constant support received during the drafting of this book. We want to deeply thank Prof. Alberto Bressan, that has been the *mentor* of the second author and scientifically the grandfather of the first. He has been a source of constant inspiration since last century and was the one initiating the whole project that gave rise to the present monograph. We are also greatly indebted with Andrei A. Agrachev for many illuminating discussions. Many other persons contributed in various ways: Bernard Bonnard, Natalia Chtcherbakova, Yacine Areski Chitour Naith–Abbas, Grégoire Charlot, Jean–Michel Coron, Mauro Garavello, Jean–Paul Gauthier, Alessia Marigo, Heinz Schättler, Ulysse Serres, Mario Sigalotti, Héctor J. Sussmann.

The webpage http://www.sissa.it/~boscain/smai-book-err.html will record errors ambiguities etc. as they come to light. Readers' contributions via e-mail addresses boscain@sissa.it, piccoli@iac.rm.cnr.it, are welcome.

Trieste, June 2003 *Ugo Boscain*
Rome, June 2003 *Benedetto Piccoli*

Basic Notation

Given a set A we indicate by $\sharp(A)$ its cardinality. If A is a topological space and B a subset of A then we indicate by $Int(B)$ the interior of B, by $Clos(B)$ the closure of B in A and by $Fr(B)$ its topological frontier. A subset B of A is said to be generic if it contains an open and dense set. Analogously a property P is said generic if the set of points satisfying P is generic.

We indicate by $\mathbb{R}$ the set of real numbers and by $\mathbb{R}^n$ the vector space of n–tuples of real numbers. $B(x,r)$ indicates the open ball centered at $x \in \mathbb{R}^n$ of radius $r > 0$.

For any function $f : A \to B$, A, B sets, the domain is indicated by $Dom(f)$, while the support by $Supp(f)$. A map $\gamma : [a,b] \to A$ is called a *curve* on A, we indicate by $In(\gamma)$ its initial point $\gamma(a)$ and $Term(\gamma)$ its terminal point $\gamma(b)$. If $[a',b'] \subset [a,b]$ we indicate by $\gamma|_{[a',b']}$ the restriction of γ to $[a',b']$.

If M is a manifold then $dim(M)$ indicates its dimension and if M has boundary then we indicate by ∂M its boundary.

List of Symbols

- $\mathcal{C}^\infty$ vector field on the plane, see Definition 15, p. 38.
- $Det(A)$, determinant of the matrix A.
- $[F,G]$, Lie bracket of the vector fields F and G. See Definition 4, p. 16.
- $F \wedge G := Det(F,G)$, see Section 2.1, p. 35,
- $\mathcal{T}$, $\mathcal{S}$: target and source, see Section 1.2, 1.2, p. 19.
- $Y := F + G, \quad X = F - G$, see p. 38 and formula (2.9), p. 41.
- Δ_A, Δ_B, see Section 2.1 and Definition 19, p. 43.
- $f(x) := -\Delta_B(x)/\Delta_A(x)$, see Definition 20, p. 44.
- $\theta^\gamma(t)$, see Definition 14, p. 37, and Section 2.7, p. 83.
- $\phi(t)$, switching function, see Definition 16, p. 40 and Section 2.7, p. 83.
- $v^\gamma(v_0, t_0; t)$, see Definition 14, p. 37.
- $\gamma^\pm$, $\gamma^\pm_{op}$, $t^\pm_{op}$, see Section 2.6, p. 58.
- $v^\pm(t)$, $\theta^\pm(t)$, $t^\pm_f$, see Section 2.8.2, p. 89.
- NTAE, Non Trivial Abnormal Extremal (i.e. abnormal extremal with at least one switching). See Section 4.3, Definition 60, p. 171.
- $X, Y, S, C, K, F, \gamma_0, \gamma_A, \gamma_k$, Frame Curves, see Section 2.6.1, p. 58.
- $\bar{C}$, Frame Curve of the Extremal Synthesis, see Section 4.1.3, p. 158.
- (X,Y), $(Y,C)_{1,2,3}$, (Y,S), $(Y,K)_{1,2,3}$, $(C,C)_{1,2}$, $(C,S)_{1,2}$, $(C,K)_{1,2}$, $(S,K)_{1,2,3}$, (K,K), $(Y,F)_1$, $(Y,F)_2$, (S,F), (C,F), (K,F), Frame Points, of the Optimal Synthesis, See Section 2.6.2, p. 60 and Section 2.6.4, p. 62.
- (X,Y), $(Y,C)_{1,2,3}$, $(Y,C)_2^{tg}$, $(Y,C)_3^{tg}$, (Y,S), $(C,C)_{1,2}$, $(C,S)_{1,2}$, $(Y,\bar{C})_1$, $(Y,\bar{C})_1^{tg}$, $(Y,\bar{C})_1^{t-o}$, $(C,\bar{C})_{1,2,3,4}$ $(\bar{C},\bar{C})_1$ $(\bar{C},S)_{1,2}$ (W,C,C) $(W,\bar{C},\bar{C})$ $(W,C,\bar{C})_{1,2}$ (S,S), Frame Points, of the Extremal Synthesis, See Chapter 4.
- Optimal Strip, see Section 3.2.1, p. 132,
- Extremal Strip, see Section 4.2, p. 167.

- $\mathcal{R}_{x_0}(T)$, reachable set from x_0, within time T. See Definition 1.2, p. 16.
- $\mathcal{R}(T)$, reachable set from the origin (or from a point x_0), within time T, for the model problem $\dot{x} = F(x) + uG(x)$. See formula (2.7), p. 39 and formula (3.2), p. 128.
- $\mathcal{R}(\infty)$, reachable set from the origin, See formula (3.3), p. 128.
- $\mathcal{N}$, see the introduction to Chapter 4, p. 153 and Definition 54, p. 155.
- N_0, see Section 4.1.4, p. 159.
- $\mathcal{Q}$, N, $\bar{N}$, see the introduction to Chapter 5, and Figure 5.1, p. 198. For N see also the introduction to Chapter 4 and Sections 4.1, 4.5, p. 189, 4.6.
- Normal, Fold, Cusp, Ribbon, Bifold, Projection Singularities, see Section 4.1.5, p. 162 and Chapter 5.
- $e^{tV}(\bar{x})$, value at time t of the solution to the Cauchy problem: $\dot{x} = V(x)$, $x(0) = \bar{x}$, see Definition 35, p. 88.
- $(e^{tV})_*$, Jacobian matrix of the map: $x \mapsto e^{tV}(x)$, see Definition 35, p. 88.
- **(P1)**,...,**(P7)**, generic conditions, see Section 2.4, p. 48.
- **(GA1)**,...,**(GA8)**, generic conditions, see Section 2.8.2, p. 89.
- **(GA9)**,...,**(GA$_\mathcal{T}$)**, generic conditions, see Section 4.1.1, p. 156.
- X–trajectory, Y–trajectory, Z–trajectory, see Definition 18, p. 41.

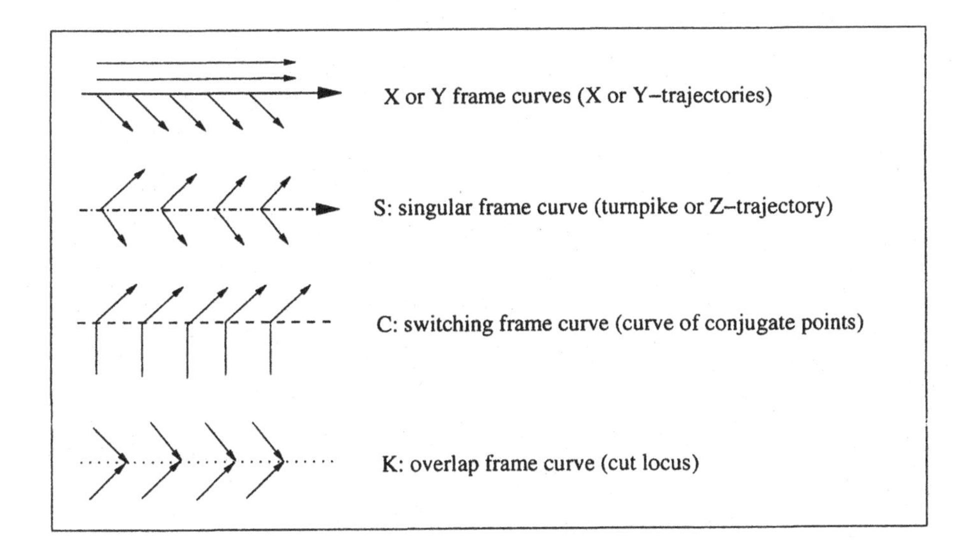

Contents

Introduction

Control Theory deals with systems that can be controlled, i.e. whose evolution can be influenced by some external agent. The birth of Control Theory can be subject of discussion, however it was after the second world war that it had a great development due to engineering applications and became a recognized mathematical research field.

At the same time, probably the most important tool in Optimal Control was proved, namely Pontryagin Maximum Principle. The goal of Optimal Control is to design trajectories that minimize some given cost and it can be viewed as a generalization of Calculus of Variations.

On the other side, another typical problem is the one of prescribing the control automatically as function of the state variables, i.e. giving a feedback, to avoid disturbances and ensure robustness of some prescribed behavior of the system. For instance, one looks for a stabilizing feedback, that is a feedback guaranteeing Lyapunov stability of a given equilibrium. Among many applications, this can be used in aerospace engineering to stabilize communications satellites.

Starting from late 60s, the use of Differential Geometry for control problems gave birth to the so called Geometric Control Theory. The development of strong mathematical tools permitted to attack problems of increasing difficulty and to suggest a systematic way towards the construction of optimal feedbacks. These *holy Grail* of control furnish the solution to both kind of problems: optimal trajectories implementation and feedback design. However, even simple optimal feedbacks are discontinuous, and the solution of the corresponding differential equation could generate not optimal trajectories. Thus optimal synthesis, i.e. collection of optimal trajectories, is the most appropriate concept of solution for Optimal Control Problems. Still the construction of optimal syntheses has been achieved only for some specific examples or class of systems in low dimensions.

The aim of this book is to develop a complete synthesis theory for minimum time on two dimensional manifolds. Beside the construction of optimal synthesis for generic smooth single-input system, we are able to operate a topological classification of the resulting non smooth flows, in the spirit of the work of Andronov-Pontryagin-Peixoto for two dimensional dynamical systems. The research encompasses a comprehensive analysis of singularities, a detailed study of the minimum time function and a deep description of the geometry underlying Pontryagin Maximum Principle.

The rest of this chapter is written in a slightly informal way to let the reader enter smoothly the subject. All details about the given concepts are better illustrated in the following chapters.

Optimal Control Problems and Syntheses

We consider control systems that can be defined as a system of differential equations depending on some parameters $u \in U \subseteq \mathbb{R}^m$:

$$\dot{x} = f(x, u), \tag{0.1}$$

where x belongs to some n–dimensional smooth manifold or, in particular, to $\mathbb{R}^n$. For each initial point x_0 there are many trajectories depending on the choice of the control parameters u.

One usually distinguishes two different ways of choosing the control:

- open loop. Choose u as function of time t,
- closed loop or Feedback. Choose u as function of space variable x.

The first problem one faces is the study of the set of points that can be reached, from x_0, using open loop controls. This is also known as the controllability problem.

If controllability to a final point x_1 is granted, one can try to reach x_1 minimizing some cost, thus defining an Optimal Control Problem:

$$\min \int_0^T L(x(t), u(t))\, dt, \qquad x(0) = x_0, \; x(T) = x_1, \tag{0.2}$$

where $L : \mathbb{R}^n \times U \to \mathbb{R}$ is the Lagrangian or running cost. To have a precise definition of the Optimal Control Problem one should specify further: the time T fixed or free, the set of admissible controls and admissible trajectories, etc. Moreover one can fix an initial (and/or a final) set, instead than the point x_0 (and x_1).

Fixing the initial point x_0 and letting the final condition x_1 vary in some domain of $\mathbb{R}^n$, we get a family of Optimal Control Problems. Similarly we can fix x_1 and let x_0 vary. One main issue is to introduce a concept of solution for this family of problems and, in this book, we focus on the concept of optimal synthesis. Roughly speaking, an optimal synthesis is a collection of optimal trajectories starting from x_0, one for each final condition x_1. As explained in next chapter, geometric techniques provide a systematic method to attack the problem of building an optimal synthesis.

For a discussion of other concepts of solution, such as feedback and value function, we refer to the bibliographical note.

We start giving some examples, which are part of the general theory developed later.

Example A. Assume to have a point of unitary mass moving on a one dimensional line and to control an external bounded force. We get the control system:

$$\ddot{x} = u, \qquad x \in \mathbb{R}, \ |u| \leq C,$$

where x is the position of the point, u is the control and C is a given positive constant. Setting $x_1 = x$, $x_2 = \dot{x}$ and, for simplicity, $C = 1$, in the phase space the system is written as:

$$\begin{cases} \dot{x}_1 = x_2 \\ \dot{x}_2 = u. \end{cases}$$

One simple problem is to drive the point to the origin with zero velocity in minimum time. From an initial position $(\bar{x}_1, \bar{x}_2)$ it is quite easy to see that the optimal strategy is to accelerate towards the origin with maximum force on some interval $[0, t]$ and then to decelerate with maximum force to reach the origin at velocity zero. The set of optimal trajectories is depicted in Figure 0.1.A: this is the simplest example of optimal synthesis for two dimensional systems. Notice that this set of trajectories can be obtained using the following feedback, see Figure 0.1.B. Define the curves $\zeta^{\pm} = \{(x_1, x_2) : \mp x_2 > 0, x_1 = \pm x_2^2\}$ and let ζ be defined as the union $\zeta^{\pm} \cup \{0\}$. We define A^+ to be the region below ζ and A^- the one above. Then the feedback is given by:

$$u(x) = \begin{cases} +1 \ if \ (x_1, x_2) \in A^+ \cup \zeta^+ \\ -1 \ if \ (x_1, x_2) \in A^- \cup \zeta^- \\ 0 \quad if \ (x_1, x_2) = (0, 0). \end{cases}$$

Notice that the feedback u is discontinuous.

Example B. The simplest model for a car-like robot is the celebrated Dubins' car. This car moves only forward at constant unitary velocity and its position is determined by the coordinates (x_1, x_2) of the center of mass and the angle θ formed by the car axis with the positive x_1 axis (see Figure 0.2). If we assume to control only the steering with a lower bound on the turning radius R, then we get the control system on $\mathbb{R}^2 \times S^1$:

$$\begin{cases} \dot{x}_1 = cos(\theta) \\ \dot{x}_2 = sin(\theta) \\ \dot{\theta} = u, \end{cases}$$

where $|u| \leq (1/R)$ and for simplicity we set $R = 1$. Consider the problem of reaching the origin of $\mathbb{R}^2$ with any orientation in minimum time, starting outside $B(0, 2)$, the ball centered at the origin of radius 2. We describe, in the

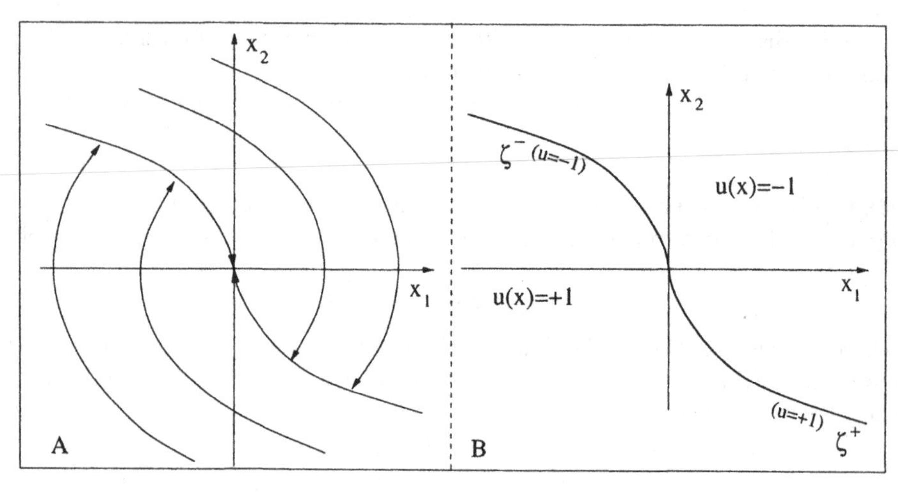

Fig. 0.1. Example A. The simplest example of optimal synthesis and corresponding feedback.

following, the optimal feedback control. Given (x_1, x_2, θ), define $\bar{\theta} \in [-\pi, \pi]$ to be the signed angle from the direction $(cos(\theta), sin(\theta))$ to the direction $(-x_1, -x_2)$, see Figure 0.2. Then, if $\bar{\theta} \notin \{0, \pm\pi\}$,

$$u(x_1, x_2, \theta) = sgn(\bar{\theta}). \tag{0.3}$$

Now if $\bar{\theta} = 0$ then we are already oriented towards the origin and we choose $u = 0$. Finally, if $\bar{\theta} \in \{\pm\pi\}$ then we can choose either $u = +1$ or $u = -1$. The set of points where $u = 0$ is a two dimensional helix in the (x_1, x_2, θ) space and trajectories starting from a point of this helix run straight to the origin lying on this set, see Figure 0.3. The set where u can be chosen to be 1 or -1 is another helix in the (x_1, x_2, θ) space, translated by π in the θ direction. The two regions between the two helices correspond one to control $+1$ and the other to control -1. If the car starts in one of these regions, then it turns until it enters the first helix and then it uses control $u = 0$ up to the origin.

Remark **1** For a complete synthesis (including initial data in $B(0, 2)$) see [118].

Example C. Consider the problem of orienting in minimum time a satellite with two orthogonal rotors. We assume to control the speed of one rotor (that we assume to be bounded), while the second rotor has constant speed. This problem can be modeled with a left invariant control system on $SO(3)$:

$$\dot{x} = x(F + uG), \quad x \in SO(3), \quad |u| \leq 1,$$

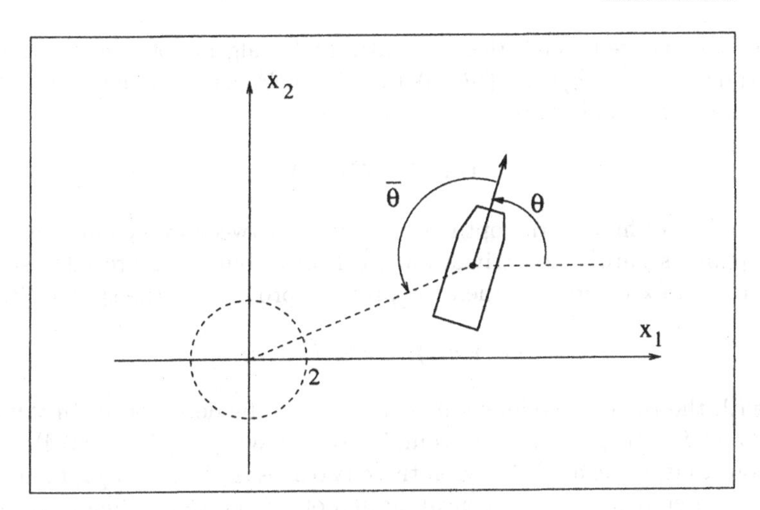

Fig. 0.2. Example B. The Dubins' Car.

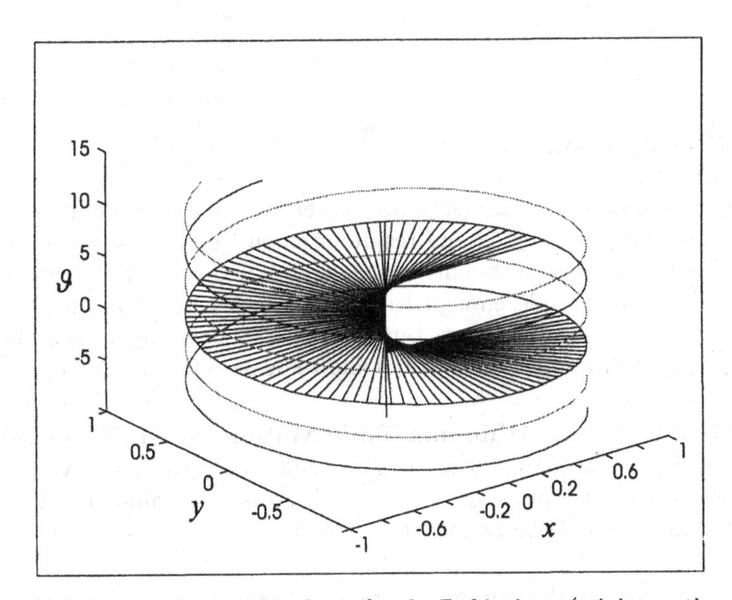

Fig. 0.3. Example B. Optimal Synthesis for the Dubins' car (minimum time, to the origin, with any orientation). The ruled helix corresponds to points where $u = 0$. The nonruled helix correspond to points where u can be chosen either $+1$ or -1. Between the two helices u is determined by formula (0.3).

where F and G are two matrices of $so(3)$, the Lie algebra of $SO(3)$. Using the isomorphism of Lie algebras $(SO(3), [\,\cdot\,,\cdot\,]) \sim (\mathbb{R}^3, \times)$, the condition that the rotors are orthogonal reads:

$$\mathrm{Trace}(F \cdot G) = 0.$$

The problem of finding the optimal trajectory between every initial and terminal point is hardly non trivial and still open, but if we are interested to orient only a fixed semiaxis then the problem projects on the sphere S^2:

$$\dot{x} = x(F + uG), \quad x \in S^2, \quad |u| \leq 1.$$

and, with the theory developed in this book, can be determined. In this case $F+G$ and $F-G$ are rotations around two fixed axes (see Figure 0.4), and it turns out that, if the angle between these two axes is less than $\pi/2$, then every optimal trajectory is a finite concatenation of arcs corresponding to constant control $+1$ or -1. The Optimal Synthesis can be obtained by the feedback shown in Figure 0.4.

The Model Problem

Even though geometric techniques are powerful, the construction of optimal syntheses is quite challenging and results are bounded to low dimensions.

We focus on a class of bidimensional Optimal Control Problems that is, at the same time, simple enough to permit the construction and complete classification of optimal syntheses, but, on the other side, sufficiently general to present a very rich variety of behaviors and to be used for several applications.

Geometric Problem: Minimum Time Motion on 2-D Manifolds
Given a smooth 2-D manifold M and two smooth vector fields X and Y on M, we want to steer a point $p \in M$ to a point $q \in M$ in minimum time using only integral curves of the two vector fields X and Y.

It may happen that the minimum time is obtained only by a trajectory γ whose velocity $\dot{\gamma}(t)$ belongs to the segment joining $X(\gamma(t))$ and $Y(\gamma(t))$ (not being an extremum of it). Thus, for existence purposes, we consider the set of velocities $\{vX(x) + (1-v)Y(x) : 0 \leq v \leq 1\}$, that does not change the value of the infimum time.

The above geometric problem can be restated as the minimum time problem for the following control system. Defining $F = \frac{Y+X}{2}$ and $G = \frac{Y-X}{2}$ this control system can be written in local coordinates as:

$$\dot{x} = F(x) + uG(x), \quad x \in M, \quad |u| \leq 1, \tag{0.4}$$

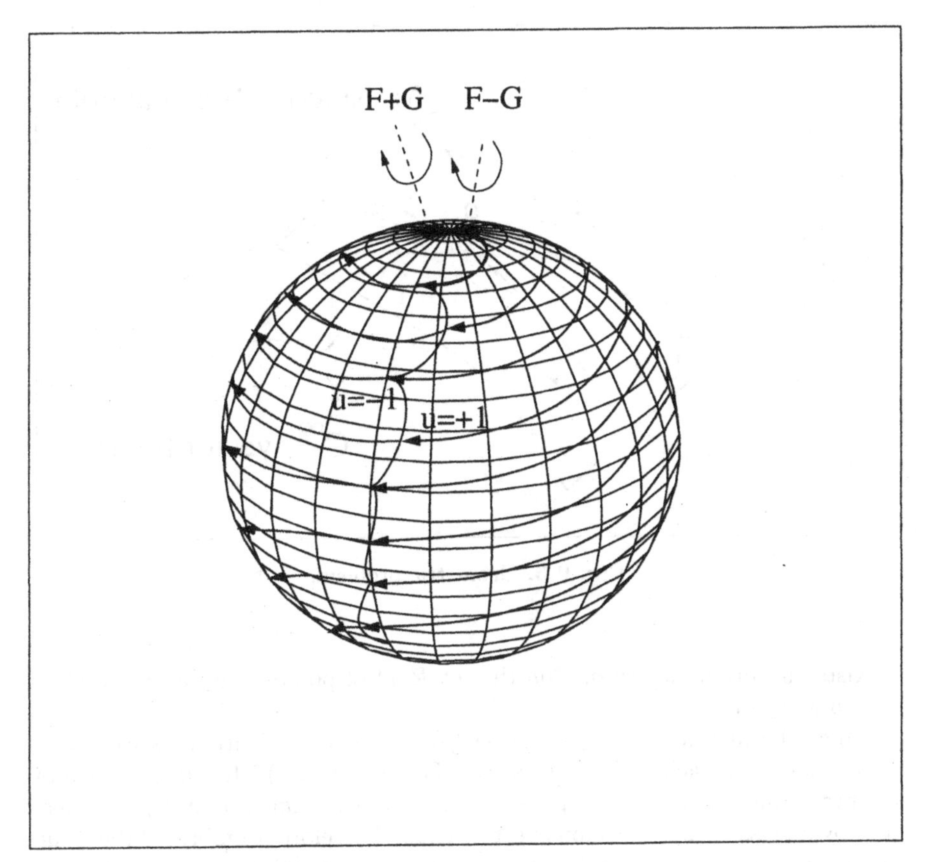

Fig. 0.4. Optimal feedback for Example C. The situation is more intricate in a neighborhood of the south pole, see [34].

where F and G are $\mathcal{C}^\infty$ vector fields on M. Our book is mainly devoted to the study of this kind of systems and, from now on, we fix the initial point p, letting q vary on M.

Therefore the problem we consider is of type (0.1)-(0.2) for the following choice. The dynamics $f(x, u)$ and the control set U are given by (0.4) and the Lagrangian L is constantly equal to 1.

The Classification Program

The geometric control approach gives good results for our model problem: one is able to prove the existence of an optimal synthesis for generic systems. More precisely, let Ξ be the set of pairs (F, G) of smooth vector fields, with the $\mathcal{C}^\infty$ topology. We prove that there exists an open dense set in Ξ for which

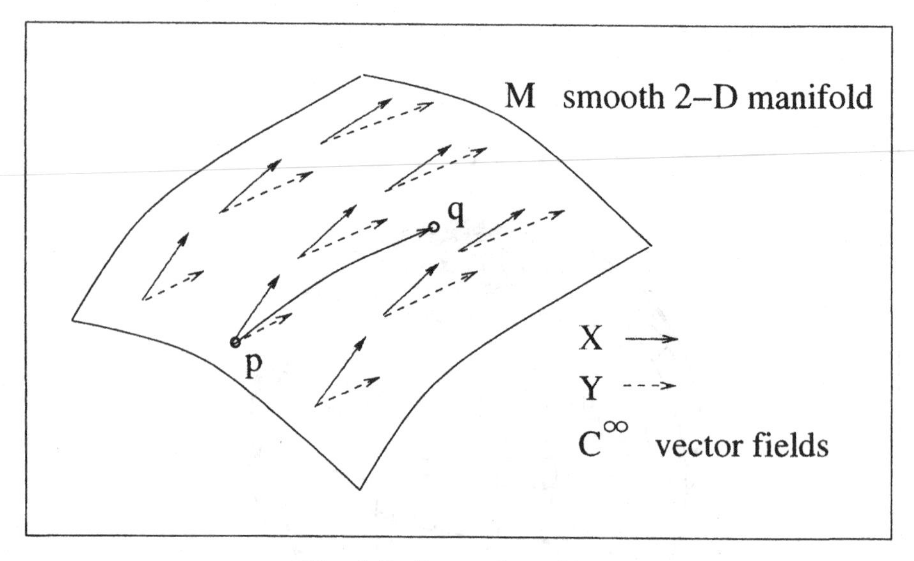

Fig. 0.5. Geometric problem.

it exists an optimal synthesis (on the set $\mathcal{R}(\tau)$ of points that can be reached from p within time τ).

Since the syntheses happen to be generated by discontinuous feedbacks $u(x)$, smooth on each stratum of a stratification of $\mathcal{R}(\tau)$, the classification of syntheses amounts to classify phase portraits of two dimensional flows, generated by a class of discontinuous vector fields. Therefore it embraces the same spirit of the topological classification program for smooth two dimensional dynamical systems, obtained with contributions of Andronov, Pontryagin and Peixoto. We can state the classification program in the following way.

Classification Program. Given $\tau > 0$, find an open dense subset Π_τ of Ξ, a set $\mathcal{G}$ of algebraic or combinatorial structures and an equivalence relation $\sim$ on optimal syntheses such that the following holds. For every couple $(F, G) \in \Pi_\tau$ there exists an optimal synthesis on $\mathcal{R}(\tau)$. Moreover the set of equivalence classes $\Pi_\tau / \sim$ can be put in bijective correspondence with the elements of the set $\mathcal{G}$.

We recall that for two dimensional smooth dynamical systems, the set $\mathcal{G}$ consists of topological graphs and the equivalence relation $\sim$ is an orbital equivalence, see [18].

We are able to complete the classification program with the following choices. The equivalence relation $\sim$ is an orbital equivalence on the flows of the discontinuous feedbacks $u(x)$, with the additional request that the singularities of the synthesis are preserved. Thus a preliminary to the classification of syntheses is the classification of generic singularities. For planar systems, the set $\mathcal{G}$ is a set of topological graphs with additional structure: names of edges, signs of two dimensional zones and lines (details are given in Chapter 2). We illustrate an example of synthesis, with relative singularities, and corresponding graph in Figures 0.6, 0.7.

For the case of a general two dimensional manifold, we need to provide more structure to graphs, giving a cyclic order to the set of edges incident at a vertex. Thanks to a theorem of Heffter, dating back to 19th century, one can individuate a minimal genus compact manifold, on which the graph can be embedded.

A key ingredient to obtain a satisfying classification is structural stability. We say that a pair (F, G) is structurally stable if a small perturbation does not change the structure of the corresponding optimal synthesis. Structural stability is guaranteed in our case by a detailed study of singularities and is ensured under generic conditions. The analysis of systems of the type (0.4) can be pushed much further as explained below.

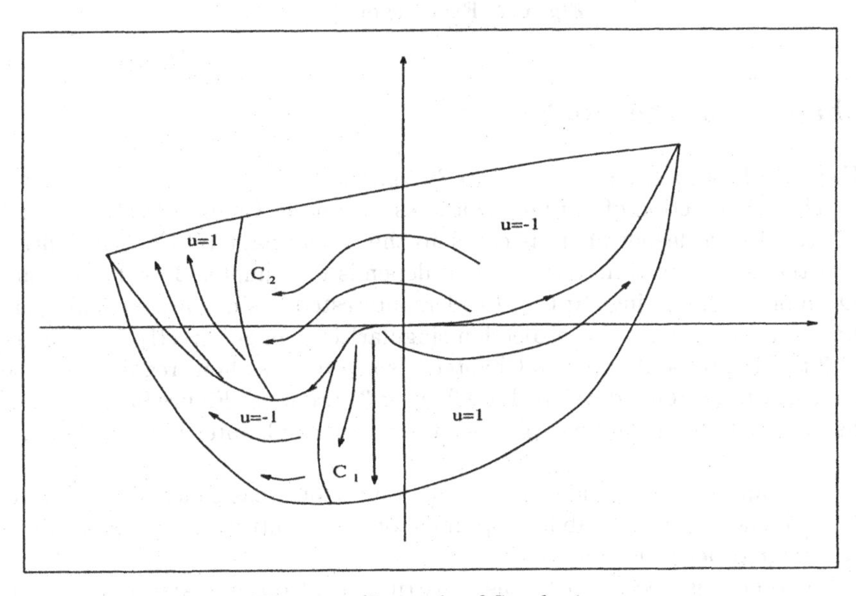

Fig. 0.6. Example of Synthesis.

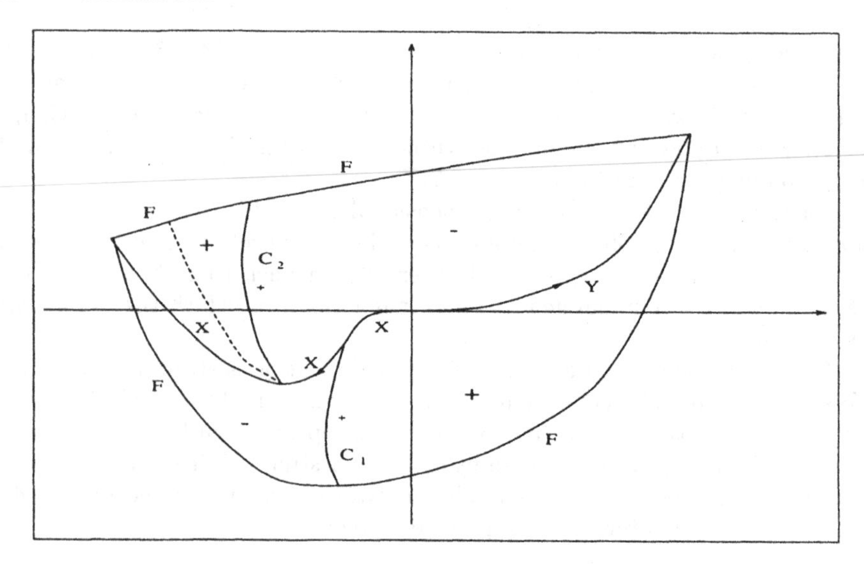

Fig. 0.7. Example of Graph.

Chapters of the Book

We give a brief outline of the chapter's contents.

The construction of optimal syntheses is done in Chapter 2. Readers not interested in a deeper analysis can skip the other chapters. Chapter 3 deals with the minimum time function and depends on Chapter 2 for the classification of synthesis singularities. Readers interested in singularities analysis in the cotangent bundle and projection singularities can go directly from Chapter 2 to Chapter 4. Anyway in Chapter 5, some results about regularity of the minimum time front (developed in Chapter 3) are used. Figure 0.8 illustrates the links between Chapters. We now describe each Chapter in more detail.

In Chapter 1 we provide an introduction to some basic facts of Geometric Control Theory: controllability, optimal control, Pontryagin Maximum Principle, high order conditions, etc.

Chapter 2 is dedicated to the construction of optimal synthesis for our model problem. First, we provide a detailed study of the structure of optimal trajectories. Using this, the existence of an optimal synthesis, under generic assumptions, is proved. A complete classification of synthesis singularities is also given, presenting explicit examples for each equivalence class. The classification program is then completed by proving structural stability of optimal syntheses and associating to each of them a topological graph. Applications are given in the last section.

In Chapter 3 we treat the problem of topological regularity of the minimum time function. We say that a continuous function, not necessarily smooth, is

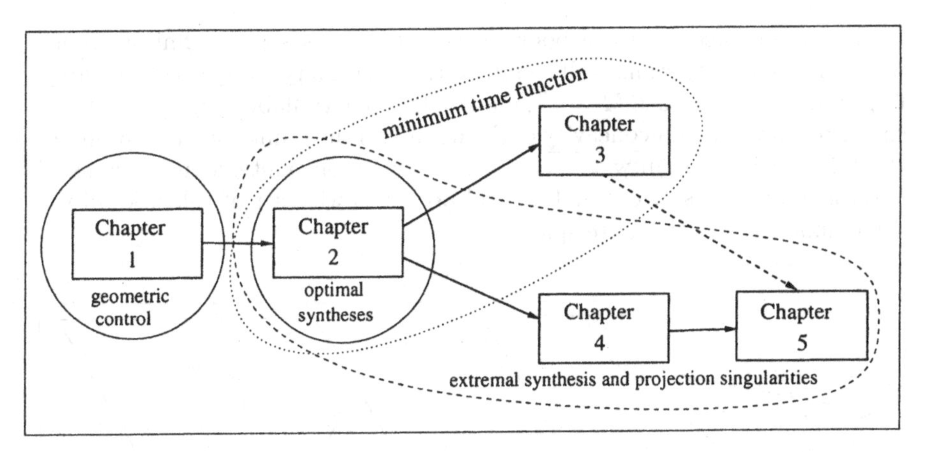

Fig. 0.8. Links between Chapters of the book

topologically a Morse function if its level sets are homeomorphic to the level sets of a Morse function. This is very useful to characterize the topological properties of reachable sets and is sufficient, for example, to derive Morse inequalities, see [101].

The most important tool for the study of optimal trajectories is the well known Pontryagin Maximum Principle (PMP), that is a first order necessary condition for optimality. For each optimal trajectory, PMP provides a lift to the cotangent bundle that is a solution to a suitable pseudo–Hamiltonian system. Chapter 4 is dedicated to analyze the set of extremals, i.e. trajectories satisfying PMP. The whole set of extremal in the cotangent bundle is called the extremal synthesis. Even if this investigation is not necessary to construct the optimal synthesis, however it permits a deeper understanding of the relationships between synthesis singularities, minimum time function and Hamiltonian singularities.

Most of the results obtained so far are key for the following Chapter 5. There are four natural spaces where the mathematical objects previously defined live: the product $T^*M \times \mathbb{R}$ of triplets $(state, costate, time)$, the cotangent bundle T^*M where extremals evolve, the product $M \times \mathbb{R}$ of couples $(state, time)$, the base space M. The set of pairs $(extremals, time) = (state, costate, time)$ is a manifold in $T^*M \times \mathbb{R}$, while the set of extremals in T^*M is only a Whitney stratified set (no more a manifold). All projection singularities between these spaces are classified under generic conditions and the links with synthesis singularities are explained. For example, we show that curves reached optimally by two different trajectories, called overlaps, are generated at projection singularities (from T^*M to M) of cusp type. On the other side, these points correspond to swallowtails in $M \times \mathbb{R}$ and to singularities of the extremal time front on M. See Figure 0.9.

Two appendices end the book. The first reports some technical proofs, while the second generalizes all results to the case of a two dimensional source. In particular, in the case $M = \mathbb{R}^2$, under a local controllability assumption, we can also prove the semiconcavity of the minimum time function. This property (that fails with a pointwise source) is typically encountered in calculus of variations problems or optimal control problems with set of velocities of the same dimension as the state space.

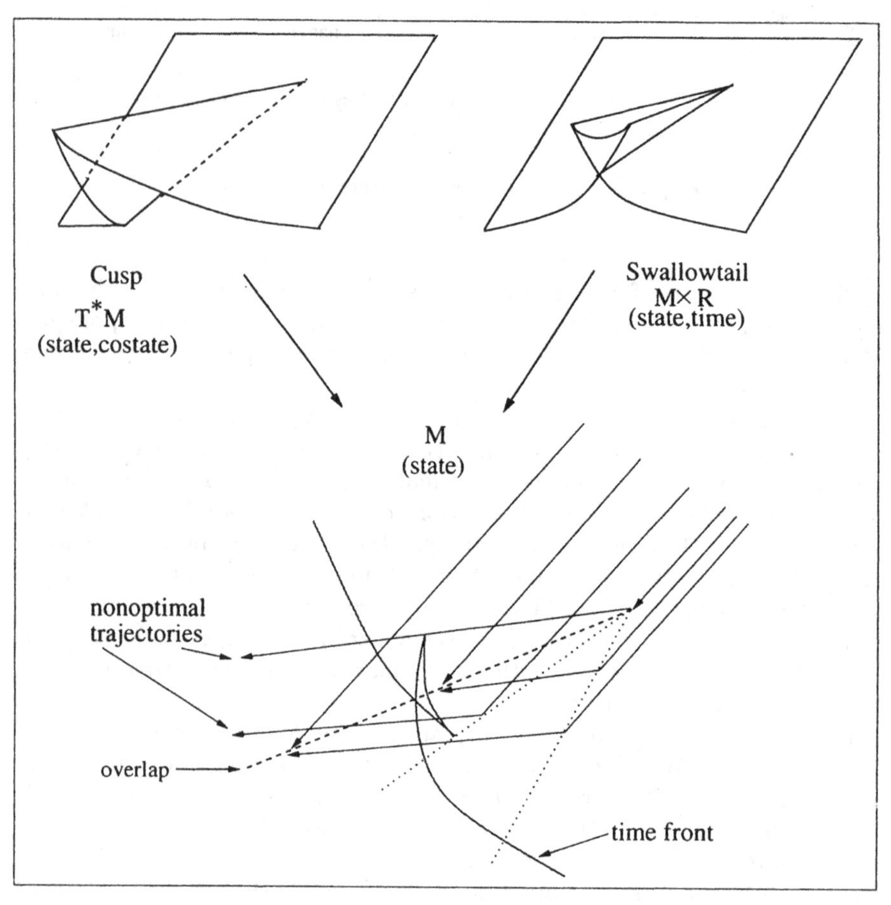

Fig. 0.9. Example of a singularity of the optimal synthesis and related projection singularities. For details see Chapter 5.

Abnormal Extremals

In the construction of both optimal and extremal synthesis, a key role is played by abnormal extremals (that are trajectories with vanishing PMP's Hamiltonian). Properties of abnormal extremals are studied all along the book. For instance we prove that they are finite concatenations of bang arc (i.e. corresponding to constant control ± 1). Moreover, the switchings (discontinuity points of the control) happen exactly when the abnormal extremal crosses the set of zeroes of the function $\Delta_A = F \wedge G$ that is when the two vector fields F and G are collinear, see Propositions 2, p. 49 and Proposition 18, p. 172. Finally, the singularities of the synthesis involving abnormal extremals have some special features (see Section 4.3, p. 171). The set of possible singularities along abnormal extremals is formed of 28 (equivalence classes of) singular points, but not all sequences of singularities can be realized. In Section 5.1.2, p. 202, we prove that all possible sequences can be classified by a set of words recognizable by an automaton and, as a consequence, that all the 28 singularities can appear for some system.

Also, the minimum time front and the extremal front have some important properties on points reached by abnormal extremals. It happens that they are always tangent to the abnormal extremals (see Theorem 28, p. 132 and Section 5.2, p. 209).

Bibliographical Note

Many engineering oriented books have been dedicated to various aspects of Control Theory. A complete list would be too long. We want to point out some texts that are more mathematically oriented, as [69, 92, 117].

Pontryagin Maximum Principle was proved in the seminal book [112]. Various generalizations were given later, see [53, 120]. As we said, Optimal Control Theory can be seen as a generalization of classical Calculus of Variations to the case of constrained velocities. The connections between the two fields are excellently illustrated in [50] and in [29, 78].

Our book focuses on a particular problem in the wide realm of Geometric Control Theory. Despite its long history, there are not so many general books dedicated specifically to Geometric Control Theory. We refer to the book of Jurdjevic [78] and the forthcoming text of Agrachev and Sachkov [2] as general references.

Synthesis theory can be dated back to the pioneering paper of Boltyanskii [27]. General results for analytic or special systems were obtained by Brunovsky [45, 46] and Sussmann [119, 121]. In most cases a synthesis is generated by a feedback that is smooth on each stratum of a stratification. The theory of stratified feedbacks were reported also in the monographs [50, 62]. For a review see also [111]. As mentioned above, discontinuous feedbacks face the problem of generating not optimal trajectories. The main issue is the definition of solution to the corresponding discontinuous ODE. Various concepts of solution are discussed in [97, 111].

The classification program for smooth two dimensional dynamical systems (to which our classification of syntheses is inspired) was developed by Andronov, Pontryagin and Peixoto, see [18, 106, 107]. A more recent improvement can be found in [104, 105].

It is well known that, under suitable assumptions, the value function, defined as the minimum of the Optimal Control Problem for each fixed terminal (initial) point, satisfies the Hamilton-Jacobi-Bellman partial differential equation in viscosity sense. There is a wide literature dedicated to the subject, see for example [23, 62, 63]. For the semiconcavity property of the value function, with set of velocities of the same dimension of the state space, see [48] and references therein.

The Dubins' Car, discussed in Example B, is one of the simplest model for a car–like robot. This problem was originally introduced by Markov in [95] and studied by Dubins in [59]. If we consider the possibility of non constant speed and admit also backward motion, then we obtain the model proposed by Reed and Shepp [113]. A family of time optimal trajectories, that is rich enough to join optimally any two points, was given by Sussmann and Tang in [128]. Then a time optimal synthesis was built by Soueres and Laumond in [118].

The optimal control problem on $SO(3)$ given in Example C, is also called *the left-invariant Dubins' problem on the unit sphere.* It is well know that every time optimal trajectory is a finite concatenation of arcs corresponding to constant control ± 1 or 0. An upper bound on the number of such arcs is given in [3], while the time optimal synthesis for the projected problem on the sphere, is determined in [34].

1

Geometric Control

This chapter provides some basic facts about Geometric Control Theory, Optimal Control and Synthesis Theory. This is a brief introduction to the theory of Geometric Control and is far from being complete: we illustrate some of the main available results of the theory, with few sketches of proofs. For a more detailed treatment of the subject, we refer the reader to the monographs [2, 78] and to the bibliographical note.

1.1 Control Systems

A control system can be viewed as a dynamical system whose dynamical laws are not entirely fixed, but depend on parameters (called controls), that can be determined by an outer agent, in order to obtain a temporal evolution with some properties. It is natural to assume that the space of all possible configurations of the system is a smooth n-dimensional manifold M. The motion of the system at each point of M can follow a number of tangent directions, depending on the choice of the control u:

$$\dot{x} = f(x,u), \quad x \in M, \quad u \in U. \tag{1.1}$$

Here the control space U can be any set. Along the book, to have simplified statements and proofs, we assume more regularity on M and U:

(H0) M is a closed submanifold of $\mathbb{R}^N$ for some N. The set U is a measurable subset of $\mathbb{R}^m$ and f is continuous, smooth with respect to x with Jacobian, with respect to x, continuous in both variables on every chart of M.

A point of view, very useful in geometric control, is to think a control system as a family of assigned vector fields on a manifold:

$$\mathcal{F} = \{F_u(\cdot) = f(\cdot,u)\}_{u \in U}.$$

We always consider smooth vector fields, on a smooth manifold M, i.e. smooth mappings $F: \ x \in M \mapsto F(x) \in T_xM$, where T_xM is the tangent space to

M at x. A vector field can be seen as an operator from the set of smooth functions on M to $\mathbb{R}$. If $x = (x_1, ..., x_n)$ is a local system of coordinates, we have:

$$F(x) = \sum_{i=1}^{n} F^i \frac{\partial}{\partial x^i}.$$

The first definition we need is of the concept of control and of trajectory of a control system.

Definition 1 *A control is a bounded measurable function $u(\cdot) : [a, b] \to U$. A trajectory of (1.1) corresponding to $u(\cdot)$ is a map $\gamma(\cdot) : [a, b] \to M$, Lipschitz continuous on every chart, such that (1.1) is satisfied for almost every $t \in [a, b]$. We write $Dom(\gamma)$, $Supp(\gamma)$ to indicate respectively the domain and the support of $\gamma(\cdot)$. The initial point of γ is denoted by $In(\gamma) = \gamma(a)$, while its terminal point $Term(\gamma) = \gamma(b)$*

Then we need the notion of reachable set from a point $x_0 \in M$.

Definition 2 *We call reachable set within time $T > 0$ the following set:*

$$\mathcal{R}_{x_0}(T) := \{x \in M : \text{ there exists } t \in [0, T] \text{ and a trajectory } \gamma : [0, t] \to M \text{ of (1.1) such that } \gamma(0) = x_0,\ \gamma(t) = x\}. \tag{1.2}$$

Computing the reachable set of a control system of the type (1.1) is one of the main issues of control theory. In particular the problem of proving that $\mathcal{R}_{x_0}(\infty)$ coincide with the whole space is the so called controllability problem. The corresponding local property is formulated as:

Definition 3 (Local Controllability) *A control system is said to be locally controllable at x_0 if for every $T > 0$ the set $\mathcal{R}_{x_0}(T)$ is a neighborhood of x_0.*

Various results were proved about controllability and local controllability. We only recall some definitions and theorems used in the sequel.

Most of the information about controllability is contained in the structure of the Lie algebra generated by the family of vector fields. We start giving the definition of Lie bracket of two vector fields.

Definition 4 (Lie Bracket) *Given two smooth vector fields X, Y on a smooth manifold M, the Lie bracket is the vector field given by:*

$$[X, Y](f) := X(Y(f)) - Y(X(f)).$$

In local coordinates:

$$[X, Y]_j = \sum_i \left(\frac{\partial Y_j}{\partial x_i} X_i - \frac{\partial X_j}{\partial x_i} Y_i \right).$$

In matrix notation, defining $\nabla Y := \left(\partial Y_j / \partial x_i \right)_{(j,i)}$ (j row, i column) and thinking to a vector field as a column vector we have $[X, Y] = \nabla Y \cdot X - \nabla X \cdot Y$.

Definition 5 (Lie Algebra of $\mathcal{F}$) *Let $\mathcal{F}$ be a family of smooth vector fields on a smooth manifold M and denote by $\chi(M)$ the set of all C^∞ vector fields on M. The Lie algebra $Lie(\mathcal{F})$ generated by $\mathcal{F}$ is the smallest Lie subalgebra of $\chi(M)$ containing $\mathcal{F}$. Moreover for every $x \in M$ we define:*

$$Lie_x(\mathcal{F}) := \{X(x) : \ X \in Lie(\mathcal{F})\}. \tag{1.3}$$

Remark **2** In general $Lie(\mathcal{F})$ is a infinite-dimensional subspace of $\chi(M)$. On the other side since all $X(x) \in T_xM$ (in formula (1.3)) we have that $Lie_x(\mathcal{F}) \subseteq T_xM$ and hence $Lie_x(\mathcal{F})$ is finite dimensional.

Remark **3** $Lie(\mathcal{F})$ is built in the following way. Define: $D_1 = Span\{\mathcal{F}\}$, $D_2 = Span\{D_1 + [D_1, D_1]\}$, $\cdots$ $D_k = Span\{D_{k-1} + [D_{k-1}, D_{k-1}]\}$. D_1 is the so called distribution generated by $\mathcal{F}$ and we have $Lie(\mathcal{F}) = \cup_{k\geq 1} D_k$. Notice that $D_{k-1} \subseteq D_k$. Moreover if $[D_n, D_n] \subseteq D_n$ for some n, then $D_k = D_n$ for every $k \geq n$.

A very important class of families of vector fields are the so called Lie bracket generating (or completely nonholonomic) systems for which:

$$Lie_x\mathcal{F} = T_xM, \quad \forall\, x \in M. \tag{1.4}$$

For instance analytic systems (i.e. with M and $\mathcal{F}$ analytic) are always Lie bracket generating on a suitable immersed analytic submanifold of M (the so called orbit of $\mathcal{F}$). This is the well know Hermann-Nagano Theorem (see for instance [78], pag. 48).

As we show later, our model problem (0.4) is Lie bracket generating under generic conditions.

If the system is symmetric, that is $\mathcal{F} = -\mathcal{F}$ (i.e. $f \in \mathcal{F} \Rightarrow -f \in \mathcal{F}$), then the controllability problem is more simple. For instance condition (1.4) with M connected implies complete controllability i.e. for each $x_0 \in M$, $\mathcal{R}_{x_0}(\infty) = M$ (this is a corollary of the well know Chow Theorem, see for instance [2]).

On the other side, if the system is not symmetric (as for our model), the controllability problem is more complicated and controllability is not guaranteed in general (by (1.4) or other simple conditions), neither locally. Anyway, important properties of the reachable set for Lie bracket generating systems are given by the following theorem (see [88] and [2]):

Theorem 1 (Krener) *Let $\mathcal{F}$ be a family of smooth vector fields on a smooth manifold M. If $\mathcal{F}$ is Lie bracket generating, then, for every $T \in]0, +\infty]$, $\mathcal{R}_{x_0}(T) \subseteq Clos(Int(\mathcal{R}_{x_0}(T))$. Here $Clos(\cdot)$ and $Int(\cdot)$ are taken with respect to the topology of M.*

Krener Theorem implies that the reachable set for Lie bracket generating systems has the following properties:

- It has nonempty interior: $Int(\mathcal{R}_{x_0}(T)) \neq \emptyset, \quad \forall\, T \in]0, +\infty]$.
- Typically it is a manifold with or without boundary of full dimension. The boundary may be not smooth, e.g. have corners or cuspidal points.

In particular it is prohibited that reachable sets are collections of sets of different dimensions as in Figure 1.1 (cfr. with the concept of stratification in Chapter 4). These phenomena happen for non Lie bracket generating systems, and it is not know if reachable sets may fail to be stratified sets (for generic smooth systems) see [72, 78, 94].

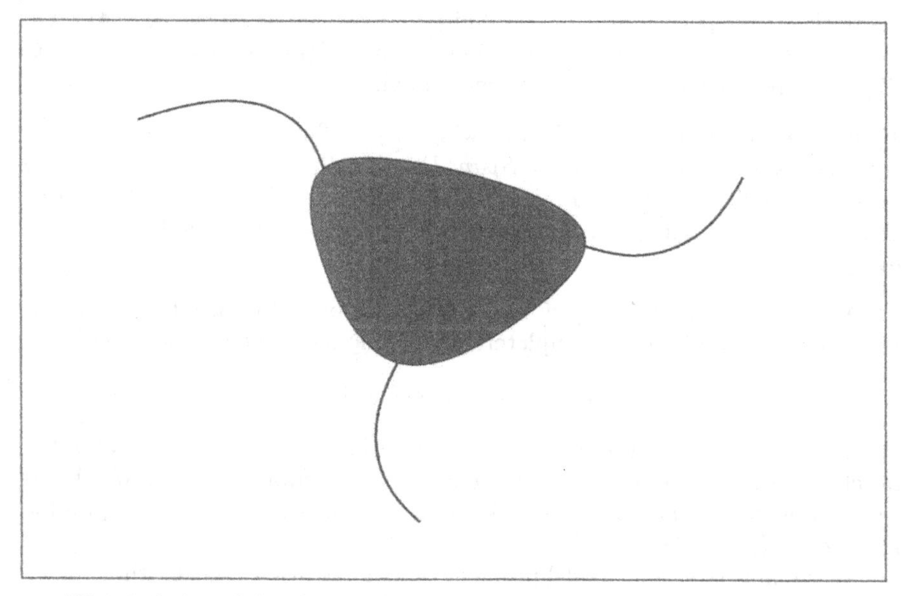

Fig. 1.1. A prohibited reachable set for a Lie bracket generating systems.

Local controllability can be detached by linearization as shown by the following important result (see [92], p. 366):

Theorem 2 *Consider the control system $\dot{x} = f(x,u)$ where x belongs to a smooth manifold M of dimension n and let $u \in U$ where U is a subset of $\mathbb{R}^m$ for some m, containing an open neighborhood of $u_0 \in \mathbb{R}^m$. Assume f of class C^1 with respect to x and u. If the following holds:*

$$\begin{aligned} & f(x_0,u_0) = 0, \\ & rank[B, AB, A^2B, ..., A^{n-1}B] = n, \\ & where\ A = (\partial f/\partial x)(x_0,u_0)\ and\ B = (\partial f/\partial u)(x_0,u_0), \end{aligned} \tag{1.5}$$

then the system is locally controllable at x_0.

Remark 4 Condition (1.5) is the well know Kalman condition that is a necessary and sufficient condition for (global) controllability of linear systems:

$$\dot{x} = Ax + Bu, \quad x \in \mathbb{R}^n, \quad A \in \mathbb{R}^{n \times n}, \quad B \in \mathbb{R}^{n \times m}, \quad u \in \mathbb{R}^m.$$

In the local controllable case we get this further property of reachable sets:

Lemma 1 *Consider the control system $\dot{x} = f(x,u)$ where x belongs to a smooth manifold M of dimension n and let $u \in U$ where U is a subset of R^m for some m. Assume f of class C^1 with respect to x and continuous with respect to u. If the control system is locally controllable at x_0 then for every $T, \varepsilon > 0$ one has:*

$$\mathcal{R}_{x_0}(T) \subseteq Int(\mathcal{R}_{x_0}(T+\varepsilon)). \tag{1.6}$$

Proof. Consider $x \in \mathcal{R}_{x_0}(T)$ and let $u_x : [0,T] \to U$ be such that the corresponding trajectory starting from x_0 reaches x at time T. Moreover, let Φ_t be the flux associated to the time varying vector field $f(\cdot, u(t))$ and notice that Φ_t is a diffeomorphism. By local controllability at x_0, $\mathcal{R}_{x_0}(\varepsilon)$ is a neighborhood of x_0. Thus $\Phi_T(\mathcal{R}_{x_0}(\varepsilon))$ is a neighborhood of x and, using $\Phi_T(\mathcal{R}_{x_0}(\varepsilon)) \subset \mathcal{R}_{x_0}(T+\varepsilon)$, we conclude. ■

1.2 Optimal Control

An Optimal Control Problem for the system (1.1), is a problem of the following type:

$$\begin{cases} \text{minimize} \quad \int_0^T L(x(t),u(t))dt + \psi(x(T)), \\ x(0) = x_0, \\ x(T) \in \mathcal{T}, \end{cases} \tag{1.7}$$

where $L : M \times U \to \mathbb{R}$ is the Lagrangian or running cost, $\psi : M \to \mathbb{R}$ is the final cost, $x_0 \in M$ is the initial condition and $\mathcal{T} \subset M$ the target. The minimization is taken on the set of all admissible trajectories of the control system (1.1) (the admissibility conditions to be specified), that start at the initial condition and end at the target in finite time T that can be fixed or free, depending on the problem. Notice that this optimal control problem is autonomous, hence we can always assume trajectories to be defined on some interval of the form $[0,T]$.

Of great importance for applications are the so called control affine systems:

$$\dot{x} = F_0 + \sum_{i=1}^{m} u_i F_i, \qquad u_i \in \mathbb{R}, \tag{1.8}$$

with quadratic cost:

$$\text{minimize} \int_0^T \sum_{i=1}^{m} u_i^2 dt, \tag{1.9}$$

or with cost equal to 1 (that is the minimum time problem) and bounded controls $|u_i| \leq 1$. The term F_0 in (1.8) is called drift term and, for the so called distributional systems (or driftless), is equal to zero. Subriemannian problems are distributional systems with quadratic cost. See also Exercise 3, p. 30. Single–input systems are control affine systems with only one control, i.e. $m = 1$.

Definition 6 *We call $\mathcal{P}_{x_1}$, $x_1 \in M$, the optimal control problem given by the dynamics (1.1) and the minimization (1.7) with $\mathcal{T} = x_1$.*

Beside (H0), (see p. 15) we make the following basic assumptions on (1.1) and (1.7).

(H1) L is continuous, smooth with respect to x with Jacobian, with respect to x, continuous in both variables on every chart of M;
(H2) ψ is a C^1 function.

We are interested in solving the family of control problems $\{\mathcal{P}_{x_1}\}_{x_1 \in M}$ and for us a solution is given by an optimal synthesis that is

Definition 7 (Optimal Synthesis) *Given $\Omega \subset M$, an optimal synthesis on Ω for the family of optimal control problems $\{\mathcal{P}_{x_1}\}_{x_1 \in \Omega}$ is a collection $\{(\gamma_{x_1}, u_{x_1}) : x_1 \in \Omega\}$ of trajectory–control pairs such that (γ_{x_1}, u_{x_1}) provides a solution to $\mathcal{P}_{x_1}$.*

Remark **5** One can invert the role of x_0 and x_1 letting x_1 be fixed and x_0 vary. Moreover, we can consider a generalization fixing an initial manifold $\mathcal{S}$, called source. In this case the initial condition reads $x_0 \in \mathcal{S}$.

Remark **6** In many cases it happens that an optimal synthesis, defined on $\mathcal{R}(T)$, is generated by a piecewise smooth feedback $u(x)$, that is a map from $\mathcal{R}(T)$ to U. This means that the optimal trajectories of the synthesis are precisely the solutions to the differential equation

$$\dot{x} = f(x, u(x)).$$

For the problem of defining a solution to this equation, when $u(x)$ is discontinuous, see the bibliographical note of the Introduction.

1.2.1 Existence

In order to prove existence for $\mathcal{P}_{x_1}$, we first give the next:

Theorem 3 *Assume f bounded, U compact and let the space of admissible controls be the set of all measurable maps $u(\cdot) : [a, b] \to U$. If the set of velocities $V(x) = \{f(x, u)\}_{u \in U}$ is convex, then the reachable set $\mathcal{R}(T)$ is compact.*

Remark **7** The fact that $\mathcal{R}(T)$ is relatively compact is an immediate consequence of the boundedness and continuity of f. One can also replace the boundedness with a linear grows condition. The convexity of $V(x)$ is the key property to guarantee that $\mathcal{R}(T)$ is closed. For a proof see [50].

From this theorem, we get immediately the following:

Theorem 4 *Assume f bounded, U compact, and let the space of admissible controls be the set of all measurable maps $u(\cdot) : [a,b] \to U$. Moreover assume that the set $\{f(x,u), L(x,u)\}_{u \in U}$ is convex. If $L \geq C > 0$ and $x_1 \in \mathcal{R}_{x_0}(\infty)$, then the problem $\mathcal{P}_{x_1}$ has a solution.*

Sketch of the Proof. There exists $T > 0$ such that every trajectory ending at x_1 after T is not optimal. Consider the augmented system obtained adding the extra variable y such that:

$$\dot{y} = L(x,u).$$

Since f is bounded, all trajectories defined on $[0,T]$ have bounded costs. Thus applying the previous Theorem to the augmented system, we get that $\mathcal{R}_{(x_0,0)}(T)$, the reachable set for the augmented system of (x,y), is compact. Therefore there exists a minimum for the function $(x,y) \to y + \psi(x)$ on the set $\mathcal{R}_{(x_0,0)}(T) \cap \{(x,y) : \; x = x_1\}$. The trajectory, reaching such a minimum, is optimal. ■

1.2.2 Pontryagin Maximum Principle

The standard tool to determine optimal trajectories is the well known Pontryagin Maximum Principle, see [78, 112], that gives a first order condition for optimality: every optimal trajectory has a lift to the cotangent bundle, formed by vector-covector pairs, satisfying a maximization condition of a suitable Hamiltonian.

Pontryagin Maximum Principle can be seen as a generalization of Weierstrass's necessary conditions for a minimum and Euler–Lagrange equations to systems with constrained velocities, called systems with nonholonomic constraints, as (1.1).

Remark **8 (Calculus of Variations)** Notice that the classical case of Calculus of Variations can be expressed as the problem (1.7) for the dynamics $\dot{x} = u$, $u \in T_xM$, fixed final time T and $\mathcal{T} = x_1 \in M$.

Pontryagin Maximum Principle can be stated in several forms depending on the following:

i) the final time is fixed or free (for fixed final time see Exercise 4, p. 31),
ii) dimension and regularity of the source and of the target,
iii) the cost contains only the running part, only the final part, or both,

iv) the source and/or the target depend on time.

Here we state a version (sufficient to treat our model problem) in which: **i)** the final time is free, **ii)** the source is zero dimensional and the target $\mathcal{T}$ is a smooth submanifold of M of any dimension, **iii)** there are both running and final cost, **iv)** the source and the target do not depend on time. Let us introduce some notation.

For every $(x, p, \lambda_0, u) \in T^*M \times \mathbb{R} \times U$ we define:

$$\mathcal{H}(x, p, \lambda_0, u) = \langle p, f(x, u) \rangle + \lambda_0 L(x, u), \tag{1.10}$$

and

$$H(x, p, \lambda_0) = \max\{\mathcal{H}(x, p, \lambda_0, u) : u \in U\}.$$

Definition 8 (Extremal Trajectories) *Consider the optimal control problem (1.1), (1.7) and assume (H0), (H1) and (H2). Let $u : [0, T] \to U$ be a control and γ a corresponding trajectory. We say that γ is an extremal trajectory if there exist a Lipschitz continuous map called covector $\lambda : t \in [0, T] \mapsto \lambda(t) \in T^*_{\gamma(t)}M$ and a constant $\lambda_0 \leq 0$, with $(\lambda(t), \lambda_0) \neq (0, 0)$ (for all $t \in [0, T]$), that satisfy:*

- *(PMP1) for a.e. $t \in [0, T]$, in a local system of coordinates, we have $\dot{\lambda} = -\frac{\partial \mathcal{H}}{\partial x}(\gamma(t), \lambda(t), \lambda_0, u(t))$;*
- *(PMP2) for a.e. $t \in [0, T]$, we have $H(\gamma(t), \lambda(t), \lambda_0) = \mathcal{H}(\gamma(t), \lambda(t), \lambda_0, u(t)) = 0$;*
- *(PMP3) for every $v \in T_{\gamma(T)}\mathcal{T}$, we have $\langle \lambda(T), v \rangle = \lambda_0 \langle \nabla\psi(\gamma(T)), v \rangle$ (transversality condition).*

In this case we say that (γ, λ) is an extremal pair.

Pontryagin Maximum Principle (briefly PMP) states the following:

Theorem 5 (Pontryagin Maximum Principle) *Consider the optimal control problem (1.1), (1.7) and assume (H0), (H1) and (H2). If $u(\cdot)$ is a control and γ a corresponding trajectory that is optimal, then γ is extremal.*

Remark **9** Notice that the dynamic $\dot{x} = f(x, u)$ and equation *(PMP1)* can be written in the pseudo-Hamiltonian form:

$$\dot{x}(t) = \frac{\partial \mathcal{H}}{\partial p}(x(t), p(t), \lambda_0, u(t)),$$

$$\dot{p}(t) = -\frac{\partial \mathcal{H}}{\partial x}(x(t), p(t), \lambda_0, u(t)).$$

Remark **10** Notice that the couple (λ, λ_0) is defined up to a positive multiplicative factor, in the sense that if the triple $(\gamma, \lambda, \lambda_0)$ represents an extremal, than the same happens for the triple $(\gamma, \alpha\lambda, \alpha\lambda_0)$, $\alpha > 0$. If $\lambda_0 \neq 0$, usually one normalizes $(\lambda(\cdot), \lambda_0)$ by $\lambda_0 = -1/2$ or $\lambda_0 = -1$.

Remark **11** If the target $\mathcal{T}$ is a point then the transversality condition (PMP3) is empty.

Remark **12** In the case in which we have also a source: $\gamma(0) \in \mathcal{S}$, we have also a transversality condition at the source:

(PMP3') For every $v \in T_{\gamma(0)}\mathcal{S}$, $\langle\lambda(0), v\rangle = 0$.

1.2.3 Abnormal Extremals and Endpoint Singular Extremals

Assume now that f is differentiable also with respect to u. The following sets of trajectories have very special features:

Definition 9 (Endpoint Singular Trajectories and Endpoint Singular Extremals) *We call endpoint singular trajectories, solutions to the following equations:*

$$\dot{x}(t) = \frac{\partial \bar{\mathcal{H}}}{\partial p}(x(t), p(t), u(t)),$$

$$\dot{p}(t) = -\frac{\partial \bar{\mathcal{H}}}{\partial x}(x(t), p(t), u(t)),$$

$$\frac{\partial \bar{\mathcal{H}}}{\partial u}(x(t), p(t), u(t)) = 0,$$

where $\bar{\mathcal{H}}(x,p,u) := \langle p, f(x,u)\rangle$, $p(t) \neq 0$ *and the constraint* $u \in U \in \mathbb{R}^m$ *is changed in* $u \in Int(U)$. *Endpoint singular trajectories that are also extremals are called endpoint singular extremals.*

Remark **13** Notice that, although endpoint singular trajectories do not depend on the cost and on the constraints on the control set, endpoint singular extremals (that for our minimization problem are the interesting ones) do depend.

The name endpoint singular trajectories comes from the fact that they are singularities of the end-point mapping that, fixed an initial point and a time T, associates to a control function $u(\cdot)$, defined on $[0,T]$, the end point of the corresponding trajectory γ:

$$\mathcal{E}^{x_0,T} : u(\cdot) \mapsto \gamma(T).$$

By singularity of the end point mapping, we mean a control at which the Fréchet derivative of $\mathcal{E}^{x_0,T}$ is not surjective. For more details see [29]. Roughly speaking, this means that the reachable set, locally, around the trajectory, does not contain a neighborhood of the end point $\gamma(T)$.

In the case of a minimum time problem for a control affine system (1.8) with $|u_i| \leq 1$, endpoint singular trajectories satisfy $< p(t), F_i(\gamma(t)) >= 0$. Endpoint singular extremals, are just endpoint singular trajectories for which there exists $\lambda_0 \leq 0$ satisfying $< p(t), F_0 > +\lambda_0 = 0$ and corresponding to admissible controls.

In the next chapters we consider control systems of the form $\dot{x} = F(x) + uG(x)$ where $|u| \leq 1$. In this case, under generic conditions, endpoint singular extremals are arcs of extremal trajectories corresponding to controls not constantly equal to $+1$ or -1.

Definition 10 (Abnormal Extremals) *We call abnormal extremals extremal trajectories for which* $\lambda_0 = 0$.

Remark **14** Abnormal extremals do not depend on the cost, but depend on the constraint $u \in U$.

Remark **15** In some problems, like in subriemannian geometry or in distributional problems for the minimum time with bounded controls, endpoint singular extremals and abnormal extremals coincide, see Exercise 7, p. 31.

In other problems (e.g. minimum time for control affine systems) the two definitions are different, but coincide for some very special class of trajectories. These trajectories are usually called singular exceptional (see [32]).

1.3 High Order Conditions

PMP is used in Synthesis Theory to attempt a finite dimensional reduction of the minimization problem, as explained in Step 2 of Section 1.4, p. 27. Of course high order conditions can be very useful for further restriction of candidate optimal trajectories.

There are several possible higher order variations. Here we present only some results used in next chapters for our model problem. We start illustrating the high order principle of Krener. Then we give a generalization proved by Bressan, finally we treat in some details envelope theory developed by Sussmann.

1.3.1 High Order Maximum Principle

In this section we consider a fixed reference trajectory $\gamma : [0,T] \to M$ corresponding to the control u, for the problem (1.7).

Definition 11 *Let* $u_1, ..., u_n$ *be controls defined on* $[0,T]$, q_1, q_2 *and* p_i, $i = 1, ..., n$, *be polynomials vanishing in zero with* $q_2(\varepsilon) \geq 0$, $p_i(\varepsilon) \geq 0$ *for* $\varepsilon \geq 0$ *small and* $q_1(\varepsilon) + q_2(\varepsilon) + \sum_{i=1}^n p_i(\varepsilon) = 0$. *Consider the following family of controls:*

$$u_{(\varepsilon,s)}(t) = \begin{cases} u(t) & \text{if } t \in [0, s + q_1(\varepsilon)] \\ u_1(t) & \text{if } t \in [s + q_1(\varepsilon), s + q_1(\varepsilon) + p_1(\varepsilon)] \\ \vdots & \\ u_n(t) & \text{if } t \in [s + q_1(\varepsilon) + \sum_{i=1}^{n-1} p_i(\varepsilon), s + q_1(\varepsilon) + \sum_{i=1}^n p_i(\varepsilon)] \\ u(t) & \text{if } t \in [s + q_1(\varepsilon) + \sum_{i=1}^n p_i(\varepsilon), T], \end{cases}$$

and let $\gamma_{(\varepsilon,s)}$ *be the corresponding trajectories, starting at* $\gamma(0)$. *A variation of order* h *at* $\tilde{t} \in]0,T[$, *where* $\tilde{t}$ *is a Lebesgue point for* $f(\gamma(t),u(t))$ *and* $L(\gamma(t),u(t))$, *is a family of trajectories* $\gamma_{(\varepsilon,s)}$, $\varepsilon \in [0,\bar{\varepsilon}]$, *such that:*

$$\left. \frac{d^j}{d\varepsilon^j}\gamma_{(\varepsilon,s)}(s)\right|_{\varepsilon=0} = 0,$$

for $j = 1,...,h-1$ *and* s *in a neighborhood of* $\tilde{t}$.

We can now state the high order principle:

Theorem 6 (Krener High Order Maximum Principle) *Let* γ *be an extremal trajectory and* $\gamma_{(\varepsilon,s)}$ *a variation of order* h *at* $\tilde{t}$. *Then there exists a covector* λ *such that* (γ,λ) *is extremal and:*

$$\lambda(\tilde{t})\left.\frac{d^h}{d\varepsilon^h}\gamma_{(\varepsilon,\tilde{t})}(\tilde{t})\right|_{\varepsilon=0} + \lambda_0 \left.\frac{d^h}{d\varepsilon^h}\int_0^{\tilde{t}} L(\gamma_{(\varepsilon,\tilde{t})}, u_{(\varepsilon,\tilde{t})})(s)\, ds\right|_{\varepsilon=0} \leq 0.$$

Using the high order maximum principle, it is possible to prove a generalized Legendre–Clebsch condition for the case of single–input systems.

Theorem 7 *Consider the minimum time problem for the system:*

$$\dot{x} = F(x) + uG(x), \qquad |u| \leq 1,$$

where F *and* G *are smooth. Assume* γ *optimal and endpoint singular extremal. Then there exists a covector* λ *such that* (γ,λ) *is extremal and for every* t:

$$\langle \lambda(t), G(\gamma(t))\rangle = 0, \qquad \langle \lambda(t), F(\gamma(t))\rangle \geq 0,$$

and

$$\langle \lambda(t), [G,[F,G]](\gamma(t))\rangle \geq 0.$$

Remark **16** The first two conditions of Theorem 7 are easily derived, as shown in the next chapter, while the third is the so called Legendre–Clebsch condition.

In some cases, it is useful to perform variations not based at a unique time $\tilde{t}$, but more generally perturbing the reference control u in L^1.

Definition 12 *A one–parameter variational family, generating a vector* v, *is a continuous map* $\varepsilon \mapsto u_\varepsilon$ *from an interval* $[0,\bar{\varepsilon}]$, $\bar{\varepsilon} > 0$, *to* $L^1([0,T],U)$ *such that:* **i)** u_ε *are uniformly bounded,* **ii)** $u_0 = u$, **iii)** *if* γ_ε *indicates the trajectory corresponding to* u_ε *and starting at* $\gamma(0)$, *it holds:*

$$\lim_{\varepsilon\to 0}\frac{\gamma_\varepsilon(T)-\gamma(T)}{\varepsilon} = v.$$

For these more general variations, we have,

Theorem 8 *Consider a minimum time problem for an affine control system* $\dot{x} = F_0(x) + \sum_{i=1}^m u_i F_i(x)$. *If* γ *is optimal and* u_ε *is a one–parameter variational family, generating a vector* v, *then there exists a covector* λ *such that* (γ,λ) *is extremal and:*

$$\lambda(T)\cdot v \leq 0.$$

1.3.2 Envelope Theory

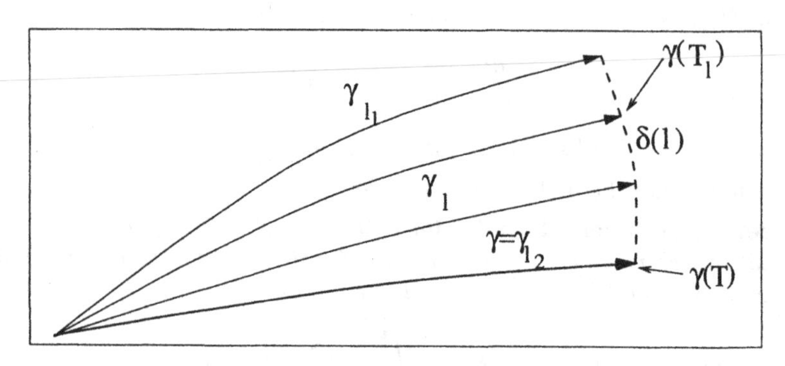

Fig. 1.2. Envelope Theory

Envelope theory is another high order technique to exclude extremal trajectories that are not locally optimal. For instance for the minimum time problem of a single input system $\dot{x} = F(x) + uG(x)$, $|u| \leq 1$, many optimal trajectories are concatenation of arcs corresponding to constant control ± 1. Using envelope theory, one can bound the number of arcs, on which $u = \pm 1$, that can form an optimal trajectory.

Recall (H0), (H1), (H2) and, for simplicity, $M = \mathbb{R}^n$ and the final cost $\psi = 0$. Let γ be an extremal trajectory defined in $[0, T]$ corresponding to control $u(\cdot)$. Consider a one-parameter family of extremals $\Lambda = \{\gamma_l : [0, T_l] \to M,\ l \in [l_1, l_2]\}$ starting at $In(\gamma)$ at time 0 with $\gamma_{l_2} = \gamma$, $T_{l_2} = T$. Assume that $\gamma_l(\cdot)$ corresponds to the control $u_l(\cdot)$ and is associated to the covector $(\lambda_l(\cdot), \lambda_l^0)$. Define the curve δ on $[l_1, l_2]$ by $\delta(l) = Term(\gamma_l)$ (see figure 1.2).

Definition 13 *We say that the curve $\delta(\cdot)$, defined above, is an envelope for γ, if the following conditions are satisfied:*

1. *The family γ_l is differentiable, that is:*
   i) the maps $l \to T_l$ and $(l, t) \to \gamma_l(t)$ are continuous on their domains of definition;
   ii) the controls $u_l(\cdot)$ are uniformly bounded. If we set $\bar{T} = \max_l T_l$ and, for some fixed $\omega \in U$, prolong u_l on $[0, \bar{T}]$ by setting $u_l(t) = \omega$ for $t \geq T_l$, then the map $l \to u_l(\cdot)$ is continuous from $[l_1, l_2]$ to $L^1([0, \bar{T}], U)$;
   iii) for every $\bar{l} \in [l_1, l_2]$ and every continuous function $\varphi : [0, \bar{T}] \to \mathbb{R}^n$ the map:

$$l \mapsto \int_0^{T_l} \varphi(t) \cdot f(\gamma_{\bar{l}}(t), u_l(t))dt,$$

   (where we set $\gamma_{\bar{l}}(t) = \gamma_{\bar{l}}(T_{\bar{l}})$ for $t \geq T_{\bar{l}}$), is differentiable at $\bar{l}$ (i.e. $l \to f(\gamma_{\bar{l}}(\cdot), u_l(\cdot))$ is weakly differentiable at $\bar{l}$ as map into the space of Borel measures.)

2. The map $l \mapsto \int_0^{T_l} L(\gamma_l(t), u_l(t))dt$ *is Lipschitz continuous.*
3. $\delta : [l_1, l_2] \to M$ *is an admissible trajectory of the system, corresponding to a control* $v(\cdot) : [l_1, l_2] \to U$.
4. the identity:

$$\mathcal{H}(\delta(l), \lambda_l(T_l), \lambda_l^0, v(l)) = 0$$

holds for almost every $l \in [l_1, l_2]$.

We have the following,

Theorem 9 (Envelope Theorem) *If* δ *is an envelope for* γ*, then the cost of* γ *is the same as the sum of the costs of* γ_{l_1} *and the envelope* δ*, i.e.:*

$$\int_0^T L(\gamma(t), u(t))\, dt = \int_0^{T_{l_1}} L(\gamma_{l_1}(t), u_{l_1}(t))\, dt + \int_{l_1}^{l_2} L(\delta(t), v(t))\, dt.$$

Envelope Theorem is then applied in this way: if one can prove that $\delta(\cdot)$ is not extremal (a condition often easily checked), and so not optimal, then γ cannot be optimal having the same cost of a trajectory (concatenation of γ_{l_1} and δ) that is not even extremal.

1.4 Geometric Control Approach to Synthesis

Geometric control provides a standard method toward the construction of an optimal synthesis for the family of problems $\{\mathcal{P}_{x_1}\}_{x_1 \in M}$, of Definition 6, p. 20.

The approach is illustrated for systems with a compact control set as for our model problem. The following scheme, consisting of four steps, elucidates the procedure for building an optimal synthesis.

Step 1. Use PMP and high order conditions to study the properties of optimal trajectories.
Step 2. Use Step 1. to obtain a finite dimensional family of candidate optimal trajectories.
Step 3. Construct a synthesis, formed of extremal trajectories, with some regularity properties.
Step 4. Prove that the regular extremal synthesis is indeed optimal.

Let us describe in more detail each Step.

Step 1.

We stated the PMP that in some cases gives many information on optimal trajectories. Beside high order maximum principle and envelopes, there are various higher order conditions that can be used to discard some extremal trajectories that are not locally optimal. These conditions come from: symplectic geometric methods, conjugate points, degree theory, etc.

Step 2. (finite dimensional reduction)

The family of trajectories and controls, on which the minimization is taken, is clearly an infinite dimensional space. Thanks to the analysis of Step 1, in some cases it is possible to narrow the class of candidate optimal trajectories to a finite dimensional family. This clearly reduces drastically the difficulty of the problem and allows a deeper analysis. More precisely, one individuates a finite number of smooth controls $u_i(x)$, such that every optimal trajectory is a finite concatenation of trajectories corresponding to the vector fields $f(x, u_i(x))$.

Step 3.

Once a finite dimensional reduction is obtained, one may construct a synthesis in the following way. Assume that on the compact set $\mathcal{R}(\tau)$, there is a bound on the number of arcs (corresponding to controls u_i) that may form, by concatenation, an optimal trajectory. Then one can construct, by induction on n, trajectories that are concatenations of n arcs and cut the not optimal ones. The latter operation produces some special sets, usually called cut loci or overlaps, reached optimally by more than one trajectory.

The above procedure is done on the base space M, however extremals admit lifts to T^*M. Thus another possibility is to construct the set of extremals in the cotangent bundle and project it on the base space. In this case, projection singularities are responsible for singularities in the synthesis.

Step 4.

Even if a finite dimensional reduction is not obtained, one can still fix a finite dimensional family of extremal trajectories and construct a synthesis on some part of the state space. If the synthesis is regular enough then there are sufficiency theorems ensuring the optimality of the synthesis.

These sufficiency theorems fit well also inside the framework of viscosity solution to the corresponding Hamilton–Jacobi–Bellman equation.

The above approach is quite powerful, but in many cases it is not known how to reach the end, namely to produce an optimal synthesis. For our model problem, we are able not only to construct an optimal synthesis under generic assumptions, but also to give a topological classification of singularities of the syntheses and of the syntheses themselves, as explained in the Introduction.

Remark **17** For general problems (not necessarily with bound on the control, e.g. with quadratic cost), Steps 1 and 4 are still used, while Step 3 is not based on a finite dimensional reduction.

Bibliographical Note

For the problem of controllability, classical results are found in the papers by Krener [88], and Lobry [93]. For the problem of controllability on Lie groups see the papers by Jurdjevic-Kupka [80, 81], Jurdjevic-Sussmann [82], Gauthier-Bornard [66], and Sachkov [114], while for controllability along a trajectory see [25].

The issue of existence for Optimal Control as well as for Calculus of Variations is a long standing research field. For a review of classical available results we refer to [50].

The terms Abnormal Extremals and Singular Extremals (in our case called Endpoint Singular Extremals) are used with several different meanings. For some results see [9, 10, 29, 30, 32, 102, 103].

The high order principle of Krener was developed in [89], while the generalization for affine control system given in Theorem 8, due to Bressan, was published in [41]. The generalization of Envelope Theory from Calculus of Variation to Optimal Control, is due to Sussmann [122, 124]. Applications of Envelope Theory can be found in [33, 90, 115, 116]. There are several other high order conditions based on different techniques: symplectic geometric methods [4, 9, 12, 13], conjugate points [49, 100], Generalized Index Theory [9] (for an application see [11]).

Geometric tools and high order conditions have been applied to various optimal control problems, for example: subriemannian geometry [1, 5, 6, 7, 24, 30, 51, 67, 76, 102, 103], analytic systems [127], systems on Lie groups [14, 15, 28, 31, 79, 131], mechanical systems [26, 77, 85], robotic applications [21, 118], etc. (The list of applications would be too long so we limit ourselves to these few examples.)

Regarding finite dimensional reductions, there is a well known single–input example with $|u| \leq 1$, due to Fuller [64], for which optimal trajectories are not concatenations of a finite number of arcs and change the control from $+1$ to -1 an infinite number of times reaching the target in finite time. This behavior is known as Fuller phenomenon. However, also for the system considered in [64] there is an optimal synthesis. Results about genericity of the Fuller phenomenon can be found in [91, 96, 132].

Fuller phenomenon is also encountered for Dubins' car with control on the steering acceleration (not on the steering velocity), see [86, 87, 123]. The construction of an optimal synthesis, for this system, is a difficult open problem.

Extremal Syntheses were considered, for example, in [38, 43, 90, 109, 118, 127]. For analytic systems, analytic stratification provides a powerful tool [119, 127], that faced some limitation, see [94].

The first sufficiency theorem, for extremal synthesis, was given by Boltyanskii in [27]. Then various generalization appeared, [45, 46, 111]. In particular the result of [111], differently from the previous ones, can be applied to systems presenting Fuller phenomena.

In **Step 3** of Section 1.4, an alternative method for the construction of the optimal synthesis is indicated. Namely, to construct all extremals in the cotangent bundle. This second method is more involved, but induces a more clear understanding of the fact that projection singularities are responsible for singularities in the optimal synthesis. This approach was used in [38, 39, 83, 84].

For sufficiency theorems, in the framework of viscosity solutions to Hamilton-Jacobi–Bellman equation (mentioned in **Step 4** of Section 1.4), see [23, 65].

Exercises

Exercise 1 Consider the control system on the plane:

$$\begin{pmatrix} \dot{x} \\ \dot{y} \end{pmatrix} = \begin{pmatrix} \sin(x) \\ 0 \end{pmatrix} + u \begin{pmatrix} G_1(x,y) \\ G_2(x,y) \end{pmatrix}, \quad u \in \mathbb{R}, \tag{1.11}$$

where $G_1, G_2 \in C^1(\mathbb{R}^2, \mathbb{R})$ and $G_1(x,y), G_2(x,y) > 0$ for every $(x,y) \in \mathbb{R}^2$. Find the set of points where Theorem 2, p. 18 permits to conclude that there is local controllability. Is it true that in all other points the system is not locally controllable?

Exercise 2 Consider the control system $\dot{x} = F(x) + uG(x)$, $u \in \mathbb{R}$ where:

$$x = \begin{pmatrix} x_1 \\ x_2 \\ x_3 \end{pmatrix}, \quad x_1^2 + x_2^2 + x_3^2 = 1,$$

$$F(x) = \begin{pmatrix} x_2 \cos(\alpha) \\ -x_1 \cos(\alpha) \\ 0 \end{pmatrix}, \quad G(x) = \begin{pmatrix} 0 \\ x_3 \sin(\alpha) \\ -x_2 \sin(\alpha) \end{pmatrix}, \quad \alpha \in]0, \pi/2[.$$

Find where the system is locally controllable and prove that it is (globally) controllable.
[Hint: to prove global controllability, find a trajectory connecting each couple of points.]

Exercise 3 Prove that, if the controls take values in the whole $\mathbb{R}^m$, then there exists no minimum for the minimum time problem of system (1.8). Moreover, show that, for a distributional problem with quadratic cost (the so called subriemannian problem), if we do not fix the final time, then there exists no minimum.

Exercise 4 Prove that if we fix the final time in the PMP, then condition (PMP2) becomes

(PMP2bis) For a.e. $t \in [0,T]$, $H(\gamma(t),\lambda(t),\lambda_0) = \mathcal{H}(\gamma(t),\lambda(t),\lambda_0,u(t)) = const \geq 0$;

and find the value of the constant for a given γ.

Exercise 5 Consider the distributional system $\dot{x} = u_1F_1(x) + u_2F_2(x)$, $x \in M$, $u_1, u_2 \in \mathbb{R}$, with quadratic cost (1.9) and fixed final time T. Prove that (λ, λ_0) can be chosen up to a multiplicative positive constant and if $\lambda_0 = -\frac{1}{2}$ and $\mathcal{H}(\gamma(t),\lambda(t),\lambda_0,u(t)) = \frac{1}{2}$ (see (PMP2bis) of the previous Exercise), then the extremal is parametrized by arclength (i.e. $u_1^2(t) + u_2^2(t) = 1$).
[Hint: Use the condition $\frac{\partial \mathcal{H}}{\partial u} = 0$ implied by the condition (PMP2bis).]

Exercise 6 Consider the control system $\dot{x} = f(x,u)$, $x \in M$ and $u \in U = \mathbb{R}^m$, and the problem (1.7) with $\mathcal{T} = \{x_1\}$, $x_1 \in M$. Prove that if γ is an endpoint singular trajectory and, for every $t \in Dom(\gamma)$, the function $u \to \langle p(t), f(\gamma(t),u)\rangle$ is strictly convex, then γ is an abnormal extremal.

Exercise 7 Consider a distributional system $\dot{x} = \sum_i u_iF_i(x)$. Prove that for the problem (1.7), with $\psi \equiv 0$, $\mathcal{T} = \{x_1\}$ and the quadratic cost $L = \sum_i u_i^2$, every endpoint singular extremal is an abnormal extremal and viceversa (what happens if $x_0 = x_1$?). Prove the same for minimum time with bounded control.

Exercise 8 Consider the problem (1.7) with $L = L(x) > 0$. Take an extremal trajectory γ and an envelope δ of γ. Assume that the equality $\dot{\gamma}_l(T_l) = f(\gamma_l(T_l), u_l(T_l))$ holds for almost every l. Prove that γ_l and δ are tangent at $\gamma_l(T_l)$ for almost every l.
[Hint: from $L(x) > 0$ we get that $\lambda(t) \neq 0$ for every t, then use the definition of envelope.]

Exercise 9 Consider the system $\dot{x} = F(x) + uG(x)$, $u \in \mathbb{R}$, $x = (x_1, x_2) \in \mathbb{R}^2$ and the optimal control problem with Lagrangian $L(x) = x_2^2$, $\psi = 0$, initial point $(0,0)$ and terminal point $(c,0)$. Assume that, for every x, we have $F_1(x) > 0$, $G_2(x) > 0$, $\Delta_A(x) := F_1(x)G_2(x) - F_2(x)G_1(x) > 0$. Prove that there exists a monotone function $\phi(t)$, such that $\gamma(t) = (\phi(t), 0)$ is the unique optimal trajectory and show that it is extremal for every covector $\lambda \equiv 0$ and $\lambda_0 < 0$.
[Hint: find a control such that the corresponding trajectory has zero cost.]

2

Time Optimal Synthesis for 2–D Systems

In this Chapter we focus on our model problem, that is on the system:

$$\dot{x} = F(x) + uG(x), \quad x \in M, \quad |u| \leq 1, \tag{2.1}$$

where M is a smooth two dimensional manifold, $F(x_0) = 0$, and consider the problem of reaching every point of M in minimum time from x_0.

The hypothesis $F(x_0) = 0$, i.e. x_0 is a stable point for F, is very natural. In fact, under generic assumptions, it guarantees local controllability. Moreover if we reverse the time, we obtain the problem of stabilizing in minimum time all the points of M to x_0. For this time-reversed problem, the stability of x_0 guarantees that once reached the origin, it is possible to stay there. The case $F(x_0) \neq 0$ is treated in Section 2.10.4, p. 122.

For simplicity, we discuss first the case $M = \mathbb{R}^2$, hence we can assume $x_0 = 0$. The necessary modification to cover the general case are treated in Section 2.9.

The PMP takes a particularly simple form for (2.1) and one can easily see that controls are always bang-bang, that is corresponding to constant control $u = +1$ or $u = -1$, unless the vector field G and the Lie bracket $[F, G]$ are parallel. Generically, this happens on a regular one dimensional submanifold and there exist extremal trajectories running this manifold with a feedback $\varphi(x)$ explicitly computed in terms of F, G and the Lie bracket $[F, G]$. These extremals are endpoint singular extremals in the sense of Section 1.2.3, p. 23, Chapter 1.

To obtain a finite dimensional reduction of the set of candidate optimal trajectories, some additional generic conditions have to be assumed (see conditions **(P1)-(P7)** of Section 2.4, p. 48), using which we can prove the existence of an optimal synthesis. Generic singularities of the synthesis are then described: the one dimensional singularities are called Frame Curves and the zero dimensional ones are called Frame Points. Frame Curves are of five types and are called, respectively, of type $Y_{1,2}, S, C, K$. The first two correspond to trajectories starting at the origin with control ± 1. These trajectories play an important role because the other optimal trajectories bifurcate from them. The third type S corresponds to endpoint singular extremals, i.e. trajectories of the feedback $\varphi(x)$. The curves of type C are formed by switching points, that are points at which optimal trajectories change the control from $+1$ to -1 or viceversa. Finally, K curves are overlaps, formed by points reached optimally by two different trajectories. In Figure 2.1 we show the optimal synthesis near Frame Curves.

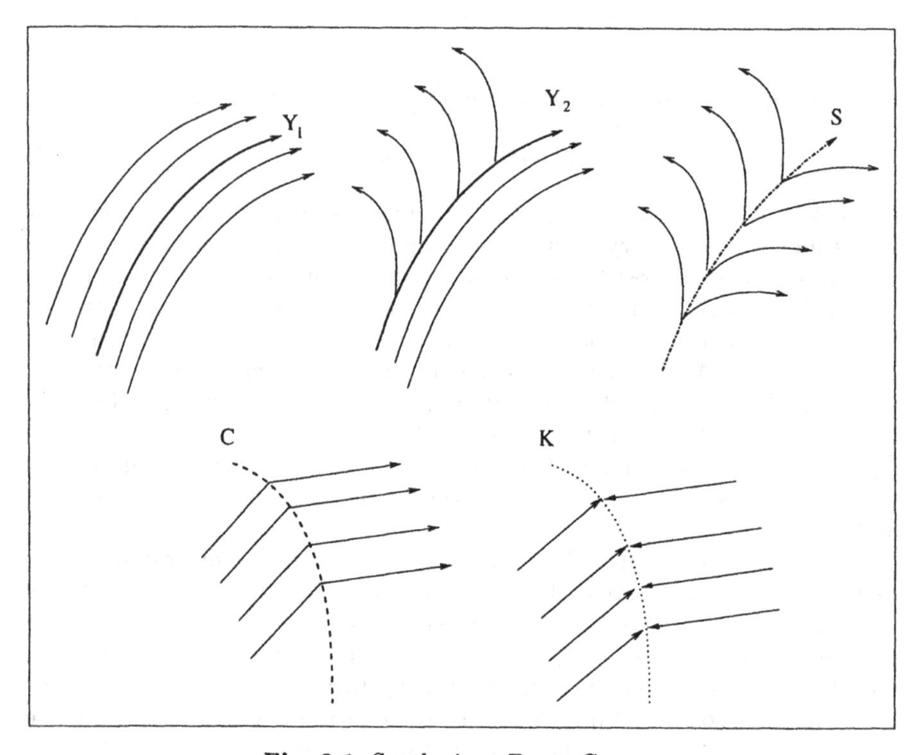

Fig. 2.1. Synthesis at Frame Curves.

Then Frame Points are described as intersections of Frame Curves. Each point is generically intersection of two curves and a detailed analysis provides 18 equivalence classes of singular points (plus 5 due to the boundary of $\mathcal{R}(\tau)$). For a picture of the synthesis near these points, see Figure 2.9, p. 61.

We then provide a detailed algorithm, with additional generic conditions, that ensures structural stability of optimal syntheses. The classification program, see the introduction, is eventually completed assigning to each optimal synthesis a topological graph with additional structure.

In Sections 2.1, 2.2, and 2.3 we introduce the basic tools of optimal synthesis theory on the plane. In Section 2.4 we prove that, under generic conditions, every trajectory has a finite number of switchings in finite time, obtaining the finite dimensional reduction of Step 2 of the geometric control approach described in Section 1.4. In Section 2.5, using a suitable algorithm (that is refined in Section 2.8) we prove the existence of an optimal synthesis under generic conditions, while in Section 2.6 we classify its singularities. In Section 2.8 we study the structural stability of syntheses and classify them by means of topological graphs. Finally in Section 2.9 we generalize our results to the case of an arbitrary two dimensional manifold, while in Section 2.10 we provide some applications and generalizations.

2.1 Introduction

The detailed analysis of the optimal synthesis requires various definitions and tools. In order to introduce them in a natural way, we first proceed giving minimal basic definitions and doing a careful description of the generic situation near the origin (we are considering the case $M = \mathbb{R}^2$ and $x_0 = 0$). This amounts to a not hard exercise, but should convince the reader of the necessity of these definitions and tools, which are better described in next sections. We hope to be excused for skipping some technical detail for sake of simplicity: details which are given with abundance in the rest of the chapter. The Pontryagin Maximum Principle (see Section 1.2.2, p. 21) in this special case states the following.

Remark **18 (Notation in the following version of the PMP)**

- $T_x^*\mathbb{R}^2 = (\mathbb{R}^2)_*$ denotes the set of row vectors.
- the duality product $< . , . >$ is now simply the matrix product and it is indicated by " $\cdot$ ".
- the Hamiltonian does not include the cost factor $\lambda_0 L$, (that in this case is just λ_0). The condition $\mathcal{H} = 0$ become then $\mathcal{H} + \lambda_0 = 0$. With this new definition of Hamiltonian, abnormal extremals are the zero levels of the $\mathcal{H}$.

Theorem 10 (Pontryagin Maximum Principle for the model problem) *Define for every* $(x, p, u) \in \mathbb{R}^2 \times (\mathbb{R}^2)_* \times [-1, 1]$*:*

$$\mathcal{H}(x, p, u) = p \cdot F(x) + u\, p \cdot G(x)$$

and:

$$H(x, p) = \max\{p \cdot F(x) + u\, p \cdot G(x) : u \in [-1, 1]\}. \tag{2.2}$$

If $\gamma : [0,a] \to \mathbb{R}^2$ *is a (time) optimal trajectory corresponding to a control* $u : [0,a] \to [-1,1]$, *then there exist a nontrivial field of covectors along* γ, *that is a function* $\lambda : [0,a] \to (\mathbb{R}^2)_*$ *never vanishing, and a constant* $\lambda_0 \le 0$ *such that for a.e.* $t \in Dom(\gamma)$*:*

i) $\dot{\lambda}(t) = -\lambda(t) \cdot (\nabla F + u(t)\nabla G)(\gamma(t))$,
ii) $\mathcal{H}(\gamma(t), \lambda(t), u(t)) + \lambda_0 = 0$,
iii) $\mathcal{H}(\gamma(t), \lambda(t), u(t)) = H(\gamma(t), \lambda(t))$.

Remark **19** Notice that, since the Lagrangian cost is constantly equal to 1, the condition $(\lambda(t), \lambda_0) \neq (0,0)$, for all $t \in [0,T]$, given in Definition 8, p. 22, now becomes $\lambda(t) \neq 0$ for all t. In fact $\lambda(t) \equiv 0$ with the condition *(PMP2)* of Definition 8 implies $\lambda_0 = 0$. To see what happens for vanishing Lagrangians cfr. Exercise 9, p. 31).

For a given pair (F, G) we want to construct a synthesis in a small neighborhood of the origin. For each extremal pair $(\gamma, \lambda) : [0,T] \to \mathbb{R}^2 \times (\mathbb{R}^2)_*$ let $\phi(t) = \lambda(t) \cdot G(\gamma(t))$ be the switching function (as in Definition 16 below). Assume that γ corresponds to a control u that is not ± 1 on some interval I of positive measure. Then, from the maximization condition of the PMP: $0 = \min_{|u|\le 1}(\lambda \cdot F + u\lambda \cdot G)$, we have $\phi = 0$ on I and thus $\dot{\phi} = 0$ on I. Now

$$\dot{\phi} = \frac{d}{dt}(\phi) = (-\lambda \cdot \nabla(F + uG)) \cdot G + \lambda \cdot (\nabla G \cdot (F + uG)) =$$

$$= \lambda \cdot (\nabla G \cdot (F + uG) - \nabla(F + uG)) \cdot G) = \lambda \cdot [F, G].$$

Hence $0 = \phi = \dot{\phi}$ results in $0 = \lambda \cdot G = \lambda \cdot [F, G]$. Since $\lambda \neq 0$ we get that G and $[F, G]$ are parallel. To avoid this situation, it is enough to assume that G and $[F, G]$ are not parallel in a neighborhood Ω of the origin. This is generic and guaranteed by the assumption:

(P1) $Det(G(0), [F,G](0)) = G_1(0)\,[F,G]_2(0) - G_2(0)\,[F,G]_1(0) \neq 0.$

Thus γ is bang-bang, i.e. the corresponding control satisfies $|u(t)| = 1$ for every t, as long as γ is in Ω. It is then natural to define the function $\Delta_B = Det(G, [F,G])$ and the condition **(P1)** reads simply $\Delta_B(0) \neq 0$.

Assume now, for example, that γ corresponds to control $+1$ on $[0, t_1]$ and then to control -1. Can it change again the control back to $+1$ remaining in Ω?

To answer to this question we need to exploit more the PMP. From the fact that γ corresponds to control $+1$ on $[0, t_1]$, we get

$$\phi(t) = \lambda(t) \cdot G(\gamma(t)) \ge 0 \qquad \forall t \in [0, t_1].$$

Consider the adjoint equation to **i)** of PMP, that is $\dot{v} = \nabla(F + uG) \cdot v$, then $\lambda(t) \cdot v(t)$ is constant and the above condition can be rewritten as:

$$\phi(t) = \lambda(0) \cdot v(G(\gamma(t)), t; 0) = \lambda(t) \cdot G(\gamma(t)) \ge 0, \tag{2.3}$$

where $v(G(\gamma(t)),t;0)$ is the solution at time 0 to $\dot{v} = \nabla(F+uG)\cdot v$ satisfying $v(t) = G(\gamma(t))$. Observe that if at t_1 we change control to -1, then it follows $\lambda(t_1)\cdot G(\gamma(t_1)) = 0 = \phi(t_1)$. In general, to check when we can change control it is enough to check if ϕ happens to be zero. Looking at (2.3) and thinking to $v(G(\gamma(t)),t;0)$ as a vector rotating in time, then the zeroes of ϕ are determined by the position of v with respect to $G(0)$. In another way, if we define θ^γ to be the angle between $G(0)$ and $v(G(\gamma(t),t;0))$, then we should check when θ^γ happens to be equal to a multiple of π (notice that $\theta^\gamma(0) = 0$ by definition). It is thus natural to give the following definition illustrated in figure 2.2.

Definition 14 *Given an extremal trajectory $\gamma : [0,T] \to \mathbb{R}^2$, let us define $v^\gamma(v_0,t_0;t)$ to be the solution to the Cauchy problem:*

$$\begin{cases} \dot{v^\gamma}(v_0,t_0;t) &= (\nabla F + u(t)\nabla G)(\gamma(t))\cdot v^\gamma(v_0,t_0;t) \\ v^\gamma(v_0,t_0;t_0) &= v_0, \end{cases} \tag{2.4}$$

where $u(\cdot)$ is the control corresponding to γ. Denote now $\bar{v}^\gamma(t) := v^\gamma(G(\gamma(t)),t;0)$ and define the function:

$$\theta^\gamma : \; Dom(\gamma) \to [-\pi,\pi], \quad \theta^\gamma(t) := arg(\bar{v}^\gamma(0),\bar{v}^\gamma(t)), \tag{2.5}$$

where arg is the angle measured counterclockwise (see figure 2.2).

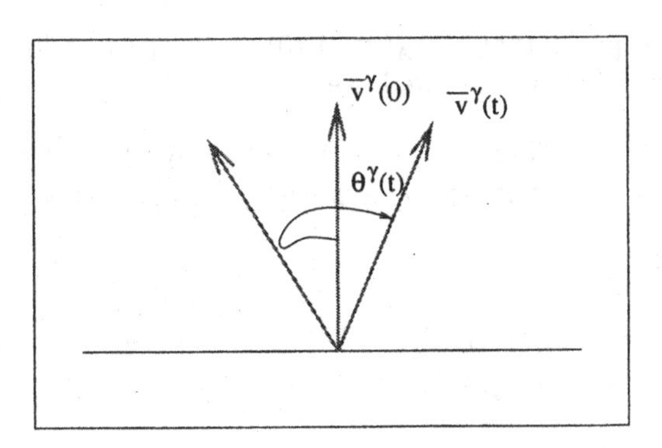

Fig. 2.2. Definition 14

Later it is proved that $sgn(\dot{\theta}^\gamma) = sgn(\Delta_B)$. Therefore if U is small enough, **(P1)** implies that θ^γ rotates always in the same direction and then t_1 is the only change of sign of u ! (To change sign again we have to rotate of another π, but this is avoided choosing U small enough.)

Finally, from the use of the PMP we obtained that all trajectories correspond to a control u that is bang-bang and changes sign only one time. The optimal synthesis near the origin is represented Figure 2.3.

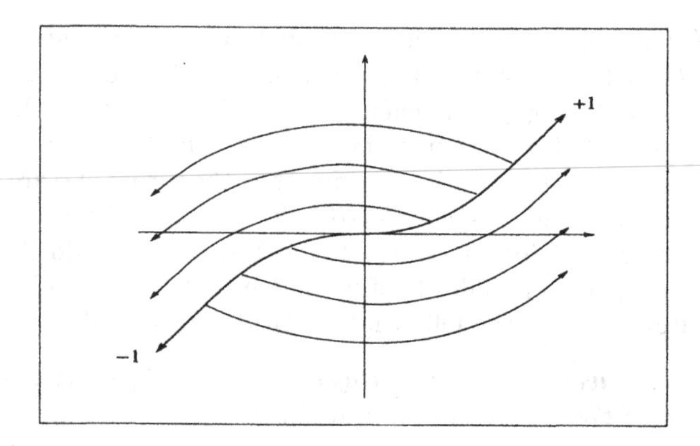

Fig. 2.3. The optimal synthesis in a neighborhood of the origin. Every point is reached starting from the origin with control ± 1 and switching at most one time.

To be more precise, the above synthesis occurs if the two vectors $X := F - G$ and $Y := F + G$ do not have the same integral trajectory through the origin. Introducing the function

$$\Delta_A := F \wedge G := Det\big(F, G\big) = F_1 G_2 - F_2 G_1,$$

this amounts to prove that $\nabla\Delta_A(0) \cdot Y(0) = \nabla\Delta_A(0) \cdot X(0) \neq 0$. Using the notation $F \wedge G := Det(F, G)$, we have:

$$\begin{aligned}(\nabla\Delta_A \cdot Y)(0) &= (\nabla(F \wedge G) \cdot Y)(0) = (\nabla(F \wedge G) \cdot G)(0) = \\ &= (\nabla F \cdot G \wedge G + F \wedge \nabla G \cdot G)(0) = (\nabla F \cdot G \wedge G)(0).\end{aligned}$$

Now

$$\begin{aligned}\Delta_B(0) &= ([F, G] \wedge G)(0) = (\nabla G \cdot F \wedge G - \nabla F \cdot G \wedge G)(0) = \\ &= (-\nabla F \cdot G \wedge G)(0).\end{aligned}$$

Hence condition **(P1)** gives the desired conclusion.

2.2 Basic Definitions

In this section we introduce the set of systems we consider, and some properties of the reachable sets.

Definition 15 ($\mathcal{C}^\infty$ vector fields) *We say that a vector field is $\mathcal{C}^\infty$ if its components admit partial derivatives of any order that are bounded on the whole plane.*

Let Ξ be the set of all couples of $\mathcal{C}^\infty$ vector fields $\Sigma = (F, G)$ such that $F(0) = 0$. From now on we endow Ξ with the <u>$\mathcal{C}^3$ topology</u>, that is the topology induced by the norm:

$$\|(F,G)\| = \sup\left\{ \left|\frac{\partial^{\alpha_1+\alpha_2}F_i(x)}{\partial x_1^{\alpha_1}\partial x_2^{\alpha_2}}\right|, \left|\frac{\partial^{\alpha_1+\alpha_2}G_i(x)}{\partial x_1^{\alpha_1}\partial x_2^{\alpha_2}}\right| : x \in \mathbb{R}^2; \right.$$
$$\left. i = 1,2;\ \alpha_1, \alpha_2 \in \mathbf{N} \cup \{0\}; \alpha_1 + \alpha_2 \le 3 \right\}. \tag{2.6}$$

We refer to $\Sigma \in \Xi$ as a control system with the meaning that we consider the corresponding control system (2.1). If $\Omega \subset \mathbb{R}^2$ is an open set then $\Sigma|_\Omega$ indicates the restriction of Σ to the set Ω.

We are interested in the <u>reachable set within time</u> $T > 0$ from the origin, that is:

$$\mathcal{R}(T) := \{x \in \mathbb{R}^2 : \text{ there exist } t \in [0,T] \text{ and a trajectory}$$
$$\gamma : [0,t] \to \mathbb{R}^2 \text{ of (2.1) such that } \gamma(0) = 0,\ \gamma(t) = x\}. \tag{2.7}$$

<u>Remark</u> **20** We restrict our analysis to vector fields with bounded derivatives on the whole plane. However, if (F,G) admits not bounded derivatives, then our analysis is still valid under the condition that the reachable set is compact.

Fix $T > 0$, the <u>minimum time function</u> $\mathbf{T}(\ .\ ) : \mathcal{R}(T) \to \mathbb{R}^+$ is by definition:

$$\mathbf{T}(x) := \min\{t \ge 0 : \text{ there exists a trajectory } \gamma(\cdot)$$
$$\text{of (2.1) s.t. } \gamma(0) = 0,\quad \gamma(t) = x\}. \tag{2.8}$$

The convexity of the set $\{F(x) + uG(x) : |u| \le 1\}$ and the bound on the derivatives of F and G imply the following (cfr. Theorem 3 of Chapter 1):

Lemma 2 *Let $\Sigma \in \Xi$ and fix $T > 0$, then the corresponding reachable set $\mathcal{R}(T)$ is compact.*

Moreover, we have:

Lemma 3 *If $\Sigma \in \Xi$ and $G, [F,G]$ are linearly independent at the origin, that is if* **(P1)** *holds, then Σ is locally controllable (see Definition 3 of Chapter 1) and the minimum time function* $\mathbf{T}$ *is continuous.*

Proof. Using the notation of Theorem 2, p. 18, $B = G(0)$ and $A = \nabla F(0)$ thus $rank[B, AB] = rank[G(0), \nabla F(0) \cdot G(0)] = rank[G(0), [F,G](0)]$. So the first statement is proved.

Now, for every sequence $x_n \to x$, we want to prove $\mathbf{T}(x_n) \to \mathbf{T}(x)$. By Lemma 1, p. 19, fixed $\varepsilon > 0$ there exists $\delta > 0$ such that $B(x,\delta) \subset \mathcal{R}(\mathbf{T}(x)+\varepsilon)$. Hence, for n sufficiently big, we get $\mathbf{T}(x_n) \le \mathbf{T}(x) + \varepsilon$. Passing to the limit: $\limsup \mathbf{T}(x_n) \le \mathbf{T}(x)$.
Assume, now, by contradiction, that $\liminf \mathbf{T}(x_n) < T < \mathbf{T}(x)$. Then, up to a subsequence, $x_n \in \mathcal{R}(T)$ and, by Lemma 2, $x \in \mathcal{R}(T)$, a contradiction. Thus we conclude. ∎

2.3 Special Curves

The aim of this section is to show the properties of optimal trajectories, following the geometric control approach illustrated in the introduction, more precisely Step 1. The use of PMP and the function θ^γ , introduced above, leads to a good characterization of the structure of optimal trajectories. This analysis permits to prove a finite dimensional reduction in next section.

Definition 16 (Switching Function) *Let $(\gamma, \lambda) : [0, \tau] \to \mathbb{R}^2 \times (\mathbb{R}^2)_*$ be an extremal pair. The corresponding switching function is defined as $\phi(t) := \lambda(t) \cdot G(\gamma(t))$. Notice that $\phi(\cdot)$ is absolutely continuous.*

From the PMP it immediately follows:

Lemma 4 *Let $(\gamma, \lambda) : [0, \tau] \to \mathbb{R}^2 \times (\mathbb{R}^2)_*$ be an extremal pair and $\phi(\cdot)$ the corresponding switching function. If $\phi(t) \neq 0$ for some $t \in]0, \tau[$, then there exists $\varepsilon > 0$ such that γ corresponds to a constant control $u = sgn(\phi)$ on $]t - \varepsilon, t + \varepsilon[$.*

Proof. There exists $\varepsilon > 0$ such that ϕ does not vanish on $]t - \varepsilon, t + \varepsilon[$. Then from condition **iii)** of PMP we get $u = \text{sgn}(\phi)$. ■

It is then natural to give the following definition:

Definition 17 (Regular Time) *Let $(\gamma, \lambda) : [0, \tau] \to \mathbb{R}^2 \times (\mathbb{R}^2)_*$ be an extremal pair and $\phi(\cdot)$ the corresponding switching function. If $\phi(t) \neq 0$ for some $t \in]0, \tau[$, we say that t is a regular time for γ.*

Reasoning as in Lemma 4 one immediately has:

Lemma 5 *Assume that ϕ has a zero at t, $\dot{\phi}(t)$ is strictly greater than zero (resp. smaller than zero) then there exists $\varepsilon > 0$ such that γ corresponds to constant control $u = +1$ on $]t - \varepsilon, t[$ and to constant control $u = -1$ on $]t, t + \varepsilon[$ (resp. to constant control $u = -1$ on $]t - \varepsilon, t[$ and to constant control $u = +1$ on $]t, t + \varepsilon[$).*

We are then interested in differentiating ϕ:

Lemma 6 *Let $(\gamma, \lambda) : [0, \tau] \to \mathbb{R}^2 \times (\mathbb{R}^2)_*$ be an extremal pair and ϕ the corresponding switching function. Then $\phi(\cdot)$ is continuously differentiable and it holds:*

$$\dot{\phi}(t) = \lambda(t) \cdot [F, G](\gamma(t)).$$

Proof. Using the PMP we have for a.e. t:

$$\begin{aligned}
\dot{\phi}(t) &= \frac{d}{dt}(\lambda(t) \cdot G(\gamma(t))) = \dot{\lambda}(t) \cdot G(\gamma(t)) + \lambda \cdot \dot{G}(\gamma(t)) \\
&= -\lambda(t)(\nabla F + u(t)\nabla G)(\gamma(t)) \cdot G(\gamma(t)) + \lambda \cdot \nabla G(\gamma(t))(F + u(t)G)(\gamma(t)) \\
&= \lambda(t) \cdot [F, G](\gamma(t)).
\end{aligned}$$

Since $\phi(\cdot)$ is absolutely continuous and $\lambda(t) \cdot [F,G](\gamma(t))$ is continuous, we deduce that ϕ is C^1. ■

Notice that if $\phi(\cdot)$ has no zeros then u is a.e. constantly equal to $+1$ or -1. Hence we are interested in determining when the control may change sign or may assume values in $]-1,+1[$. For this purpose we give the following definition:

Definition 18 • *If $u_1 : [a,b] \to [-1,1]$ and $u_2 : [b,c] \to [-1,1]$ are controls, their concatenation $u_2 * u_1$ is the control:*

$$(u_2 * u_1)(t) := \begin{cases} u_1(t) \text{ for } t \in [a,b] \\ u_2(t) \text{ for } t \in]b,c]. \end{cases}$$

If $\gamma_1 : [a,b] \to \mathbb{R}^2$ and $\gamma_2 : [b,c] \to \mathbb{R}^2$ are trajectories of $\Sigma \in \Xi$, corresponding respectively to u_1 and u_2, such that $\gamma_1(b) = \gamma_2(b)$, then the concatenation $\gamma_2 * \gamma_1$ is the trajectory:

$$(\gamma_2 * \gamma_1)(t) := \begin{cases} \gamma_1(t) \text{ for } t \in [a,b] \\ \gamma_2(t) \text{ for } t \in [b,c]. \end{cases}$$

- *Define the vector fields:*

$$X := F - G, \quad Y := F + G. \tag{2.9}$$

  We say that γ is a X-trajectory on the interval $[a,b]$ if on $[a,b]$ it corresponds to constant control -1. Similarly we define Y-trajectories.
- *If a trajectory γ is a concatenation of an X-trajectory and a Y-trajectory, then we say that γ is a $Y * X$-trajectory (The X-trajectory comes first). Similarly we define trajectories of kind $X * Y$, $X * Y * X$ and so on. A bang trajectory is a trajectory corresponding to constant control +1 or to constant control -1. A bang–bang trajectory is a trajectory obtained as a finite concatenation of X- and Y-trajectories. The times at which the control changes sign are called switching times. Switching times are particular cases of non regular times.*
- *An extremal trajectory γ defined on $[c,d]$ is said to be singular or a Z-trajectory if the switching function ϕ vanishes on $[c,d]$.*

Remark **21** A singular extremal in the sense above is also an endpoint singular extremal in the sense of Section 1.2.3, p. 23. In fact for these trajectories the Hamiltonian is independent from the control. In the following we use the term singular trajectories with the same meaning of endpoint singular extremal.

Remark **22** On any interval where ϕ has no zeroes (respectively finitely many zeroes) the corresponding control is bang (respectively bang-bang).

Singular trajectories are studied in details below. Notice that if an extremal trajectory γ is a singular trajectory on $[c,d]$, then every $t \in [c,d]$ is not a regular time. In Figure 2.4 we give an example to illustrate the relationship between an extremal trajectory and the corresponding switching function. The control φ, corresponding to the singular arc, is computed below, see Lemma 10, p. 46.

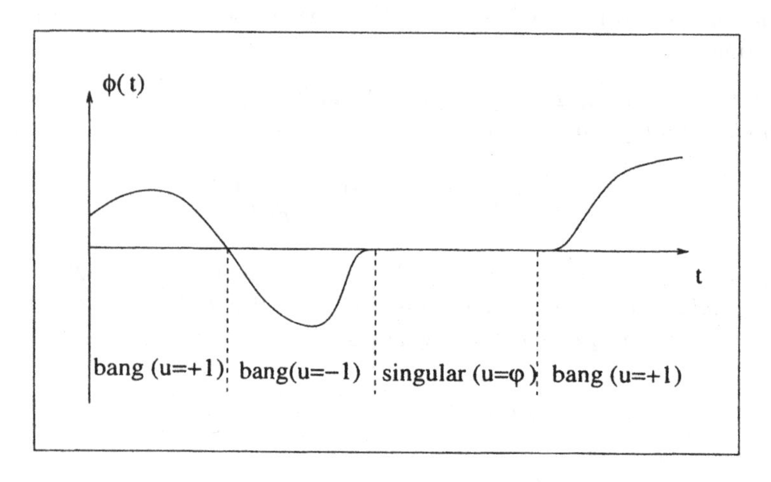

Fig. 2.4. The switching function

Recall Definition 14, p. 37. We have the following:

Lemma 7 *For fixed* t_0, t, *the map* $f_{t_0,t} : v_0 \mapsto v^\gamma(v_0, t_0; t)$ *is linear and injective. If* $\gamma : [a, b] \to \mathbb{R}^2$ *is an extremal trajectory of (2.1), corresponding to a constant control* $\bar{u}$, *then for every* $t, t_0 \in [a, b]$ *it holds:* $v^\gamma((F + \bar{u}G)(\gamma(t_0)), t_0; t) = (F + \bar{u}G)(\gamma(t))$. *Moreover, if* (γ, λ) *is an extremal pair then* $\lambda(t) \cdot v^\gamma(v_0, t_0; t)$ *is constant.*

Proof. The first assertion follows from the fact that v^γ is the solution to a linear ODE. The second claim is consequence of:

$$\frac{d}{dt}(F + \bar{u}G)(\gamma(t)) = \nabla(F + \bar{u}G)(\gamma(t)) \cdot (F + \bar{u}G)(\gamma(t)),$$

ensuring that $F + \bar{u}G$ solves the same equation as v^γ. Finally for a.e. t:

$$\frac{d}{dt}(\lambda(t) \cdot v^\gamma(v_0, t_0; t)) =$$

$$-\lambda(t) \cdot \nabla(F + \bar{u}G)(\gamma(t)) \cdot v^\gamma(v_0, t_0; t) + \lambda(t) \cdot \nabla(F + \bar{u}G)(\gamma(t)) \cdot v^\gamma(v_0, t_0; t) = 0,$$

moreover λ, v^γ are absolutely continuous, hence the last claim holds true. ∎

In the following a key role is played by the functions:

Definition 19 *For each $x \in \mathbb{R}^2$, define:*

$$\Delta_A(x) = Det\big(F(x), G(x)\big) = F_1(x)G_2(x) - F_2(x)G_1(x),$$
$$\Delta_B(x) = Det\big(G(x), [F,G](x)\big) = G_1(x)[F,G]_2(x) - G_2(x)[F,G]_1(x).$$

The first function is useful for studying abnormal extremals while the second for detecting singular trajectories. Next Lemma shows the relation between θ^γ, Δ_B and the switching times of an extremal trajectory.

Lemma 8 *Let $\gamma : [0,T] \to \mathbb{R}^2$ be an extremal trajectory, $t \in Dom(\gamma)$ such that $G(\gamma(t))$ is different from zero, and consider the corresponding function θ^γ of Definition 14, p. 37. Then θ^γ is C^1 in a neighborhood of t and:*

$$sgn(\dot{\theta}^\gamma(t)) = sgn(\Delta_B(\gamma(t))).$$

Moreover if $t_1, t_2 \in Dom(\gamma)$ are such that $G(\gamma(t_i)) \neq 0$, $\frac{d}{dt}\theta^\gamma(t_i) \neq 0$, t_1 switching time, then t_2 is a switching time if and only if $\theta^\gamma(t_1) = \theta^\gamma(t_2) + \alpha\pi$ where $\alpha \in \{0, \pm 1\}$.

Proof. Using the notation $F \wedge G := Det(F,G)$, one has:

$$\frac{d}{dt}\theta^\gamma(t) = \frac{\bar{v}^\gamma(t) \wedge \dot{\bar{v}}^\gamma(t)}{\|\bar{v}^\gamma(t)\|^2}. \tag{2.10}$$

Let $M(t)$ be the fundamental matrix solution to the system (2.4) for $t_0 = 0$. Then for almost every t at which $G(\gamma(t))$ does not vanish, one has

$$\begin{aligned}\frac{d}{dt}v^\gamma(G(\gamma(t)),\ t;\ 0) &= \lim_{\varepsilon\to 0}\frac{v^\gamma(G(\gamma(t+\varepsilon)),\ t+\varepsilon;\ 0) - v^\gamma(G(\gamma(t)), t;\ 0)}{\varepsilon}\\ &= M(t)\cdot\lim_{\varepsilon\to 0}\frac{v^\gamma(G(\gamma(t+\varepsilon),\ t+\varepsilon;\ t) - G(\gamma(t))}{\varepsilon}\\ &= M(t)\cdot\big[F + u(t)G,\ G\big](\gamma(t)) = M(t)\cdot[F,G](\gamma(t)).\end{aligned}$$

Since $M(t)$ preserves orientation, by (2.10) we have

$$\begin{aligned}sgn(\dot{\theta}^\gamma(t)) &= sgn\Big(M(t)G(\gamma(t))\ \wedge\ M(t)[F,G](\gamma(t))\Big)\\ &= sgn\Big(G(\gamma(t))\ \wedge\ [F,G](\gamma(t))\Big) = sgn(\Delta_B(\gamma(t))),\end{aligned} \tag{2.11}$$

proving the first statement.

Now t_2 is a switching time if and only if $\lambda(\gamma(t_2))\cdot G(\gamma(t_2)) = 0$. By Lemma 7, p. 42, this happens only when $v^\gamma(G(\gamma(t_1), t_1; 0)$ and $v^\gamma(G(\gamma(t_2), t_2; 0)$ are parallel. By definition of θ^γ we conclude. ∎

Definition 20 *A point $x \in \mathbb{R}^2$ is called an* <u>ordinary point</u> *if $x \notin \Delta_A^{-1}(0) \cup \Delta_B^{-1}(0)$. If x is an ordinary point, then $F(x), G(x)$ form a basis of $\mathbb{R}^2$ and we define the scalar functions f, g to be the coefficients of the linear combination: $[F,G](x) = f(x)F(x) + g(x)G(x)$.*

The relation between f, Δ_A and Δ_B is given by:

Lemma 9 *Let x an ordinary point then:*

$$f(x) = -\frac{\Delta_B(x)}{\Delta_A(x)}. \tag{2.12}$$

Proof. We have:

$$\begin{aligned}\Delta_B(x) &= G(x) \wedge [F,G](x) = G(x) \wedge (f(x)F(x) + g(x)G(x)) \\ &= f(x)\, G(x) \wedge F(x) = -f(x)\Delta_A(x).\end{aligned}$$

■

On a set of ordinary points the structure of optimal trajectories is particularly simple:

Theorem 11 *Let $\Omega \in \mathbb{R}^2$ be an open set such that every $x \in \Omega$ is an ordinary point. Then all extremal trajectories γ of $\Sigma|_\Omega$ are bang-bang with at most one switching. Moreover if $f > 0$ throughout Ω then γ is a X, Y or $Y * X$–trajectory. If $f < 0$ throughout Ω then γ is an X, Y or $X * Y$–trajectory.*

Proof. Assume $f > 0$ in Ω, the opposite case being similar. Let (γ, λ) be an extremal pair such that γ is contained in Ω. Let $\bar{t}$ be such that $\lambda(\bar{t}) \cdot G(\gamma(\bar{t})) = 0$, then:

$$\dot{\phi}(\bar{t}) = \lambda(\bar{t}) \cdot [F,G](\gamma(\bar{t})) = \lambda(\bar{t}) \cdot (fF + gG)(\gamma(\bar{t})) = f(\gamma(\bar{t}))\ \lambda(\bar{t}) \cdot F(\gamma(\bar{t})).$$

Now from PMP, we have $H(\gamma(\bar{t}), \lambda(\bar{t})) = \lambda(\bar{t}) \cdot F(\gamma(\bar{t})) \geq 0$. Hence $\dot{\phi} > 0$, since $F(\gamma(\bar{t}))$, and $G(\gamma(\bar{t}))$ are independent. This proves that ϕ has at most one zero with positive derivative at the switching time and gives the desired conclusion. ■

We are now interested in understanding what happens at points that are not ordinary.

Definition 21 *A point $x \in \mathbb{R}^2$ is called an* <u>non ordinary point</u> *if $x \in \Delta_A^{-1}(0) \cup \Delta_B^{-1}(0)$.*

The following proposition describes some basic properties of abnormal extremals in relation to the set $\Delta_A^{-1}(0)$.

Proposition 1 *Let $\gamma : [0, \tau] \to \mathbb{R}^2$ be an extremal trajectory for the control problem (2.1), $\lambda : [0, \tau] \to (\mathbb{R}^2)_*$ the corresponding covector and $t_0 \in]0, \tau[$ a non regular time with $G(\gamma(t_0)) \neq 0$. Then the following conditions are equivalent:*

(a) γ is an abnormal extremal;
(b) $\gamma(t_0) \in \Delta_A^{-1}(0)$;
(c) $\gamma(t) \in \Delta_A^{-1}(0)$ for every non regular time $t \in]0, \tau[$.

Proof. Proof that **(a)** implies **(c)**. For each non regular time $t \in]0, \tau[$ we have $\lambda(t) \cdot G(\gamma(t)) = 0$ and there exists a sequence $t'_m \nearrow t$ such that

$$\mathcal{H}(\gamma(t'_m), \lambda(t'_m), u(t'_m)) = \lambda(t'_m) \cdot (F + u(t'_m)G)(\gamma(t'_m)) = 0.$$

Hence $\lambda(t'_m) \cdot F(\gamma(t'_m)) \to 0$ and $\lambda(t) \cdot F(\gamma(t)) = 0$. If $G(\gamma(t)) \neq 0$, then we can conclude that $F(\gamma(t))$ and $G(\gamma(t))$ are parallel (possibly $F(\gamma(t)) = 0$), $\lambda(t)$ being not equal to 0. It follows **(c)**
Proof that **(b)** implies **(a)**. Assume now that $\Delta_A(\gamma(t_0)) = 0$. We have $\lambda(t_0) \cdot G(\gamma(t_0)) = 0$ and from $\Delta_A(\gamma(t_0)) = 0$ we get $\lambda(t_0) \cdot F(\gamma(t_0)) = 0$. If λ_0 is the constant of the PMP, there exists a sequence $t'_m \nearrow t_0$ such that:

$$-\lambda_0 = \mathcal{H}(\gamma(t'_m), \lambda(t'_m), u(t'_m)) = \lambda(t'_m) \cdot (F + u(t'_m)G)(\gamma(t'_m)) \to 0.$$

Hence we can conclude that $\mathcal{H}(\gamma(t), \lambda(t), u(t)) = 0$ almost everywhere.
(c)⇒(b) is obvious. This concludes the proof. ■

In the following we study some properties of singular trajectories in relation to non ordinary points on which $\Delta_B = 0$.

Definition 22 • *An non ordinary arc is a C^2 one-dimensional connected embedded submanifold S of $\mathbb{R}^2$ with the property that every $x \in S$ is a non ordinary point.*

• *A non ordinary arc is said isolated if there exists a set Ω satisfying the following conditions:*

(C1) Ω is an open connected subset of $\mathbb{R}^2$.
(C2) S is a relatively closed subset of Ω.
(C3) If $x \in \Omega \setminus S$ then x is an ordinary point.
(C4) The set $\Omega \setminus S$ has exactly two connected components.

• *A turnpike (resp. anti-turnpike) is an isolated non ordinary arc that satisfies the following conditions:*

(S1) For every $x \in S$ the vectors $X(x)$ and $Y(x)$ are not tangent to S and point to opposite sides of S.
(S2) For every $x \in S$ one has $\Delta_B(x) = 0$ and $\Delta_A(x) \neq 0$.

(S3) Let Ω be an open set which satisfies **(C1)–(C4)** *above and $\Delta_A \neq 0$ on Ω. If Ω_X and Ω_Y are the connected components of $\Omega \setminus S$ labeled in such a way $X(x)$ points into Ω_X and $Y(x)$ points into Ω_Y, then the function f satisfies*

$$f(x) > 0 \text{ (resp. } f(x) < 0) \text{ on } \Omega_Y$$
$$f(x) < 0 \text{ (resp. } f(x) > 0) \text{ on } \Omega_X$$

The following Lemmas describes the relation between turnpikes, anti-turnpikes and singular trajectories. In Lemma 10 we compute the control corresponding to a trajectory whose support is a turnpike or an anti-turnpike. In Lemma 11 we prove that if this control is admissible (that is the turnpike or the anti-turnpike is regular, see Definition 23) then the corresponding trajectory is extremal and singular. In Lemma 12 we show that anti-turnpike are locally not optimal.

Lemma 10 *Let S be a turnpike or an anti-turnpike and $\gamma : [c, d] \to \mathbb{R}^2$ a trajectory of $\Sigma \in \Xi$ such that $\gamma(c) = x_0 \in S$. Then $\gamma(t) \in S$ for every $t \in [c, d]$ iff γ corresponds to the feedback control (called singular):*

$$\varphi(x) = -\frac{\nabla\Delta_B(x) \cdot F(x)}{\nabla\Delta_B(x) \cdot G(x)}, \tag{2.13}$$

Proof. Assume that $\gamma([c, d])) \subset S$ and let u be the corresponding control, that is $\dot{\gamma}(t) = F(\gamma(t)) + u(t)G(\gamma(t))$, for almost every t. From $\Delta_B(\gamma(t)) = 0$, for a.e. t we have:

$$0 = \frac{d}{dt}\Delta_B(\gamma(t)) = \nabla\Delta_B \cdot (F(\gamma(t)) + u(t)G(\gamma(t))).$$

This means that at the point $x = \gamma(t)$ we have to use control $\varphi(x)$ given by (2.13). ■

Definition 23 (regular turnpike or anti-turnpike) *We say that a turnpike or anti-turnpike S is regular if $|\varphi(x)| < 1$ for every $x \in S$.*

Lemma 11 *Let $(\gamma, \lambda) : [0, \bar{t}] \to \mathbb{R}^2$ be an extremal pair that verifies $\gamma(\bar{t}) = x$, $x \in S$ where S is a turnpike or an anti-turnpike, and $\lambda(\bar{t}) \cdot G(\gamma(\bar{t})) = 0$. Moreover let $\gamma' : [0, t'] \to \mathbb{R}^2$ $(t' > \bar{t})$ be a trajectory such that:*

- $\gamma'|_{[0,\bar{t}]} = \gamma$,
- $\gamma'([\bar{t}, t']) \subset S$.

Then γ' is extremal. Moreover if ϕ' is the switching function corresponding to γ' then $\phi'|_{[\bar{t},t']} \equiv 0$.

Proof. Define λ' to be the covector along γ' that coincide with λ on $[0,\bar{t}]$ and ϕ' the corresponding switching function. Define $\tilde{\gamma}(t) = \gamma'(t-\bar{t})$ for every $t \in [\bar{t}, t']$ and $\tilde{\theta} = \theta^{\tilde{\gamma}}$ (see Definition 14). From Lemma 8 we get that $\frac{d}{dt}\tilde{\theta}(t) = 0$ for every $t \in [\bar{t}, t']$. Since $\tilde{\theta}(0) = 0$, we conclude $\tilde{\theta} \equiv 0$. It follows, from Lemma 7, that for every $t \in [\bar{t}, t']$:

$$\phi'(t) = \lambda'(t) \cdot G(\tilde{\gamma}(t-\bar{t})) = \lambda'(\bar{t}) \cdot v^{\tilde{\gamma}}(G(\tilde{\gamma}(t-\bar{t})), t-\bar{t}; 0) = 0.$$

Hence γ' is extremal and $\phi' \equiv 0$ on $[\bar{t}, t']$. ∎

Lemma 12 *Let S be an anti-turnpike and $\gamma : [c,d] \to \mathbb{R}^2$ be an extremal trajectory such that $\gamma([c,d]) \subset S$. Then γ is not optimal.*

This Lemma can be proved using the generalized Legendre–Clebsch condition given in Theorem 7, see Exercise 14. Here we give a nice alternative proof, see [126], that uses Stokes' theorem.

Proof. Choose an open set Ω containing $\gamma([c,d])$ such that $\Delta_A \neq 0$ on Ω and define the differential form ω on Ω by $\omega(F) = 1$, $\omega(G) = 0$. Let $\gamma_1 : [c, d_1] \to \Omega$ be any trajectory such that $\gamma_1(c) = \gamma(c)$, $\gamma_1(d_1) = \gamma(d)$, $\gamma_1(t) \notin S$ for every $t \in]c, d_1[$. Notice that $d - c = \int_\gamma \omega$, $d_1 - c = \int_{\gamma_1} \omega$. Hence $d - d_1 = \int_{\gamma_1^{-1} * \gamma} \omega$ where γ_1^{-1} is γ_1 run backward. Assume that $\gamma_1^{-1} * \gamma$ is oriented counterclockwise, being similar the opposite case. Then by Stokes' Theorem, $d - d_1 = \int_A d\omega$ where A is the region enclosed by $\gamma_1^{-1} * \gamma$. Now $d\omega(F,G) = F \cdot \nabla\omega(G) - G \cdot \nabla\omega(F) - \omega([F,G]) = -f$, by definition of ω. Since $d\omega(F,G) = \Delta_A d\omega(\partial_x, \partial_y)$, we get

$$d - d_1 = \int_A \left(-\frac{f}{\Delta_A} \right) dx\, dy.$$

If Δ_A is positive (resp. negative) then Y (resp. X) points to the side of S where γ_1 is contained and, by definition of anti-turnpike, f is negative (resp. positive) on A. We conclude that $d > d_1$ and γ is non optimal. ∎

One finally gets:

Theorem 12 *Let $\gamma : [0,\bar{t}] \to \mathbb{R}^2$ be an optimal trajectory that it is singular on some interval $[c,d] \subset [0,\bar{t}]$. Then, under generic conditions, $Supp(\gamma|_{[c,d]})$ is contained in a regular turnpike S.*

Proof. From $\phi \equiv 0$ on $[c,d]$ it follows $\dot{\phi} \equiv 0$ on $[c,d]$. By Lemma 6, $Supp(\gamma|_{[c,d]}) \subset \Delta_B^{-1}(0)$. Under generic conditions, $\Delta_B^{-1}(0) \cap \mathcal{R}(\bar{t})$ is formed by a finite number of turnpikes, anti-turnpikes and isolated points (at intersections with $\Delta_A^{-1}(0)$). By Lemma 12 we conclude. ∎

2.4 Bound on the Number of Arcs

The aim of this section is to prove, given $\tau > 0$, the existence of generic conditions on F, G ensuring that every time optimal trajectory in $R(\tau)$ is a finite concatenation of $X-$, $Y-$ and $Z-$trajectories; more precisely for each Σ in a generic subset of Ξ there exists $N(\Sigma)$ that bounds the number of these trajectories.

First we focus on the concept of genericity.

Definition 24 (Generic) *A subset of Ξ is said to be generic if it contains an open and dense subset of Ξ. A condition for $\Sigma = (F, G) \in \Xi$ is a logic proposition involving the components of the vector fields (F, G), their derivatives or set and functions that can be defined using them. Given a condition P for $\Sigma \in \Xi$ we write $P(\Sigma)$ if the system satisfies the condition P. A condition P is said to be generic if $\{\Sigma \in \Xi : P(\Sigma)\}$ is generic. If $P_1, \ldots, P_n$ are generic conditions then it is easy to verify that $\{\Sigma \in \Xi : P_1(\Sigma), \ldots, P_n(\Sigma)\}$ is generic.*

We now give a finite number of generic conditions $P_1, \ldots, P_n$ that ensure the genericity of the set for which there exist an optimal synthesis. These condition are essential to prove that every optimal trajectory is a finite concatenation of bang and singular arcs.

(P1) The vectors $G(0)$ and $[F, G](0)$ are linearly independent, i.e. $\Delta_B(0) \neq 0$.
(P2) Zero is a regular value for Δ_A and Δ_B i.e. $\Delta_A(x) = 0$ implies $\nabla \Delta_A(x) \neq 0$ and similarly for Δ_B.
(P3) The set $\Delta_A^{-1}(0) \cap \Delta_B^{-1}(0)$ is locally finite.

Let Tan_A be the set of points $x \in \Delta_A^{-1}(0)$ such that $X(x)$ or $Y(x)$ is tangent to $\Delta_A^{-1}(0)$. Define Tan_B in the same way using Δ_B rather than Δ_A.

(P4) Tan_A and Tan_B are locally finite sets.

Let $Bad := (\Delta_A^{-1}(0) \cap \Delta_B^{-1}(0)) \cup Tan_A \cup Tan_B$.

(P5) Bad is locally finite.

Notice that **(P5)** is a consequence of **(P3)** and **(P4)**.

(P6) If $x \in Bad$, $G(x) = 0$ then $F(x) \cdot \nabla(\Delta_A)(x) \neq 0$.
(P7) If $x \in Bad$, $G(x) \neq 0, x \in (\Delta_A^{-1}(0) \cap \Delta_B^{-1}(0)) \cap Tan_A$, then $x \notin Tan_B$, $\partial_y(X \cdot \nabla \Delta_A)|_{y=x} \neq 0$, $X(x) \neq 0$, $Y(x) \neq 0$.

Remark **23** Notice that **(P6)** follows from **(P3)** if $\Delta_A^{-1}(0)$ and $\Delta_B^{-1}(0)$ are transversal, see Lemma 13 C.

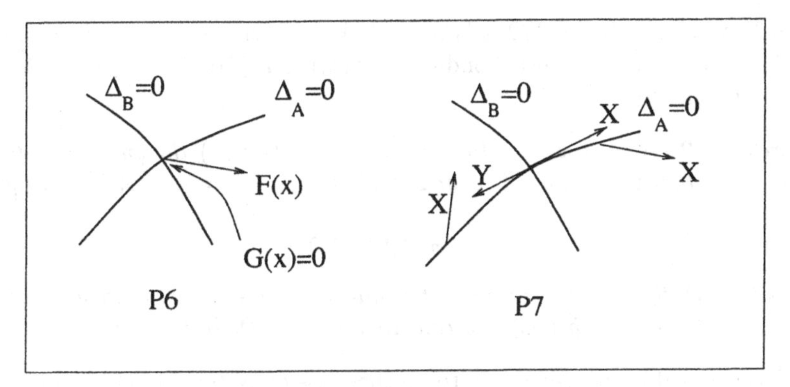

Fig. 2.5. Generic Conditions.

The generic conditions **(P6)**, **(P7)** are showed in Figure 2.5.

From Lemma 5, Proposition 1 we get that, under generic conditions, abnormal extremals are bang-bang trajectories with switchings happening on the set of zeroes of the function Δ_A (that is where the two vector fields F and G are collinear):

Proposition 2 *Let $\gamma : [0, \tau] \to \mathbb{R}^2$ be an optimal abnormal extremal for the control problem (2.1), then the generic conditions* **(P3)**, **(P6)**, **(P7)**, *imply the following:*

- *γ is bang–bang,*
- *if $t_0 \in]0, \tau[$ is a switching time then $\gamma(t_0) \in \Delta_A^{-1}(0)$.*

Proof. Assume by contradiction that γ is not bang–bang. Then there exists a set $I \subset [0, \tau]$ with an accumulation point such that the switching function ϕ vanishes on I. Then reasoning as in Proposition 1, $\lambda \cdot F(\gamma) \equiv 0$ on I and $\Delta_A \equiv 0$ on I. Let $\bar{t}$ be an accumulation point of I so that there exists a monotone sequence $t_i \in I$ with $t_i \to \bar{t}$ for $i \to \infty$. We get $0 = \dot{\phi}(\bar{t}) = \lambda(\bar{t}) \cdot [F, G](\gamma(\bar{t}))$, thus $\Delta_B(\gamma(\bar{t})) = 0$. By optimality $\gamma(t_i) \neq \gamma(\bar{t})$, thus from **(P3)** for i sufficiently big $G(\gamma(t_i)) \neq 0$ and $\Delta_B(\gamma(t_i)) \neq 0$. Then t_i are switching times and $X(\gamma(\bar{t}))$ and $Y(\gamma(\bar{t}))$ are tangent to $\Delta_A^{-1}(0)$. From **(P6)** it follows $G(\gamma(t_0)) \neq 0$ thus we are in the situation of the picture of **(P7)**. Define $b_i := |\gamma(t_i) - \gamma(\bar{t})|$, then we easily obtain the estimate $b_{i+1} \geq b_i - Cb_i^2$. for some $C > 0$. Then the series b_i diverges by comparison with the series $\big(i(C+1)\big)^{-1}$. Since $|t_{i+1} - t_i| > C'b_i$, for some $C' > 0$, we obtain a contradiction.

The second claim is proved as in Proposition 1. ■

In Proposition 18, p. 172 it is proved also the viceversa i.e. for an abnormal extremal, under generic conditions, $\gamma(t_0) \in \Delta_A^{-1}(0)$ implies that t_0 is a switching time for γ.

Definition 25 *Given a trajectory γ, we denote by $n(\gamma)$ the smallest integer, if any, such that there exist X, Y or Z trajectories γ_i $i = 1, \dots, n(\gamma)$ verifying:*

$$\gamma = \gamma_{n(\gamma)} * \cdots * \gamma_1.$$

We call $n(\gamma)$ the number of arcs of γ and by abuse of notation a switching time is a time at which two arcs concatenate, cfr. Definition 18.

Given $\tau > 0$ let us define Π_τ to be the class of systems having an a priori bound on the number of arcs of optimal trajectories:

$$\Pi_\tau = \{\Sigma \in \Xi : \exists N(\Sigma) \ s.t. \ \forall \gamma \text{ optimal }, Supp(\gamma) \subset \mathcal{R}(\tau), n(\gamma) \leq N(\Sigma)\}.$$

From Theorem 11 we have a bound on the number of arcs of time optimal trajectories in a neighborhood of an ordinary point. In the following Section we prove:

Theorem 13 *Under generic conditions on $F, G \in C^3$, for every x there exist Ω_x, neighborhood of x, and $N_x \in \mathbb{N}$ such that if γ is optimal and $Supp(\gamma) \subset \Omega_x$ then:*

$$n(\gamma) \leq N_x.$$

It follows:

Corollary 1 *For every $\tau > 0$ the set Π_τ is a generic subset of Ξ.*

Proof. Using Theorem 13, for each $x \in \mathcal{R}(\tau)$ we can select an neighborhood Ω_x such that every optimal trajectory remaining in Ω_x is the concatenation of at most N_x bang or singular arcs. Choose $\varepsilon_x > 0$ such that $B(x, 2\varepsilon_x) \subset \Omega_x$. Since $\mathcal{R}(\tau) \subset \cup_{x \in \mathcal{R}(\tau)} B(x, \varepsilon_x)$, by compactness of $\mathcal{R}(\tau)$ (see Lemma 2, p. 39), we can extract a finite subcover $B(x_i, \varepsilon_i)$, $i = 1, \dots, m$, $\varepsilon_i = \varepsilon_{x_i}$. Consider an optimal trajectory $\gamma : [0, \tau] \mapsto \mathbb{R}^2$, $\gamma(0) = 0$. Define:

$$\varepsilon = \min_{i=1,\dots,m} \varepsilon_i, \qquad N = \max_{i=1,\dots,m} N_{x_i}.$$

Choose i_1 such that $0 \in B(x_{i_1}, \varepsilon_{i_1})$. Let t_1 be either the first time such that $\gamma(t_1) \notin B(x_{i_1}, 2\varepsilon_{i_1})$ or $t_1 = \tau$ if γ remains in $B(x_{i_1}, 2\varepsilon_{i_1})$. In the former case, there exists $i_2 \neq i_1$ such that $\gamma(t_1) \in B(x_{i_2}, \varepsilon_{i_2})$. Let t_2 be either the first time for which $\gamma(t_2) \notin B(x_{i_2}, 2\varepsilon_{i_2})$ or $t_2 = \tau$ if γ remains in $B(x_{i_2}, 2\varepsilon_{i_2})$. We proceed in the same way defining a set of increasing times $\{t_0 = 0, t_1, \dots, t_\nu = \tau\}$. If $M = \max\{|F(x)| + |G(x)| : x \in \mathcal{R}(\tau)\}$ denotes the maximum speed of trajectories inside $\mathcal{R}(\tau)$, it is clear that $t_j - t_{j-1} \geq \varepsilon / M$. Therefore:

$$\nu \leq \frac{M \, \tau}{\varepsilon}. \tag{2.14}$$

By definition, for each t_j, $j = 1, \ldots, \nu$, we have that $\{\gamma(t) : t \in [t_{j-1}, t_j[\}$ is contained in $B(x_{i_j}, 2\varepsilon_{i_j})$, hence $n(\gamma|_{[t_{j-1},t_j[}) \leq N$. From (2.14) we obtain:

$$n(\gamma) \leq N(\Sigma) = N\,\frac{M\,\tau}{\varepsilon}.$$

■

2.4.1 Proof of Theorem 13

To prove Theorem 13, p. 50 we need two Lemmas.

Lemma 13 *Let* $x \in Bad(\tau)$ *then:*

A $G(x) \neq 0$, $x \in (\Delta_A^{-1}(0) \cap \Delta_B^{-1}(0)) \quad \Rightarrow \quad x \in Tan_A$;
B $x \in Tan_A$, $X(x)$ *or* $Y(x) \neq 0 \quad \Rightarrow \quad x \in (\Delta_A^{-1}(0) \cap \Delta_B^{-1}(0))$.
C $G(x) = 0 \quad \Rightarrow \quad F(x) \cdot \nabla\Delta_B(x) = 0$.

Proof. The proof of **A** is as follows: being $G(x) \neq 0$ we can choose a local system of coordinates such that $G \equiv (1,0)$, then from $x \in \Delta_A^{-1}(0)$ we have $\alpha F(x) = G(x)$ ($\alpha \in \mathbb{R}$) and:

$$\nabla(\Delta_A)(x) = \nabla(F_1G_2 - G_1F_2)(x) = -\nabla F_2(x)$$

$$[F,G] = -\nabla F \cdot G.$$

From $\Delta_B(x) = 0$ we have $[F,G](x), G(x)$ are collinear and then:

$$0 = (\partial_1 F_2)(x)\,G_1(x) + (\partial_2 F_2)(x)\,G_2(x) = (\partial_1 F_2)(x)$$

finally:

$$\nabla(\Delta_A)(x) \cdot G(x) = 0.$$

We conclude that $x \in Tan_A$.
Let us prove **B**. From **(P6)**, we have $G(x) \neq 0$ and we can choose a local system of coordinates such that $G \equiv (1,0)$. From $x \in Tan_A$ it follows $x \in \Delta_A^{-1}(0)$, hence $F(x) = \alpha G \quad (\alpha \in \mathbb{R})$. Similarly to above, we have:

$$\Delta_B(x) = -\partial_1 F_2(x). \tag{2.15}$$

Assume $X(x) \neq 0$ being the other case similar. Since $X(x)$ is tangent to $\Delta_A^{-1}(0)$, $\nabla\Delta_A(x) \cdot X(x) = (\alpha - 1)\nabla\Delta_A(x) \cdot G = 0$. From $X(x) \neq 0$ we have that $\alpha \neq 1$, hence $\nabla\Delta_A \cdot G = 0$. This implies $\partial_1 F_2 = 0$ and using (2.15) we obtain $\Delta_B(x) = 0$, i.e. $x \in \Delta_B^{-1}(0)$.
Let us prove **C**. If $F(x) = 0$ there is nothing to prove, otherwise choose a local system of coordinates such that $F \equiv (1,0)$. Then $[F,G] = \nabla G \cdot F = (\partial_1 G_1, \partial_1 G_2)$. Then $\Delta_B = G_1\partial_1 G_2 - G_2\partial_1 G_1$, hence $F \cdot \nabla\Delta_B = G_1\partial_1^2 G_2 - G_2\partial_1^2 G_1$. From $G(x) = 0$ we conclude. ■

We now use envelope theory to prove a bound on the number of arcs in a neighborhood of anti-turnpikes.

Lemma 14 *Let S be an anti-turnpike and $x \in S$. Then, under the generic condition* **(P2)**, *there exists a neighborhood Ω of x such that every optimal trajectory γ contained in Ω satisfies $n(\gamma) \leq 2$.*

Proof. From Lemma 12, p. 47, we already know that taking Ω sufficiently small, γ has no singular arc. We prove that $Y * X * Y$ trajectories are not optimal, in a small neighborhood of x, being similar the proof of nonoptimalty of $X * Y * X$.

Take a local system of coordinates such that $x = (0,0)$, $X \equiv (1,0)$, $\Delta_B^{-1}(0) = \{(x,y) : x = 0\}$ and let $Y = (\alpha, \beta)$. By definition of anti-turnpike $\alpha < 0$ and without loss of generality we can assume $\beta > 0$ on Ω.

Recalling Lemma 8, p. 43, if γ_y is the X–trajectory passing through $(0,y)$ at time 0, then for ε sufficiently small, we can define $\psi : [-\varepsilon, 0] \times [-\varepsilon, \varepsilon] \to \mathbb{R}^+$ such that

$$\theta^{\gamma_y}(G(\gamma_y(x)), x; 0) = \theta^{\gamma_y}(G(\gamma_y(\psi(x,y))), \psi(x,y); 0). \tag{2.16}$$

We have that given $(x,y) \in]-\varepsilon, 0] \times [-\varepsilon, \varepsilon]$, $\psi(x,y)$ is the unique positive number with this property. Moreover, by **(P2)**, ψ is smooth, $\partial_x \psi < 0$, $\psi(0,y) = 0$ hence $\partial_y \psi(0,0) = 0$. Possibly restricting Ω, we can assume that:

$$|\partial_y \psi(x,y) \beta(x,y)| < \min_{\Omega} \partial_x \psi(x,y) \alpha(x,y). \tag{2.17}$$

We also define $\Psi(x,y) = (\psi(x,y), y)$. Assume $\gamma = \gamma_1 * \gamma_2 * \gamma_3$ where γ_1 and γ_3 are Y–trajectories and γ_2 is a X–trajectory. If γ_2 does not cross $\Delta_B^{-1}(0)$ then γ is not optimal by Theorem 11. We thus assume the opposite and refer to Figure 2.6. Let $z_{12} = \gamma_1(t_{12}) = \gamma_2(t_{12})$ and $z_{23} = \gamma_2(t_{23}) = \gamma_3(t_{23})$ be the switching points of γ and define $\bar{\gamma}(t) = \Psi(\gamma_3(t_{23} + t))$, for t in a left neighborhood of 0. We want to prove that $\bar{\gamma}$ is an envelope for $\gamma_2 * \gamma_3$.
For this, define, for $\sigma < 0$ sufficiently close to 0, the extremal trajectory γ_σ in the following way. We start from the initial point of γ_3 and reach $\gamma_3(t_{23} + \sigma)$ with control $+1$, then we switch to control -1 and reach the point $\Psi(\gamma_3(t_{23} + \sigma))$. Denote also by T_σ the final time of γ_σ and $\gamma_3(t_{23} + \sigma) = z_\sigma = (x_\sigma, y_\sigma)$, thus $\gamma_\sigma(T_\sigma) = \Psi(z_\sigma)$. We have to verify the conditions of envelope given in Definition 13. Condition 1-i) is immediately verified. Let u_σ be the control associated to γ_σ, thus u_σ is bang–bang with one switching and assumption 1-ii) is promptly checked. Now, for every continuous function $\varphi : [0, \bar{T}] \to \mathbb{R}^2$, we have (see Exercise 15):

$$\lim_{\varepsilon \to 0} \frac{1}{\varepsilon} \left(\int_0^{T_\sigma + \varepsilon} \varphi(t) \cdot f(\gamma_\sigma(t), u_{\sigma+\varepsilon}(t)) dt - \int_0^{T_\sigma} \varphi(t) f(\gamma_\sigma(t), u_\sigma(t)) dt \right) =$$
$$= 2G(z_\sigma) + [1 + (\nabla \psi \cdot Y)(z_\sigma) - \alpha(z_\sigma)] (F - G)(\Psi(z_\sigma)), \tag{2.18}$$

thus condition 1-iii) is verified. Since $L \equiv 1$, condition 2) follows from the continuity of ψ.

To verify condition 3) we compute:

$$\frac{d}{dt}\bar{\gamma}(t) = D\Psi \cdot \frac{d}{dt}\gamma_3(t_{23}+t) = \begin{pmatrix} \partial_x\psi & \partial_y\psi \\ 0 & 1 \end{pmatrix} \begin{pmatrix} \alpha \\ \beta \end{pmatrix} = \begin{pmatrix} \partial_x\psi\ \alpha + \partial_y\psi\ \beta \\ \beta \end{pmatrix}.$$

Now, $\beta > 0$ and from (2.17), $\partial_x\psi\ \alpha + \partial_y\psi\ \beta > 0$, hence, up to a reparametrization of $\bar{\gamma}$, $\frac{d}{dt}\bar{\gamma}(t)$ is a strict convex combination of $Y = (\alpha, \beta)$ and $X = (1, 0)$. Therefore we may assume that $\bar{\gamma}$ is a trajectory. Now, to verify condition 4), denote by λ_σ the covector associated to γ_σ. From (2.16), we have that T_σ is a switching point of γ_σ, thus $\lambda_\sigma(T_\sigma) \cdot G(\Psi(z_\sigma)) = 0$. Using the maximization condition of PMP for γ_σ we thus conclude.
Therefore, Theorem 9 ensures that γ and $\bar{\gamma} * \gamma_\sigma$ steers the same initial point to z_{12} taking the same amount of time. But $\bar{\gamma}$ is not bang-bang and hence not optimal because of Theorem 11. We conclude that γ is not optimal as well. ■

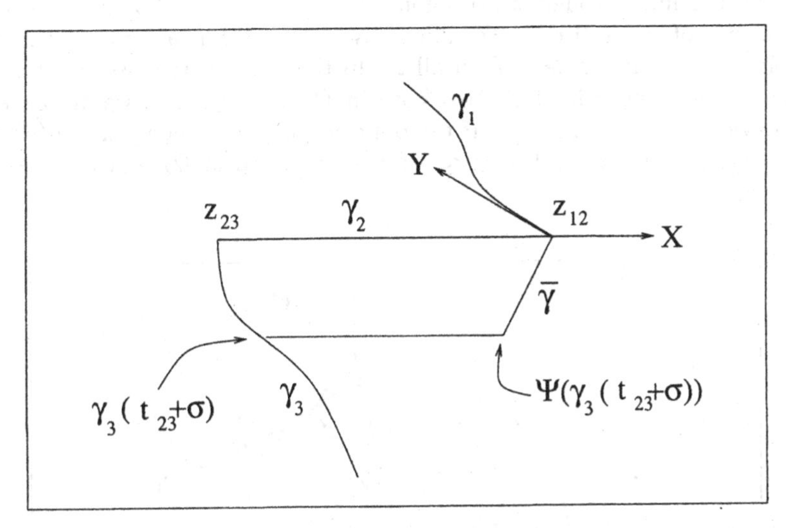

Fig. 2.6. Proof of Lemma 14

Proof of Theorem 13. We assume the generic conditions **(P1)**–**(P7)**. If x is an ordinary point, the conclusion is given by Theorem 11, p. 44. From now on we assume that x is a nonordinary point.

In case of $x \in Bad(\tau)$, using **(P5)**, we always assume that x is the only bad point in Ω_x.

Assume first $\Delta_B(x) \neq 0$ and let γ be an optimal trajectory contained in a open neighborhood Ω_x of x such that $\Delta_B(y) \neq 0$ for every $y \in \Omega_x$. From Lemma 8, p. 43 we have that θ^γ is monotonic. Since F, G are smooth, choosing Ω_x sufficiently small, we can assume that $TV(\theta^\gamma) < \pi$ where TV is the total variation. From the second statement of Lemma 8, p. 43 we get $n(\gamma) \leq 2$.

From now on we assume $\Delta_B(x) = 0$ and distinguish two cases:

a) $\Delta_A(x) \neq 0$,
b) $\Delta_A(x) = 0$.

Let us first treat case **a)**.

Assume $x \notin Tan_B$. If X, and Y point to the same side of $\Delta_B^{-1}(0)$ then every optimal trajectory moves from one side of $\Delta_B^{-1}(0)$ to the other side. Since $f = -\Delta_A/\Delta_B$ does not vanish outside $\Delta_B^{-1}(0)$, from Theorem 11, p. 44 we get $N_x = 3$. Assume now that X, Y point to opposite side. If $\Delta_B^{-1}(0)$ is locally a turnpike then, from Theorem 11, p. 44 we get $N_x \leq 3$. If $\Delta_B^{-1}(0)$ is locally a anti-turnpike then, from Lemma 14, p. 52 we get $N_x = 2$.

Consider now $x \in Tan_B$. We refer to Figure 2.7. Since $\Delta_A(x) \neq 0$, $X(x)$ and $Y(x)$ do not vanish and are not parallel. Without loss of generality we assume $X(x) \cdot \nabla \Delta_B(x) = 0$, $Y(x) \cdot \nabla \Delta_B(x) \neq 0$, hence $F(x) \cdot \nabla \Delta_B(x) \neq 0$. Let ℓ be the line through x orthogonal to $F(x)$ and Ω_1, Ω_2 the connected components of $\Omega_x \setminus \ell$. Taking Ω sufficiently small, we may assume that X and Y point to the same side of ℓ on all Ω. In the not generic case in which X points to the same side of $\Delta_B^{-1}(0)$ both in Ω_1 and Ω_2, then we are back to the previous case and $N_x \leq 3$. If the opposite happens, then either $n(\gamma) \leq 2$ for γ contained in Ω_1 and $n(\gamma) \leq 3$ for γ contained in Ω_2 or viceversa. We conclude $N_x \leq 5$.

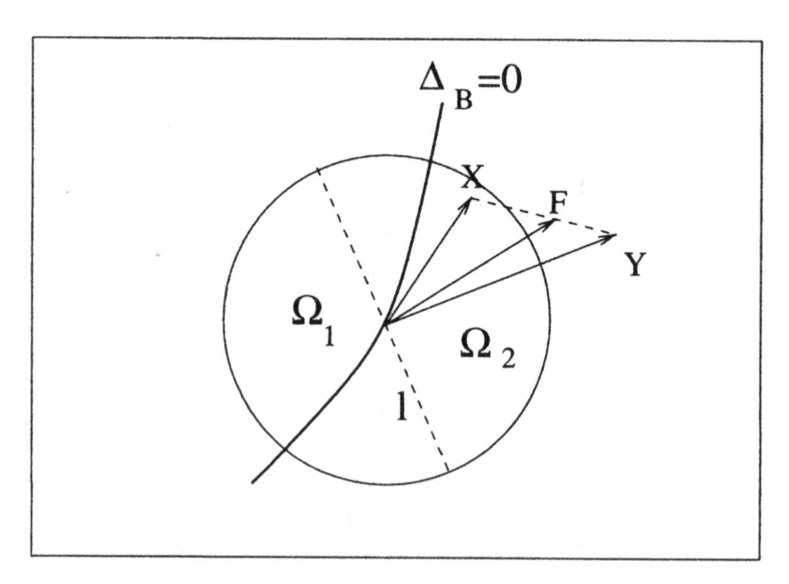

Fig. 2.7. Proof of Theorem 13

Let us now treat case **b)**. Assume first $G(x) = 0$, then by **(P6)**, F is not tangent to $\Delta_A^{-1}(0)$, while by **C** of Lemma 13 F is tangent to $\Delta_B^{-1}(0)$. Let ℓ be the line through x orthogonal to $F(x)$ and Ω_1, Ω_2 the connected components of $\Omega_x \setminus \ell$. Since $G(x) = 0$, taking Ω_x sufficiently small, we have that X and Y point to the same side of $\Delta_A^{-1}(0)$ on all Ω_x. Then $n(\gamma) \leq 3$ for γ contained in Ω_1 and for γ contained in Ω_2. We conclude $N_x \leq 6$.

Assume now that $G(x) \neq 0$, hence, possibly shrinking Ω_x, G does not vanish on Ω_x. If $X(x)$ and $Y(x)$ have the same versus then we conclude as in the previous cases, hence we assume the opposite. From Lemma 13, p. 51 **A** and assumption **(P7)**, we are in the situation of figure 2.5 (case P7). We can assume that X and Y do not vanish on Ω_x. Let γ be an optimal trajectory contained in Ω_x. If γ switches at x then by Proposition 1, p. 45, γ has no more switchings.

Assume now that γ switches at a point $y \neq x$, $y \in \Delta_A^{-1}(0)$. then by Proposition 1, p. 45, γ switches at every crossing of $\Delta_A^{-1}(0)$. Assume that γ has at least two switchings then we can apply the same reasoning of Lemma 14, p. 52 obtaining the non optimality of γ. Indeed let x_1, x_2 be two consecutive switching points of γ and Ω_i, $i = 1, ..., 4$, be the connected components of $\Omega_x \setminus (\Delta_A^{-1}(0) \cup \Delta_B^{-1}(0))$, labeled as in Figure 2.8.

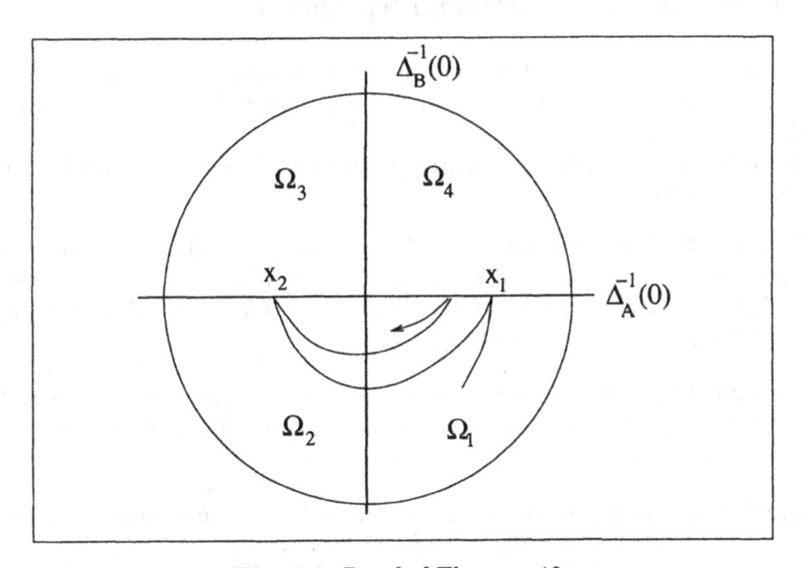

Fig. 2.8. Proof of Theorem 13

Let $x_1 = \gamma(t_1)$ and consider the trajectory γ_ε switching at point $\gamma(t_1 - \varepsilon)$, then γ_ε switches at a point $y_\varepsilon \in \Omega_2$. It is easy to check that:

$$\left.\frac{d}{d\varepsilon} y_\varepsilon\right|_{\varepsilon=0}$$

belongs to the positive cone generated by $X(x_2)$ and $Y(x_2)$. Hence up to a reparametrization, $\varepsilon \to y_\varepsilon$ is a trajectory and we can reason as in Lemma 14, p. 52 to construct an envelope for γ.

Assume now that γ has two switchings on Ω_x outside $\Delta_A^{-1}(0) \cup \Delta_B^{-1}(0)$. From **(P2)**, referring to figure 2.8, γ switches either on $\Omega_1 \cup \Omega_3$ or on $\Omega_2 \cup \Omega_4$. Then again we can reason as in Lemma 14 to construct an envelope for γ.

If γ contains a singular arc, then reasoning on θ^γ as for the case $\Delta_B(x) \neq 0$, we easily get $n(\gamma) \leq 3$. We thus conclude.

■

2.5 Existence of an Optimal Synthesis

In this section we complete the construction of an optimal synthesis according to Step 3. and 4. of the geometric control approach illustrated in Section 1.4, p. 27.

We introduce the definition of stratification of $\mathcal{R}(\tau)$, $\tau > 0$, and of optimal synthesis on $\mathcal{R}(\tau)$.

Definition 26 (stratification) *A stratification of $\mathcal{R}(\tau)$, $\tau > 0$, is a finite collection $\{M_i\}$ of connected embedded C^1 submanifolds of $\mathbb{R}^2$, called strata, such that the following holds. If $M_j \cap Clos(M_k) \neq \emptyset$ with $j \neq k$ then $M_j \subset Clos(M_k)$ and $dim(M_j) < dim(M_k)$.*

Remark **24** The given definition of stratification is quite rough, indeed the optimal synthesis is more regular as showed below. More refined definitions of stratification are introduced in next chapters.

Definition 27 (regular optimal synthesis) *A regular optimal synthesis for $\Sigma \in \Xi$ on $\mathcal{R}(\tau)$ is a collection of trajectory-control pairs $\{(\gamma_x, u_x) : x \in \mathcal{R}(\tau)\}$ satisfying the following properties:*

1. *For every $x \in \mathcal{R}(\tau)$, $\gamma_x : [0, t_x] \to \mathbb{R}^2$ steers the origin to x in minimum time.*
2. *If $y = \gamma_x(t)$ for some $t \in Dom(\gamma_x)$ then γ_y is the restriction to $[0, t]$ of γ_x.*
3. *There exists a stratification of $\mathcal{R}(\tau)$ such that $u(x) = u_x(t_x)$ is smooth on each stratum (assuming each u_x left continuous).*

Given a system $\Sigma \in \Pi_\tau$, since there is a finite dimensional reduction of the space of optimal trajectories, it is possible to construct a synthesis Γ for Σ following the classical idea of constructing extremal trajectories and deleting those trajectories that are not optimal. By construction this synthesis is optimal.

We describe an algorithm $\mathcal{A}$ by induction. At step N, we construct precisely those trajectories which are concatenation of N bang or singular arcs and satisfy the PMP. The end points of the arcs forming these trajectories, corresponding to the switching times of the control, are determined by certain nonlinear equations. Under generic conditions, such equations can be solved by the implicit function theorem, thus determining a smooth switching locus. Eventually the algorithm will partition the reachable set $\mathcal{R}(\tau)$ into finitely many open regions (where the optimal feedback is either $u = +1$ or $u = -1$), separated by boundary curves and points, called Frame Curves (in the following FCs) and Frame Points (in the following FPs), respectively. At each step it may happen that distinct extremal trajectories reach the same point at different times. It is therefore necessary to delete from the synthesis those trajectories which are not globally optimal. This procedure usually produces new "overlap curves", consisting of points reached in minimum time by two distinct trajectories: one ending with the control value $u = +1$, the other with $u = -1$.

If at step N the algorithm $\mathcal{A}$ does not construct any new trajectory, then we say that $\mathcal{A}$ stops at step N (for Σ at time τ) or that $\mathcal{A}$ succeeds for Σ. From Theorem 13 it is clear that, under generic assumptions, there exists $N(\Sigma)$ such that $\mathcal{A}$ stops before step $N(\Sigma)$ and every γ constructed by $\mathcal{A}$ is optimal. By definition $Fr(\mathcal{R}(\tau))$, the topological frontier of $\mathcal{R}(\tau)$, is a frame curve and its intersections with other frame curves are frame points.

If $\mathcal{A}$ stops, then for each $x \in \mathcal{R}(\tau)$ there exists a set of constructed trajectories that reach x. Define $\Gamma_x := \{\gamma : \gamma \text{ is constructed by } \mathcal{A}, Term(\gamma) = x\}$. We want to select for each $x \in \mathcal{R}(\tau)$, a trajectory from Γ_x to form a synthesis. Define K_k to be the set of points $x \in \mathcal{R}(\tau)$ reached at least by one constructed trajectory γ, satisfying $n(\gamma) \leq k$. Note that K_k is compact for every k and $K_{N(\Sigma)} = \mathcal{R}(\tau)$. We proceed by induction on k. Given $x \in K_k \setminus K_{k-1}$, we consider the optimal trajectories $\gamma \in \Gamma_x$ formed by k arcs, for which the following holds. If $y = \gamma(t)$ is the initial point of the last arc of γ, then $\gamma|_{[0,t]}$ has been selected from Γ_y by induction. Finally if there is more than one such trajectory, then we select one, say, according to the preference order X, Y, Z on the type of the last arc.

In this way, at step $N(\Sigma)$ we have constructed a synthesis for Σ at time τ. We use the symbol $\Gamma_\mathcal{A}(\Sigma, \tau)$ to denote this synthesis and call it the synthesis generated by the algorithm $\mathcal{A}$. If τ is fixed we also use the symbol $\Gamma_\mathcal{A}(\Sigma)$.

Theorem 14 *Consider $\Sigma \in \Xi$ and $\tau > 0$. If $\mathcal{A}$ stops for Σ at time τ then $\Gamma_\mathcal{A}(\Sigma, \tau)$ is an optimal synthesis*

Remark **25** A more detailed algorithm is described in Section 2.8.2, p. 89. The latter provides generic conditions under which the structural stability of the synthesis is guaranteed.

2.6 Classification of Synthesis Singularities

The aim of this Section is to provide a generic classification of all Frame Curves and Frame Points that appear in optimal syntheses built by the algorithm described in the previous Section.

Let us introduce some more notation. We call $\gamma^\pm : [0,\tau] \to \mathcal{R}(\tau)$ the trajectories exiting the origin with constant control ± 1. We let $t^\pm_{op}$ the last times at which $\gamma^\pm$ are optimal and define $\gamma^\pm_{op} = \gamma^\pm|_{[0,t^\pm_{op}]}$.

Definition 28 (conjugate times) *Let t_0, t belong to the domain of an extremal trajectory. If $v^\gamma(G(\gamma(t), t; t_0)$ and $G(\gamma(t_0))$ are linearly dependent (i.e. either if at least one vanish or if $\theta^\gamma(t) = \theta^\gamma(t_0) \pm a$, $a \in \{0, \pm\pi\}$), we say that t_0 and t_1 are conjugate along γ.*

From the proof of Lemma 8, p. 43 we get

Lemma 15 *Let $(\gamma, \lambda) : [0,\tau] \to \mathbb{R}^2$ be an extremal pair and t_1, t_2 consequent switching times. Then t_1 and t_2 are conjugate along γ.*

Definition 29 *Let D_1, D_2 be two connected C^2 embedded submanifolds of $\mathbb{R}^2$. We say that they are conjugate along the X-trajectories, if there exists a diffeomorphism $\psi : D_1 \to D_2$ satisfying the following. For every $y \in D_1$ let γ_y be the X-trajectory trough y at time 0. Then there exists $t(y)$ depending continuously on y such that $\gamma_y(t(y)) = \psi(y) \in D_2$ and $0, t(y)$ are conjugate along γ_y. Conjugate curves along Y – trajectories are defined similarly.*

2.6.1 Frame Curves

Definition 30 *Given $x \in \mathcal{R}(\tau)$ we denote by γ_x, u_x the trajectory of $\Gamma_A(\Sigma,\tau)$ and the corresponding control such that $\gamma_x(t_x) = x$. Also, we denote by u_A the feedback control $u_x(t_x)$.*

Definition 31 (Frame Curves and Points) *A one-dimensional connected embedded C^2 submanifold with boundary D of $\mathcal{R}(\tau)$ is called a Frame Curve if u_A is discontinuous at each point of of D and D is maximal. Frame Points are defined as intersection of Frame Curves.*

Under generic conditions, the Algorithm constructs only six types of frame curves:

1. the trajectory γ^-_{op} starting from zero and corresponding to constant control -1. We say that this curve is of kind X;

2. the trajectory γ_{op}^{+} starting from zero and corresponding to constant control $+1$. We say that this curve is of kind Y;
3. singular trajectories that are trajectories corresponding to the singular control φ. We say that these are FCs of kind S;
4. conjugate curves to other FCs. These curves are called C frame curves or switching curves, since the optimal control changes sign crossing them;
5. overlap curves, formed by points reached optimally by two distinct trajectories. We call these curves K frame curves;
6. the topological frontier of the reachable set. We call this curve a F frame curve.

Our aim is to topologically classify all frame curves in the following sense.

Definition 32 (Local Topological Equivalence of Syntheses) *Consider two systems Σ_1, Σ_2, a time $\tau > 0$ and two open sets $U_1 \subset \mathcal{R}_1(\tau)$ and $U_2 \subset \mathcal{R}_2(\tau)$; here $\mathcal{R}_1$ and $\mathcal{R}_2$ denote the reachable sets of Σ_1 and Σ_2 respectively. Assume that $\mathcal{A}$ succeeds for Σ_1 and for Σ_2 at time τ. We say that $\Gamma_1 = \Gamma_{\mathcal{A}}(\Sigma_1, \tau)|U_1$ (the restriction of $\Gamma_{\mathcal{A}}(\Sigma_1, \tau)$ to U_1) and $\Gamma_2 = \Gamma_{\mathcal{A}}(\Sigma_2, \tau)|U_2$ are equivalent if there exists an homeomorphism $\varphi : U_1 \mapsto U_2$ such that:*

(E1) φ induces a bijection on Γ_i: $\{\varphi(\gamma_x(t)) : t \in Dom(\gamma_x)\} \cap U_1 = \{\gamma_{\varphi(x)}(t) : t \in Dom(\gamma_{\varphi(x)})\} \cap U_2$ for every $x \in U_1$; if the two sets are oriented for increasing t then φ preserves the orientation;
(E2) φ induces a bijection on frame curves, i.e. for each FC D_1 of Γ_1 we have that $\varphi(D_1)$ is a FC of Γ_2 of the same type and viceversa, assuming that the types $X-$, $Y-$ are equivalent.

In this case we write $\Gamma_1|U_1 \equiv \Gamma_2|U_2$.

Remark **26** Note that in the definition of equivalence there is no request about the time evolution along γ_x, in fact there is no condition of the type $\varphi(\gamma_x(t)) = \gamma_{\varphi(x)}(t)$. It is necessary to give a not too strict definition of equivalence to have a discrete set of equivalence classes. The same problem occurs in the definition of equivalence for a singular point of a dynamical system. In this case the orbital equivalence was introduced, see [18].

Definition 33 (Topological Equivalence of FCs and FPs) *Given x_1, x_2 we say that Γ_1 at x_1 and Γ_2 at x_2 are equivalent, or we write $\Gamma_1|x_1 \equiv \Gamma_2|x_2$, if there exist U_1, U_2 neighborhoods of x_1, x_2, respectively, such that $\Gamma_1|U_1 \equiv \Gamma_2|U_2$. We say that two FC's D_i of Γ_i, $i = 1, 2$, are equivalent if for each $y_1 \in D_1 \setminus \partial D_1$, $y_2 \in D_2 \setminus \partial D_2$ we have that $\Gamma_1|y_1 \equiv \Gamma_2|y_2$. Similarly two Frame Points x_i of Γ_i, $i = 1, 2$, are equivalent if $\Gamma_1|x_1 \equiv \Gamma_2|x_2$.*

Definition 34 *A Frame Curve D is simple if $D \setminus \partial D$ does not contain any frame point.*

It is clear that a classification of FPs and simple FCs, immediately provides a classification of all FCs. Recall Figure 2.1, p. 34, where the optimal synthesis near each simple Frame Curve is shown (except the F curve for which the synthesis is trivial: only one side is covered by trajectories corresponding to the same bang control). Explicit examples of FCs are given in Section 2.6.4. We have the following Theorem, proved in Appendix A.

Theorem 15 *Consider $\Sigma \in \Xi$ and $\tau > 0$. If A succeeds for Σ at time τ and D is a simple FC of $\Gamma_A(\Sigma, \tau)$, then, under generic conditions, D is of one of the following six types Y_1, Y_2, S, C, K, F.*

2.6.2 Frame Points

In this section we give a description of the local structure of Γ_A in a neighborhood of a Frame point. More precisely, only generic frame points are considered. Therefore, all frame points are intersections of no more than two FCs. Indeed an intersection of three or more FCs can be destroyed by an arbitrary small perturbation of the system.

Consider a frame point x and two FCs D_1 and D_2 such that $\{x\} = D_1 \cap D_2$. Then there are four possible cases:

(FP0) $x \in D_1 \setminus \partial D_1$, $x \in D_2 \setminus \partial D_2$.
(FP1) $x \in D_1 \setminus \partial D_1$, $x \in \partial D_2$.
(FP2) $x \in \partial D_1$, $x \in D_2 \setminus \partial D_2$.
(FP3) $x \in \partial D_1$, $x \in \partial D_2$.

It is easy to check that, using Theorem 15, $(FP0)$ can never occur generically. Thus we have to examine the other three possibilities. The classification of generic Frame Points is based on the types of the two intersecting curves, hence if $x = D_1 \cap D_2$, we say that x is a (D_1, D_2) Frame Point.

The shape of the optimal synthesis near each Frame Point is shown in Figure 2.9, except for FPs belonging to $\partial\mathcal{R}(\tau)$. Explicit examples of FPs are given in Section 2.6.4. Next Theorem is proved in Appendix A.

Theorem 16 *Consider $\Sigma \in \Xi$ and $\tau > 0$. If A succeeds at time τ for Σ and x is a frame point, then, under generic conditions, x is one of the following 23 types (X,Y), $(Y,C)_{1,2,3}$, (Y,S), $(Y,K)_{1,2,3}$, $(C,C)_{1,2}$, $(C,S)_{1,2}$, $(C,K)_{1,2}$, $(S,K)_{1,2,3}$, (K,K), $(Y,F)_1$, $(Y,F)_2$, (S,F), (C,F), (K,F).*

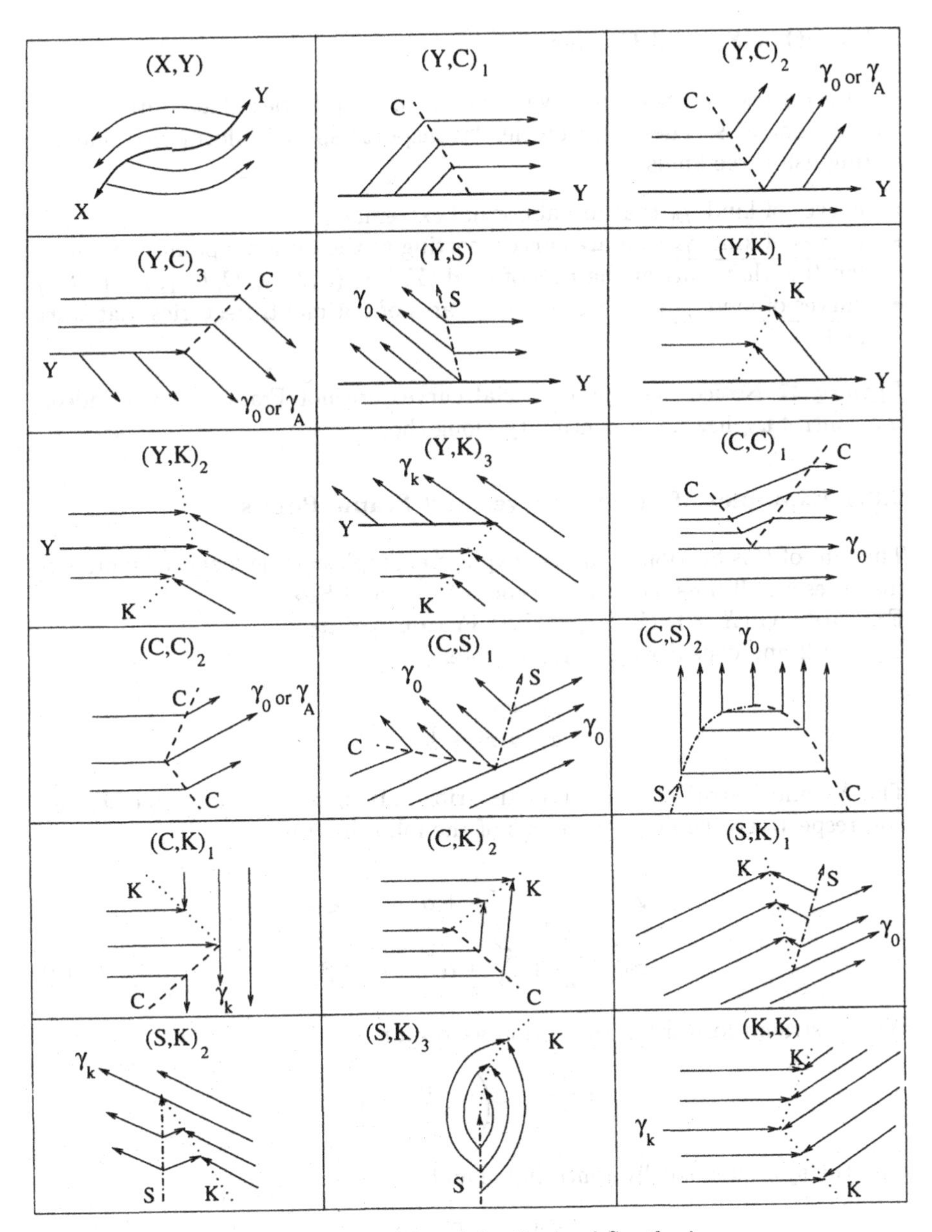

Fig. 2.9. The FPs of the Optimal Synthesis.

2.6.3 Other Special Curves

For future use, we need to give a name to trajectories that start at FPs and are not FCs. These trajectories "transport" special information and we distinguish three kinds:

- curves of kind γ_A that are abnormal extremals,
- curves of kind γ_k that are curves starting at the terminal point of an overlap (i.e. that start at the FPs of kind $(Y,K)_3$ $(S,K)_2, (C,K)_1$, see below)
- curves of kind γ_0 that are the other arcs of optimal trajectories that start at FPs.

Remark **27** Notice that these special curves are not Frame Curves, indeed the control $u_{\mathcal{A}}$ has no discontinuity along them.

2.6.4 Examples of Frame Curves and Frame Points

The aim of this Section is to give explicit examples (some local) of syntheses that present all possible generic types of FCs and FPs.
Example 1: (X,Y), $(Y,K)_1$, (Y,S) **Frame Points.**
Let $\tau > 2$ and consider the control system:

$$\begin{cases} \dot{x}_1 = u \\ \dot{x}_2 = x_1 + \frac{1}{2}x_1^2 \end{cases} .$$

The $X-$ and $Y-$trajectories can be described giving x_2 as a function of x_1 and are, respectively, cubic polynomials of the following type:

$$\begin{aligned} x_2 &= -\frac{x_1^3}{6} - \frac{x_1^2}{2} + \alpha \qquad \alpha \in \mathbb{R} \\ x_2 &= \frac{x_1^3}{6} + \frac{x_1^2}{2} + \alpha \qquad \alpha \in \mathbb{R}. \end{aligned} \tag{2.19}$$

With a straightforward computation we obtain:

$$[F,G] = \begin{pmatrix} 0 \\ -1 - x_1 \end{pmatrix}$$

then the system is locally controllable and:

$$\Delta_B(x) = \det \begin{pmatrix} 1 & 0 \\ 0 & -1 - x_1 \end{pmatrix} = -1 - x_1 . \tag{2.20}$$

From equation (2.20) it follows that every turnpike is subset of $\{(x_1,x_2) \in \mathbb{R}^2 : x_1 = -1\}$. Indeed, at the second step the algorithm $\mathcal{A}$ constructs the turnpike:

$$S = \left\{ (x_1,x_2) : x_1 = -1, x_2 \le -\frac{1}{3} \right\}.$$

Given b, consider the trajectories $\gamma_1 : [0,b] \mapsto \mathbb{R}^2$ for which there exists $t_0 \in [0,b]$ such that $\gamma_1|[0,t_0]$ is a Y–trajectory and $\gamma_1|[t_0,b]$ is an X–trajectory, and the trajectories $\gamma_2 : [0,b] \mapsto \mathbb{R}^2$, $b > 2$, for which there exists $t_1 \in [2,b]$ such that $\gamma_2|[0,t_1]$ is an X–trajectory and $\gamma_2|[t_1,b]$ is a Y–trajectory. For every $b > 2$, these trajectories cross each other in the region of the plane above the cubic (2.19) with $\alpha = 0$ and determine an overlap curve K that originates from the point $(-2, -\frac{2}{3})$. We use the symbols $x^{+-}(b,t_0)$ and $x^{-+}(b,t_1)$ to indicate, respectively, the terminal points of γ_1 and γ_2 above. Explicitly we have:

$$x_1^{+-} = 2t_0 - b \qquad x_2^{+-} = -\frac{(2t_0-b)^3}{6} - \frac{(2t_0-b)^2}{2} + t_0^2 + \frac{t_0^3}{3} \tag{2.21}$$

$$x_1^{-+} = b - 2t_1 \qquad x_2^{-+} = \frac{(b-2t_1)^3}{6} + \frac{(b-2t_1)^2}{2} - t_1^2 + \frac{t_1^3}{3}. \tag{2.22}$$

Now the equation:

$$x^{+-}(b,t_0) = x^{-+}(b,t_1), \tag{2.23}$$

as b varies in $[2,+\infty[$, describes the set K. From (2.21), (2.22) and (2.23) it follows:

$$t_0 = b - t_1 \qquad t_1\Big(-2t_1^2 + (2+3b)t_1 + (-b^2-2b)\Big) = 0.$$

Solving for t_1 we obtain three solutions:

$$t_1 = 0, \qquad t_1 = b, \qquad t_1 = 1 + \frac{b}{2}. \tag{2.24}$$

The first two of (2.24) are trivial, while the third determines a point of K, so that:

$$K = \left\{(x_1,x_2) : x_1 = -2, x_2 \geq -\frac{2}{3}\right\}.$$

The set $\mathcal{R}(\tau)$ is portrayed in Fig. 2.10

Example 2: $(C,C)_2$, $(Y,C)_3$ Frame Points.
Consider $\tau > \pi$ and the control system:

$$\begin{cases} \dot{x}_1 = x_2 \\ \dot{x}_2 = -x_1 + u \end{cases}.$$

This example, called the forced pendulum is accurately described also in [92], pp.11–14 and in [69], p.80.
The X– and Y–trajectories are circles centered at $(-1,0)$ and at $(1,0)$, respectively. The algorithm $\mathcal{A}$ constructs $\gamma^{\pm}$ only up to time π, indeed after this time they are not extremal.

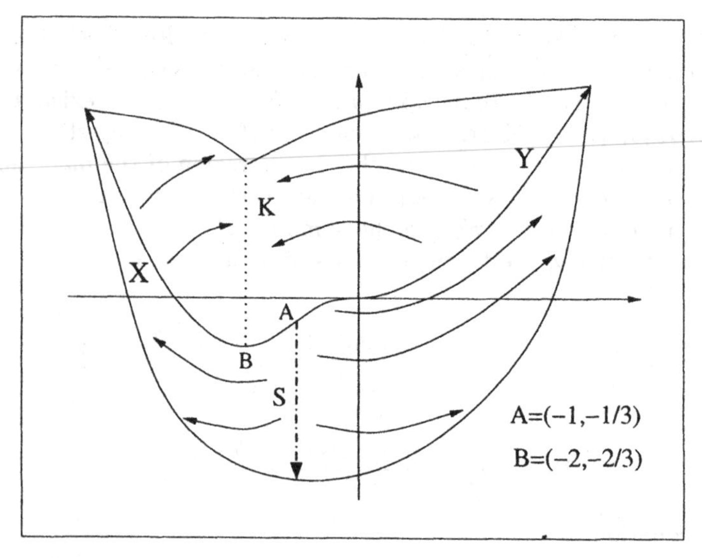

Fig. 2.10. Example 1, (X,Y), $(Y,K)_1$, (Y,S) Frame Points

At step n+1 the following switching curves originate:

(a) All the semicircles of radius 1 centered at $(2n+1,0)$ and contained in the half plane $\{(x_1,x_2) : x_2 \geq 0\}$.
(b) All the semicircles of radius 1 centered at $(-2n-1,0)$ and contained in the half plane $\{(x_1,x_2) : x_2 \leq 0\}$.

Along the switching curves described in a) the constructed trajectories arrive as Y–trajectories and leave as X–trajectories, i.e. the controls switch from $+1$ to -1. The opposite happens along the switching curves described in b). The set $\mathcal{R}(3\pi)$ is represented in Figure 2.11.

Remark **28** Notice that in this example if $T \geq \pi$ then $\partial\mathcal{R}(T)$ is smooth. In Chapter 3 the smoothness properties of $\partial\mathcal{R}(T)$ are studied in details.

Example 3: $(Y,C)_1$ $(Y,C)_2$ **Frame Points.**
Consider $\tau > \frac{1}{3}ln(4)$ and the system Σ:

$$\begin{cases} \dot{x}_1 = 3x_1 + u \\ \dot{x}_2 = x_1^2 + x_1 \end{cases} . \tag{2.25}$$

Since $\Sigma \in \Xi$ and

$$[F,G] = \begin{pmatrix} -3 \\ -2x_1 - 1 \end{pmatrix},$$

the system is locally controllable.

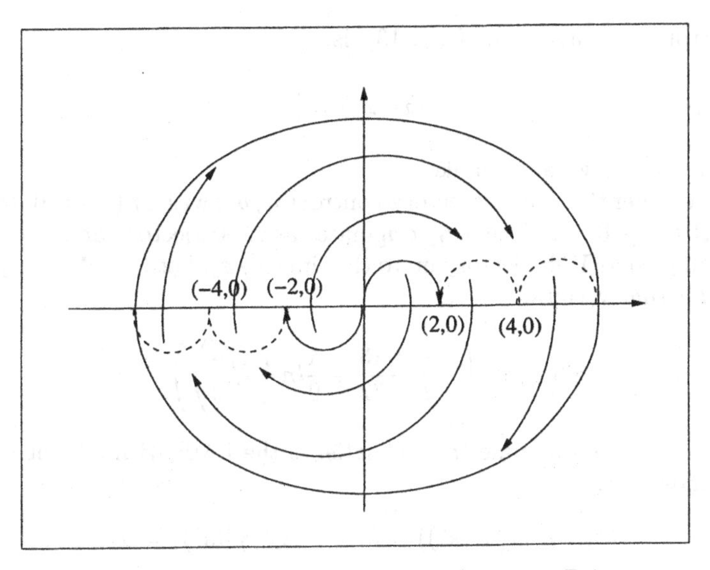

Fig. 2.11. Example 2, $(C,C)_2$, $(Y,C)_3$ Frame Points.

The X–trajectory passing through the point (x_1^0, x_2^0) at time 0 is:

$$x_1(t) = \left(x_1^0 - \frac{1}{3}\right) e^{3t} + \frac{1}{3}$$

$$\begin{aligned} x_2(t) = &\frac{1}{6}\left(x_1^0 - \frac{1}{3}\right)^2 e^{6t} + \frac{5}{9}\left(x_1^0 - \frac{1}{3}\right) e^{3t} + \frac{4}{9}t + x_2^0 \\ &- \frac{1}{6}\left(x_1^0 - \frac{1}{3}\right)^2 - \frac{5}{9}\left(x_1^0 - \frac{1}{3}\right). \end{aligned}$$

While the Y–trajectory passing through the point (x_1^0, x_2^0) at time 0 is:

$$x_1(t) = \left(x_1^0 + \frac{1}{3}\right) e^{3t} - \frac{1}{3}$$

$$\begin{aligned} x_2(t) = &\frac{1}{6}\left(x_1^0 + \frac{1}{3}\right)^2 e^{6t} + \frac{1}{9}\left(x_1^0 + \frac{1}{3}\right) e^{3t} - \frac{2}{9}t + x_2^0 \\ &- \frac{1}{6}\left(x_1^0 + \frac{1}{3}\right)^2 - \frac{1}{9}\left(x_1^0 + \frac{1}{3}\right). \end{aligned}$$

The equation for turnpikes is:

$$0 = \Delta_B(x_1, x_2) = -(2x_1 + 1), \tag{2.26}$$

hence every turnpike is subset of $S = \{(x_1, x_2) : x_1 = -\frac{1}{2}\}$.

The control φ to stay on S, cfr. (2.13), is:

$$\varphi(x_1, x_2) = \frac{3}{2}$$

then there is no regular turnpike.

Now consider the pairs trajectory-control (γ_s, u_s) with $In(\gamma_s) = 0$, $Dom(\gamma_s) = [0, s + \varepsilon_s]$ $(\varepsilon_s \geq 0)$, such that γ_s originates as X–trajectory and switches at time s going on as Y–trajectory up to the time $s+\varepsilon_s$. Let $(\gamma^*, u^*) = (\gamma_{s^*}, u_{s^*})$ be the pair that verifies:

$$\gamma^*(s^*) = \left(-\frac{1}{2}, -\frac{13}{72} + \frac{4}{9} ln \left(\sqrt[3]{\frac{5}{2}}\right)\right). \tag{2.27}$$

Define $\varepsilon^* = \varepsilon_{s^*}$ and assume that γ^* satisfies the PMP with adjoint variable λ^*. We know that:

$$\lambda^*(s^*) \cdot G(\gamma^*(s^*)) = 0, \qquad \Delta_B(\gamma^*(s^*)) = 0 \tag{2.28}$$

then:

$$\frac{d}{dt}\left(\lambda^*(t) \cdot G(\gamma^*(t))\right)\Big|_{t=s^*} = \lambda^*(s^*) \cdot [F, G](\gamma^*(s^*)) = 0. \tag{2.29}$$

From (2.28) and (2.29) and straightforward calculations we have at $\gamma^*(s^*)$ the situation of Fig. 2.12.

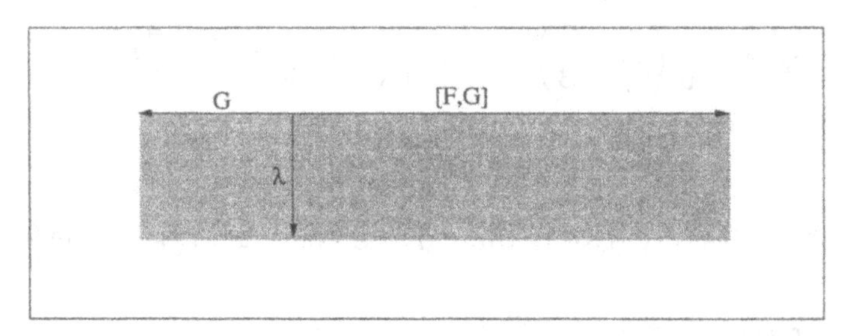

Fig. 2.12. Example 3

Now from (2.26) it follows:

$$\forall t \in [s^*, s^* + \varepsilon^*] \qquad \Delta_B(\gamma^*(t)) > 0 \tag{2.30}$$

thus, for each $t \in [s^*, s^* + \varepsilon^*]$, the couple of vectors $(G(t), [F, G](t))$ is a positive oriented basis of $\mathbb{R}^2$. Since $u^*(t)|[s^*, s^* + \varepsilon^*] \equiv 1$, it follows that the two functions $\lambda^*(t) \cdot G(\gamma^*(t))$ and $\lambda^*(t) \cdot [F, G](\gamma^*(t))$ are positive in a right neighborhood of s^*. Therefore G and $[F, G]$ lie in the darkened region of Fig. 2.12. But this is prohibited by (2.30), hence it follows that $\varepsilon^* = 0$.

Similarly, if the trajectories γ_s, $s \in [ln(\sqrt[3]{2}), s^*]$,satisfy the PMP, then ε_s is small. More precisely the algorithm $\mathcal{A}$ constructs trajectories that have a second switching point and these switching points form a switching curve C_1 originating at (2.27).

The above geometric reasoning is very general, however in this case we can compute explicit calculations. Suppose that γ_s satisfies the PMP with adjoint variable λ^s. The equation for λ^s is:

$$\dot{\lambda}^s(t) = -\lambda^s(t) \cdot \left(\nabla F(\gamma_s(t)) + u_s(t)\ \nabla G(\gamma_s(t))\right) = -\lambda^s(t) \cdot \left(\nabla F(\gamma_s(t))\right) \quad (2.31)$$

Let x_1^s be the first component of γ_s. For time $t \geq s$, the explicit form of (2.31) is:

$$(\dot{\lambda}_1^s, \dot{\lambda}_2^s)(t) = \left(-3\lambda_1^s(t) - \lambda_2^s(t)\left[2\left(x_1^s(s) + \frac{1}{3}\right)\ e^{3(t-s)} + \frac{1}{3}\right], 0\right). \quad (2.32)$$

Denote by ϕ_s the switching function along $(\gamma_s, u_s, \lambda^s)$. The solution to (2.32) with initial condition:

$$\lambda^s(s) \cdot G(\gamma_s(s)) = 0$$

is:

$$\lambda_2^s(t) \equiv \lambda_2^s(0)$$

$$\phi_s(t) = \lambda_1^s(t) = \lambda_2^s(t)\left[\frac{1}{3}\left(x_1^s(s) + \frac{2}{3}\right)\ e^{-3(t-s)} - \frac{1}{3}\left(x_1^s(s) + \frac{1}{3}\right)\ e^{3(t-s)} - \frac{1}{9}\right]$$

Now, the equation $\phi_s(t) = 0$ has two solutions:

$$t_1^s = s \qquad t_2^s = s + ln\left(\sqrt[3]{\frac{-3x_1^s(s) - 2}{3x_1^s(s) + 1}}\right)$$

thus

$$x_1^s(t_2^s) = -x_1^s(s) - 1$$

gives the first component of switching points belonging to C_1.

The point $\gamma^-(ln\ \sqrt[3]{4})$ is conjugate to the origin along γ^-. Consider the trajectories γ_r, $In(\gamma_r) = 0$, $Dom(\gamma_r) = [0, b_r]$ $(b_r \geq r)$, that originate as Y–trajectories and have a switching at time r going on as X–trajectories. Again we can make direct calculations and obtain the existence of a second switching time (if $r < ln\ \sqrt[3]{2}$):

$$t_r = r + ln\left(\sqrt[3]{-1 - \frac{10}{6\ a_r}}\right) \quad (2.33)$$

where $a_r = \left(x_1^r(r) - \frac{1}{3}\right)$ and x_1^r is first component of γ_r. These switching points form another switching curve C_2 that intersects γ^- at $\gamma^-(ln\sqrt[3]{4})$. Then from (2.33) it follows:

$$x_1^r(t_r) = -x_1^r(r) - 1.$$

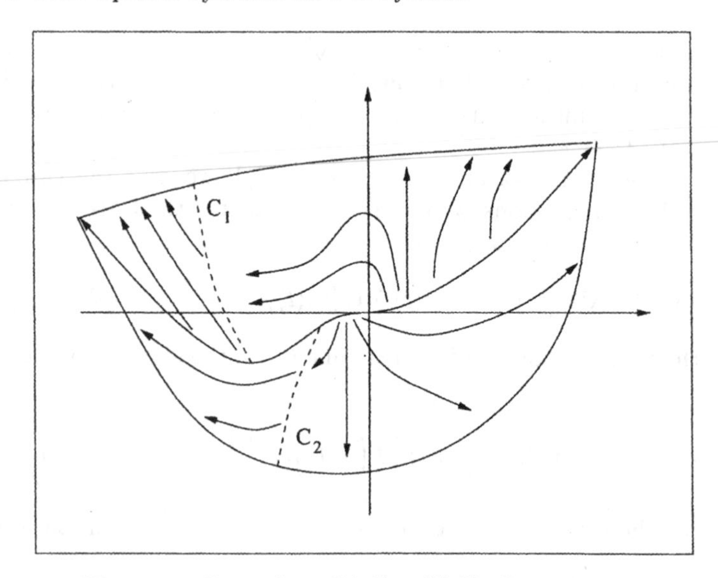

Fig. 2.13. Example 3, $(Y,C)_1$, $(Y,C)_2$ Frame Points

In Fig. 2.13, the reachable set $\mathcal{R}(\tau)$ is represented.

Example 4: $(Y,K)_2$ Frame Point.
Consider ε, $0<\varepsilon<1$, $\tau>\frac{\pi}{\sqrt{1-\varepsilon}}$ and the system Σ:

$$\begin{cases} \dot{x}_1 = \varepsilon x_2 + u x_2 \\ \dot{x}_2 = u(1-x_1) \end{cases} . \tag{2.34}$$

It is easy to check that:

$$[F,G] = \begin{pmatrix} -\varepsilon(1-x_1) \\ -\varepsilon x_2 \end{pmatrix} . \tag{2.35}$$

From (2.35) and Lemma 3, p. 39 we have that the system is locally controllable.
The X–trajectory passing through the point (x_1^0, x_2^0) at time 0 is:

$$x_1(t) = (x_1^0 - 1)\cos(\sqrt{1-\varepsilon}\; t) + x_2^0\sqrt{1-\varepsilon}\sin(\sqrt{1-\varepsilon}\; t) + 1 \tag{2.36}$$

$$x_2(t) = \frac{(x_1^0 - 1)}{\sqrt{1-\varepsilon}}\sin(\sqrt{1-\varepsilon}\; t) - x_2^0\cos(\sqrt{1-\varepsilon}\; t). \tag{2.37}$$

The Y–trajectory passing through the point (x_1^0, x_2^0) at time 0 is:

$$x_1(t) = (x_1^0 - 1)\cos(\sqrt{1+\varepsilon}\; t) + x_2^0\sqrt{1+\varepsilon}\sin(\sqrt{1+\varepsilon}\; t) + 1, \tag{2.38}$$

$$x_2(t) = -\frac{(x_1^0 - 1)}{\sqrt{1+\varepsilon}} \sin(\sqrt{1+\varepsilon}\ t) + x_2^0 \cos(\sqrt{1+\varepsilon}\ t). \tag{2.39}$$

The equation for turnpikes is:

$$0 = \Delta_B(x_1, x_2) = -\varepsilon x_2^2 + \varepsilon(1 - x_1)^2. \tag{2.40}$$

Hence every turnpike is a subset of $S = \{(x_1, x_2) : x_2 = \pm(1 - x_1)\}$. Using (2.38)-(2.40) it is easy to verify that the trajectory γ^+ intersects the set S in a point (x_1^+, x_2^+) of the first quadrant. The algorithm $\mathcal{A}$ constructs the turnpike $S_1 = \{(x_1, x_2) : x_2 = 1 - x_1, x_1^+ \leq x_1 < 1\}$. The singular control φ_S^1 along the turnpike S_1, cfr. (2.13), is:

$$\varphi_S^1(x_1, x_2) = -\frac{\varepsilon x_2}{1 - x_1 + x_2} > -1. \tag{2.41}$$

From (2.41) we have:

$$\dot{x}_1(\varphi_S^1) = \frac{\varepsilon}{2}\,(1 - x_1)$$

hence the point $(1, 0)$ is not reached in finite time by a singular trajectory. Similarly using (2.36), (2.37) it is easy to verify that the trajectory γ^- intersects the set S in a point of the fourth quadrant:

$$(x_1^-, x_2^-) \doteq \gamma^-\left(\frac{1}{\sqrt{1-\varepsilon}} \arccos\left(\sqrt{\frac{1}{2-\varepsilon}}\right)\right).$$

Hence the algorithm $\mathcal{A}$ constructs the turnpike $S_2 = \{(x_1, x_2) : x_2 = x_1 - 1, x_1 \leq x_1^-\}$. Indeed, the control φ_S^2, cfr. (2.13), is:

$$\varphi_S^2(x_1, x_2) = \frac{\varepsilon x_2}{1 - x_1 - x_2}.$$

The trajectories $\gamma^\pm$ are very close to the circle A of center $(1, 0)$ and radius 1; γ^+ runs clockwise and γ^- counterclockwise. From (2.36)-(2.39) we have that γ^+ lies inside A, γ^- outside, and:

$$\gamma^+ \cap \gamma^- \cap A = \{(0, 0), (2, 0)\}.$$

However the two trajectories $\gamma^\pm$ do not meet each other at $(2, 0)$, indeed:

$$(2, 0) = \gamma^+\left(\frac{\pi}{\sqrt{1+\varepsilon}}\right) = \gamma^-\left(\frac{\pi}{\sqrt{1-\varepsilon}}\right).$$

But the $X * Y$ and $Y * X$ trajectories constructed by the algorithm give rise to an overlap curve K and $\gamma^\pm$ end on it. In Fig. 2.14, $\mathcal{R}(\tau)$ is represented.

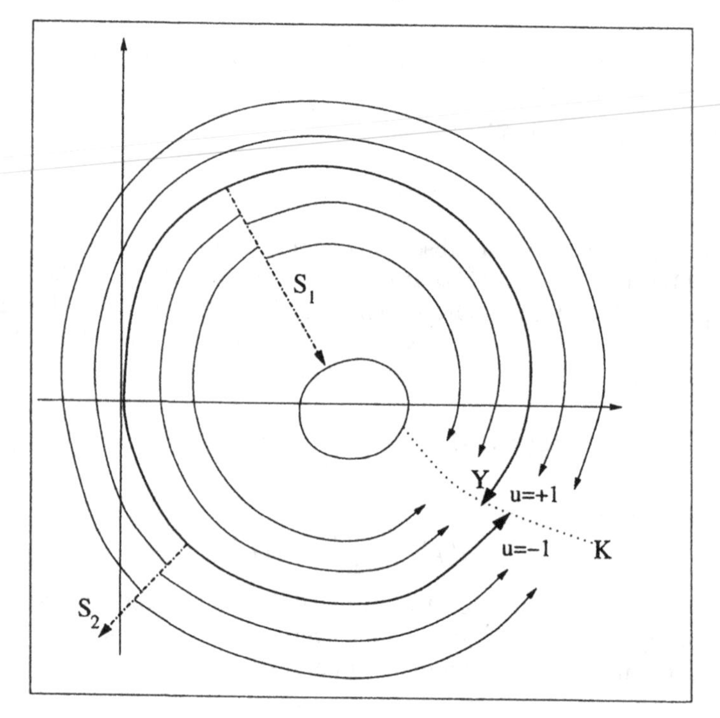

Fig. 2.14. Example 4, $(Y, K)_2$ Frame Point

Example 5: $(Y, K)_3$ **Frame Point.**
Consider the system (cfr. Example 1):

$$\begin{cases} \dot{x}_1 = u \\ \dot{x}_2 = \frac{1}{2}x_1^2 + x_1 \end{cases} \tag{2.42}$$

and the two embedded submanifold:

$$M_1 = \left\{ (x_1, x_2) : x_1 = 0, -\frac{2}{3} \leq x_2 \leq 0 \right\}$$

$$M_2 = \left\{ (x_1, x_2) : x_1 = \sqrt{2}, -\frac{2}{3} \leq x_2 \leq \frac{2}{3} \right\}.$$

We assume that for each $y \in M_1$ there exists a Y–trajectory $\gamma_1(y)$ that verifies $\gamma_1(y)(0) = y$. Moreover, for each $x \in M_2$ there exists an X–trajectory $\gamma_2(x)$ that arises from x at time 0. Finally, an X–trajectory originates from each point of the Y–trajectory $\gamma_1(0)$, i.e. $\gamma_1(0)$ is the trajectory γ^+ of a given system.
At the point $(1, \frac{2}{3})$ the trajectories $\gamma_1(0)$, $\gamma_2((\sqrt{2}, 0))$ meet each other. After this point $\gamma_1(0)$ is not constructed by $\mathcal{A}$ because the trajectories

$\gamma_2((\sqrt{2},c))$, $c \geq 0$, achieve a better performance. The trajectories $\gamma_1((0,-c))$ and $\gamma_2((\sqrt{2},-c))$ meeting each other give raise to an overlap curve:

$$K = \left\{(x_1,x_2) : x_1 = -1, 0 \leq x_2 \leq \frac{2}{3}\right\}.$$

In Fig. 2.15, this local example is portrayed.

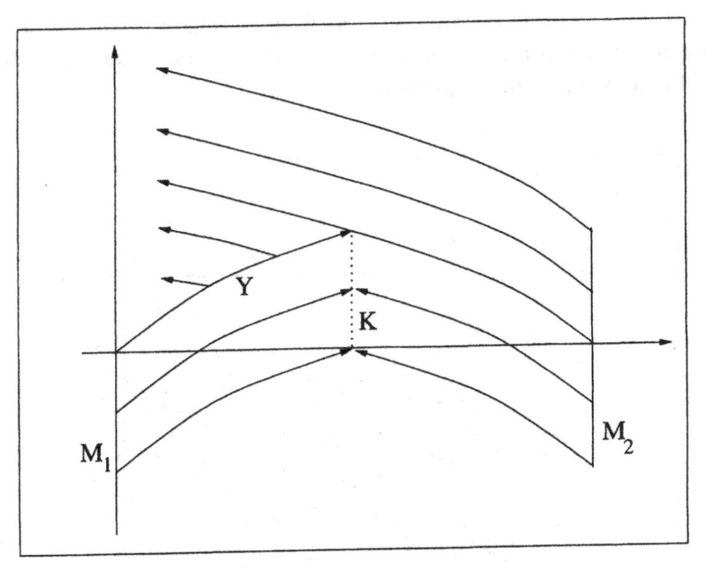

Fig. 2.15. Example 5, $(Y,K)_3$ Frame Point

Example 6: $(C,C)_1$ Frame Point.
Consider the system (2.25) of Example 3 and the manifold:

$$M = \{(x_1,x_2) : x_1 = 0, |x_2| \leq 1\}.$$

We assume that from every point $(0,x_2) \in M$ an X–trajectory $\gamma(x_2)$ arises, with initial time 0, and with adjoint variable satisfying:

$$[\lambda_1(x_2)](0) = -\frac{9}{36} - \frac{1}{36}\, sgn(x_2)\, x_2 \qquad [\lambda_2(x_2)](0) = -1.$$

With simple calculations we obtain the solutions to the equation $[\phi(x_2)](t) = 0$, where $\phi(x_2)$ is the switching function along $(\gamma(x_2), -1, \lambda(x_2))$:

$$t^{\pm}(x_2) = ln\left(\sqrt[3]{\frac{5}{2} \pm \frac{1}{2}\sqrt{9 + 36\,[\lambda_1(x_2)](0)}}\right).$$

Hence, the trajectories $\gamma(x_2)$, $x_2 \leq 0$, have a switching at time $t^-(x_2)$, while the trajectories $\gamma(x_2)$, $x_2 > 0$, do not switch. These switching points form a switching curve C_1 having the point (2.27) as endpoint.
Now the equation $\phi(x_2) = 0$ has another solution after the time $t^-(x_2)$, namely:

$$t'(x_2) = t^-(x_2) + ln\left(\sqrt[3]{\frac{-3\,p_1(x_2) - 2}{3\,p_1(x_2) + 1}}\right)$$

where $p_1(x_2)$ is the first coordinate of the first switching point of $\gamma(x_2)$. These switching points form another switching curve C_2 that meet C_1 at the point (2.27). This local example is portrayed in Fig. 2.16.

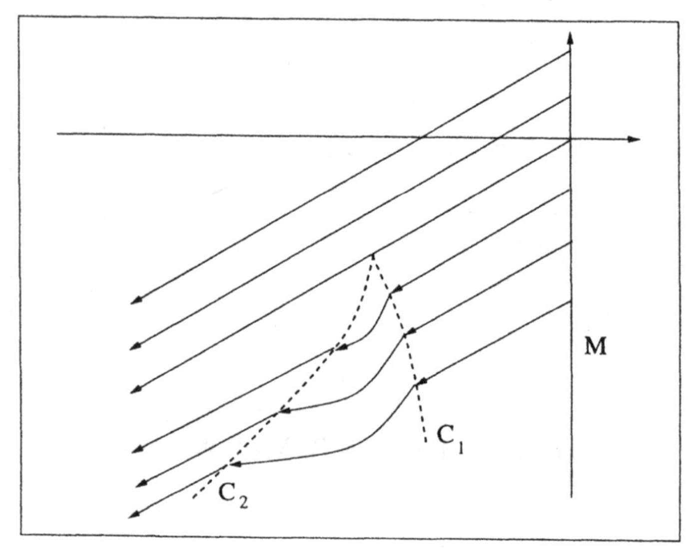

Fig. 2.16. Example 6, $(C,C)_1$ Frame Point

Example 7: $(C,S)_1$ **Frame Point.**
Consider the system (2.42) of Example 5 and the manifold:

$$M = \left\{(x_1, x_2) : x_1 = 0, -\frac{1}{3} \leq x_2 \leq \frac{1}{3}\right\}.$$

We assume that from each $(x_1, x_2) \in M$ there arises, with initial time 0, an X–trajectory $\gamma(x_1, x_2) = \gamma(x_2)$ with adjoint variable $\lambda(x_2)$ that satisfies:

$$[\lambda_1(x_2)](0) = \frac{-1 - 4\,sgn(x_2)\,x_2^2}{2} \qquad [\lambda_2(x_2)](0) = -1,$$

where $sgn(x) = x \mid x \mid^{-1}$ if $x \neq 0$ and $sgn(0) = 0$. Now, the switching function along $(\gamma(x_2), -1, \lambda(x_2))$ is:

$$[\phi(x_2)](t) = \lambda_1(t) = -\frac{t^2}{2} + t - \frac{1 + 4\, sgn(x_2)\, x_2^2}{2}.$$

If $x_2 \leq 0$ the equation $[\phi(x_2)](t) = 0$ has the following solutions:

$$t_1(x_2) = 1 + 2 \mid x_2 \mid \qquad t_2(x_2) = 1 - 2 \mid x_2 \mid$$

otherwise there is no solution. Then every trajectory $\gamma(x_2)$, $x_2 \leq 0$, switches at the point:

$$[\gamma(x_2)](t_2) = \left(2 \mid x_2 \mid -1, -\frac{(2 \mid x_2 \mid -1)^3}{6} - \frac{(2 \mid x_2 \mid -1)^2}{2} + x_2\right).$$

These switching points form a switching curve C.
The trajectory $\gamma(0)$ crosses the set $\{(x_1, x_2) : \Delta_B(x_1, x_2) = 0\} = \{(x_1, x_2) : x_1 = -1\}$ at a switching point, hence the algorithm $\mathcal{A}$ constructs the turnpike:

$$S = \left\{(x_1, x_2) : x_1 = -1, x_2 \leq -\frac{1}{3}\right\}.$$

This local example is represented in Fig. 2.17.

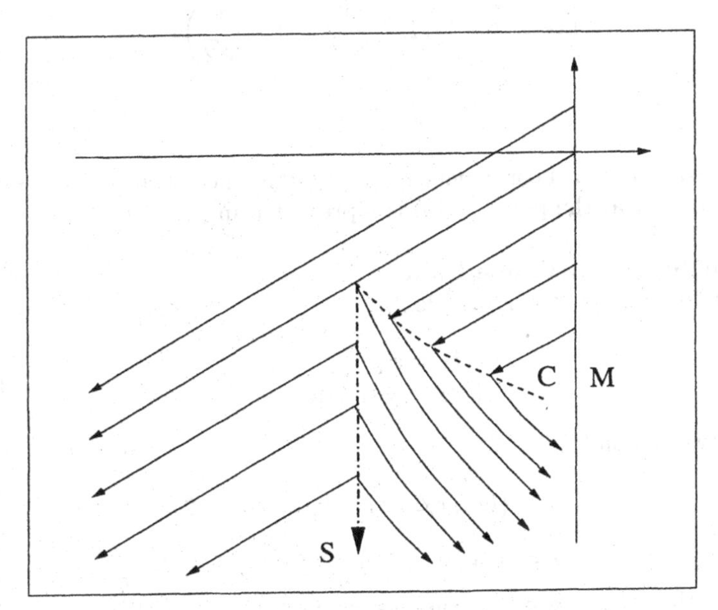

Fig. 2.17. Example 7, $(C, S)_1$ Frame Point

Example 8: $(C,S)_2$ Frame Point.
Let $\tau > \frac{7}{3} + \sqrt[3]{4}$ and consider the system:

$$\begin{cases} \dot{x}_1 = u \\ \dot{x}_2 = \Big(x_1 + \psi(x_2)\Big) + \frac{1}{2}\Big(x_1 + \psi(x_2)\Big)^2 \end{cases} \tag{2.43}$$

where:

$$\psi(x_2) = \begin{cases} 0 & x_2 > -1 \\ (x_2+1)^4 & x_2 \leq -1. \end{cases}$$

Observe that for $x_2 > -1$ the system is the same as the first example. There is a turnpike S that lies on the line $x_1 = -1$ between the points $(-1, -\frac{1}{3})$ and $(-1, -1)$. Moreover, for $x_2 \leq -1$, S is represented by the equations:

$$x_1 + (x_2+1)^4 + 1 = 0 \qquad x_2 \leq -1. \tag{2.44}$$

Recalling (2.13), from (2.43)-(2.44) we have that the control φ is:

$$\begin{array}{ll} \varphi(x_1, x_2) = 0, & \text{if } x_2 \geq -1, \\ \varphi(x_1, x_2) = 2(x_2+1)^3, & \text{if } x_2 \leq -1. \end{array} \tag{2.45}$$

By (2.45), the turnpike S is regular up to the point:

$$(\bar{x}_1, \bar{x}_2) = \left(-1 - \frac{1}{2\sqrt[3]{2}}, -1 - \frac{1}{\sqrt[3]{2}}\right) \tag{2.46}$$

indeed:

$$\varphi(\bar{x}_1, \bar{x}_2) = -1.$$

Hence, the algorithm $\mathcal{A}$ constructs a turnpike that ends at the the point (2.46). The set $\mathcal{R}(\tau)$ near the point (2.46) is represented in Fig. 2.18.

Example 9: $(C,K)_1$ Frame Point.
Consider the system (cfr. Example 2):

$$\begin{cases} \dot{x}_1 = x_2 \\ \dot{x}_2 = -x_1 + u \end{cases} \tag{2.47}$$

and the two manifolds:

$$M_1 = \{(x_1, x_2) : x_1 = 0, 1 \leq x_2 \leq 2\}$$

$$M_2 = \{(x_1, x_2) : 1 \leq x_1 \leq 2, x_2 = 3\}.$$

The algorithm $\mathcal{A}$ succeeds for the system (2.47) at time 4. A Y–trajectory $\gamma'(x) \in \Gamma_{\mathcal{A}}(\Sigma, 4)$, with an associate adjoint variable $\lambda'(x)$, passes through each point $x \in M_1$. We suppose that from each $x \in M_1$ a Y–trajectory $\gamma(x)$ with adjoint variable $\lambda(x)$ arises at time 0. Moreover, $(\gamma(x), \lambda(x))$ is obtained from $(\gamma'(x), \lambda'(x))$ shifting the time.

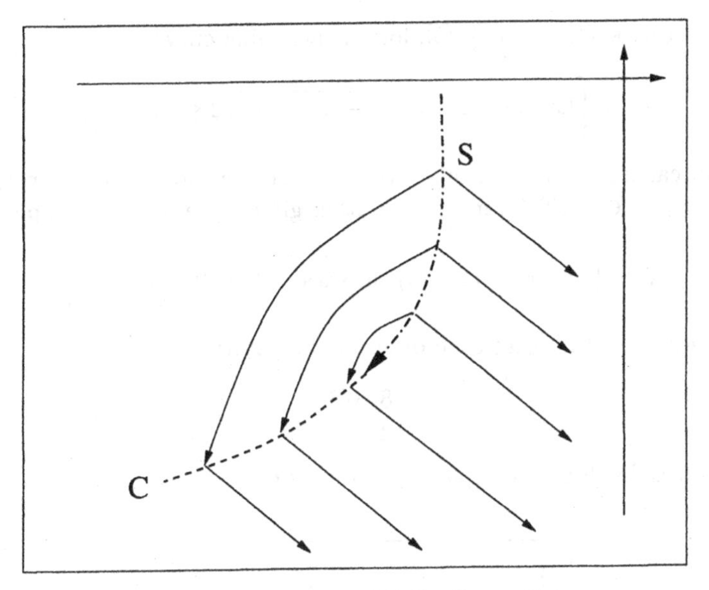

Fig. 2.18. Example 8, $(C, S)_2$ Frame Point

Consider the line given by the equation:

$$x_2 = (\sqrt{2} - 3)\ x_1 + 8 - \sqrt{8}. \tag{2.48}$$

For each $s \in [1, \frac{2}{3}\sqrt{6}]$ the trajectory $\gamma^s = \gamma\big((0, s)\big)$ intersects the line (2.48) in a point, say (x_1^s, x_2^s).
Let $r(s) \in [1, 2]$ be such that the X–trajectory passing through $(r(s), 3)$ intersects the line (2.48) in (x_1^s, x_2^s). We assume that an X–trajectory γ_r with initial time $t(r)$ originates from each $(r, 3) \in M_2$. If $r = r(s)$ for some s let define, denoting by d the Euclidean distance:

$$t_s \doteq t\big(r(s)\big) = 2\ \arcsin\left(\frac{d\big((0, s), (x_1^s, x_2^s)\big)}{2\sqrt{1 + s^2}}\right) - 2\ \arcsin\left(\frac{d\big((r(s), 3), (x_1^s, x_2^s)\big)}{2\sqrt{(r(s) + 1)^2 + 9}}\right) \tag{2.49}$$

otherwise:

$$t(r) \doteq \max\,\{t_s : s \in [1, 2]\}.$$

We associate to every γ_r an adjoint variable λ_r verifying:

$$\lambda_1^r\big(t(r)\big) = -1 \qquad \lambda_2^r\big(t(r)\big) = 0. \tag{2.50}$$

The equation (2.50) implies that $\lambda^r\big(t(r)\big) \cdot G\big((r, 3)\big) = 0$ then to satisfy the PMP the trajectory γ_r is an X–trajectory for a time interval of length π.

The trajectories γ^s, $s \in [1, \frac{2}{3}\sqrt{6}]$, form a switching curve:

$$C = \left\{ (x_1, x_2) : x_2 = \sqrt{1 - (x_1 - 3)^2}, 2 \le x_1 \le \frac{8}{3} \right\}.$$

By direct calculations, one can verify that (2.49) ensures that the trajectories γ^s and $\gamma_{r(s)}$, $s \in [\frac{2}{3}\sqrt{6}, 2]$, meet each other giving rise to an overlap curve:

$$K = \left\{ (x_1, x_2) : (x_1, x_2) \text{ satisfies (2.48)}\ , 2 \le x_1 \le \frac{8}{3} \right\}.$$

The curves K and C meet each other at the point:

$$(\frac{8}{3}, \frac{\sqrt{8}}{3}).$$

In Fig. 2.19, this local example is portrayed.

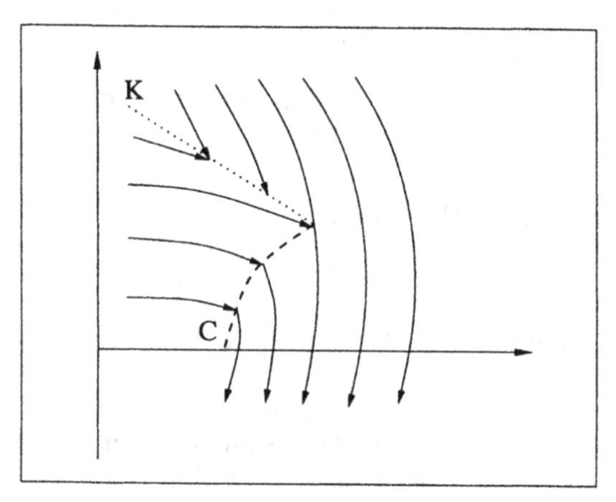

Fig. 2.19. Example 9, $(C, K)_1$ Frame Point

Example 10: $(C, K)_2$ Frame Point, with $\Delta_B = 0$.
Consider the system:

$$\begin{cases} \dot{x}_1 = 3\, x_1 - u \\ \dot{x}_2 = x_1^2 + x_1 \end{cases} \tag{2.51}$$

that is obtained from the system (2.25) replacing G with $-G$, and the manifold:

$$M = \{(x_1, x_2) : x_1 = 0, |x_2| < \varepsilon\}.$$

We assume that from every point $(0, x_2) \in M$ a Y–trajectory $\gamma(x_2)$ with initial time $t_0(x_2)$ arises.

Moreover, $\gamma(x_2)$ admits an adjoint variable satisfying:

$$[\lambda_1(x_2)](0) = -\frac{9}{36} + 25\, sgn(x_2)\, x_2 \qquad [\lambda_2(x_2)](0) = -1.$$

With simple calculations we obtain the solutions to the equation $[\phi(x_2)](t) = 0$, where $\phi(x_2)$ is the switching function along $(\gamma(x_2), +1, \lambda(x_2))$:

$$t^{\pm}(x_2) = t_0(x_2) + ln\left(\sqrt[3]{\frac{5}{2} \pm \frac{1}{2}\sqrt{9 + 36\,[\lambda_1(x_2)](0)}}\right). \tag{2.52}$$

Hence the trajectories $\gamma(x_2)$, $x_2 \geq 0$, have a switching at time $t^-(x_2)$, while the trajectories $\gamma(x_2)$, $x_2 < 0$, do not switch. Let $x^-(x_2) \doteq [\gamma(x_2)](t^-(x_2))$, $x_2 \geq 0$. From (2.52) we have:

$$x_1^- = -\frac{1}{2} + 5\, x_2^2 \tag{2.53}$$

and these switching points form a switching curve C_1 originating from (2.27).

As for Example 6, the equation $\phi(x_2) = 0$, where $\phi(x_2)$ denotes again the switching function along $\gamma(x_2)$, after the time $t^-(x_2)$ has another solution:

$$t'(x_2) = t^-(x_2) + ln\left(\sqrt[3]{\frac{-3\, x_1^-(x_2) - 2}{3\, x_1^-(x_2) + 1}}\right). \tag{2.54}$$

These switching points give raise to another switching curve C_2 that meets C_1 at the point (2.27).
It is easy to verify that the X–trajectories leaving from C_1 cross the trajectory $\gamma_0 \doteq \gamma(0)$ before reaching the switching curve C_2. Hence, we can define $P(x_2)$, $x_2 \geq 0$, to be the point at which $\gamma(x_2)$ meets γ_0.
Let $r(x_2)$, $x_2 \geq 0$, be such that the trajectory $\gamma(r(x_2))$ meets $\gamma(x_2)$ at the point $Q(x_2) \doteq [\gamma(x_2)](t'(x_2))$, i.e. they meet each other on C_2.
Now let $t_1(x_2), t_2(x_2)$ be the time in which, respectively, $\gamma(x_2), \gamma_0$ reaches $P(x_2)$ and let $t_3(x_2), t_4(x_2)$ be the time in which, respectively, $\gamma(x_2), \gamma(r(x_2))$ reaches $Q(x_2)$. If x_2 is sufficiently small, from (2.53), (2.54) we have that:

$$t_4 - t_3 < t_2 - t_1,$$

then, taking ε sufficiently small, we can define:

$$\begin{cases} t_0(x_2) = 0 & \text{if } \ x_2 < 0, \\ t_0(x_2) = \frac{(t_2 - t_1) + (t_4 - t_3)}{2} & \text{if } \ x_2 \geq 0. \end{cases}$$

With this choice, the trajectories $\gamma(x_2)$, $x_2 \geq 0$, and $\gamma(x_2)$, $x_2 < 0$, meet each other forming an overlap curve K that meets C_1 at (2.27). The curve C_2 is deleted by the algorithm.
In Fig. 2.20, this local example is portrayed.

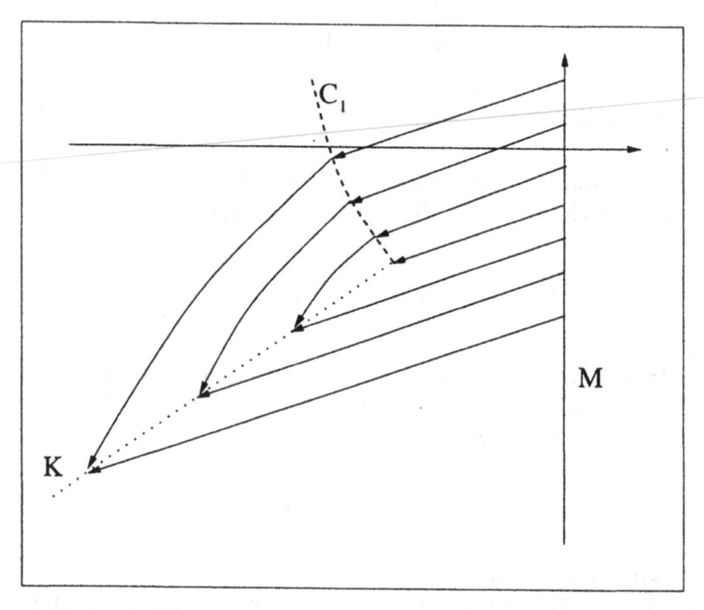

Fig. 2.20. Example 10: $(C,K)_2$ Frame Point, with $\Delta_B = 0$

Example 11: $(C,K)_2$ Frame Point, with $\Delta_B \neq 0$.
Consider the system:

$$\begin{cases} \dot{x}_1 = \frac{x_1+1}{2} + u\,\frac{x_1-1}{2} \\ \dot{x}_2 = \frac{x_2}{2} + u\,\frac{x_2}{2} \end{cases} \tag{2.55}$$

and the manifold:

$$M = \{(x_1, x_2) : x_1 = 0, 0 < x_2 < 1\}.$$

Assume that from every point $(0, x_2) \in M$, with initial time 0, an X–trajectory $\gamma(x_2)$ originates with adjoint variable satisfying:

$$[\lambda_1(x_2)](0) = \frac{x_2}{1 - \sqrt{1 - x_2^2}} \qquad [\lambda_2(x_2)](0) = -1. \tag{2.56}$$

It is easy to verify, from (2.55), (2.56), that every $\gamma(x_2)$ switches at time:

$$t(x_2) = 2 - \sqrt{1 - x_2^2},$$

and the corresponding switching points form a switching curve:

$$C = \{(x_1, \psi(x_1)) : 1 < x_1 < 2\}, \qquad \psi(x_1) \doteq \sqrt{1 - (2 - x_1)^2},$$

that is an arc of circle.

Observe that for ε small, $Y(\varepsilon, \psi(\varepsilon))$ points to the right of C and $Y(2 - \varepsilon, \psi(2-\varepsilon))$ points to the left of C. Then there exists $(\bar{x}_1, \bar{x}_2) \in C$ such that $Y(\bar{x}_1, \bar{x}_2)$ is tangent to C. Define $C' \doteq \{(x_1, \psi(x_1)) \in C : x_1 \geq \bar{x}_1\}$. The trajectories $\gamma(x_2)$ that reach C' meet other trajectories $\gamma(x_2)$ giving rise to an overlap curve K. It is possible to move along C' with a trajectory of the system, hence we can construct an envelope for the curves $\gamma(x_2)|[0, t(x_2)]$ that reach C'; see Section 1.3.2, p. 26 for envelope theory. Hence the subset C' of C is removed by the algorithm.
In Fig. 2.21 this local example is represented.

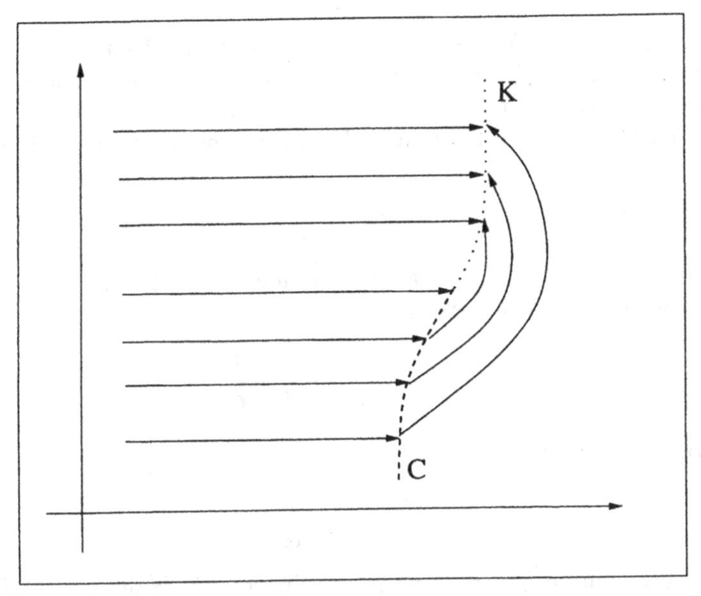

Fig. 2.21. Example 11, $(C, K)_2$ Frame Point with $\Delta_B \neq 0$

Example 12: $(S, K)_1$ Frame Point.
Consider the same system and the same manifold of Example 7 and define S in the same way. We assume that from each $(0, x_2) \in M$ an X–trajectory $\gamma(x_2)$ arises with initial time:

$$t_0(x_2) = -\frac{2}{3}\, x_2$$

and with adjoint variable satisfying:

$$[\lambda_1(x_2)](0) = \frac{-1 - \alpha\, sgn(x_2)\, x_2^2}{2} \qquad [\lambda_2(x_2)](0) = -1,$$

where $\alpha > 0$ and sgn is defined as in Example 7. There is again a switching curve C.
The X–trajectories starting from $(0, x_2)$, $x_2 \leq 0$, reach $(-1, -\frac{1}{3} - x_2) \in S$ at time:

$$1 + \frac{2}{3}\,|x_2|.$$

On the other hand the $Z * X$–trajectories, concatenations of an X- and a Z-trajectory, starting from 0 reach the same point at time:

$$1 + \frac{|x_2|}{2}.$$

Therefore the $Y * Z * X$–trajectories, concatenations of an X-, a Z- and a Y–trajectory, starting from the origin and the X–trajectories starting from $(0, x_2)$, $x_2 \leq 0$, give rise to an overlap curve K having $(-1, -\frac{1}{3})$ as endpoint. Let $(s, k(s))$ be a parametrization of K in a neighborhood of $(-1, -\frac{1}{3})$ and define:

$$\beta \doteq \left.\frac{dk(s)}{ds}\right|_{s=-1^+}.$$

Now let $(c_1(x_2), c_2(x_2))$ be a parametrization of the switching curve C. After straightforward calculations we have:

$$\left.\frac{dc_2}{dc_1}\right|_{c_1=-1^+} = -\frac{3}{2} - \frac{1}{\alpha}.$$

Thus if α is sufficiently small:

$$\beta \geq -\frac{3}{2} - \frac{1}{\alpha}$$

and the overlap curve K arises. Therefore the curve C is deleted by the algorithm.
This local example is portrayed in Fig. 2.22.

Example 13: $(S, K)_2$, (K, K) Frame Points.
Consider $0 < \varepsilon << 1$, $\tau > 1$ and the system Σ:

$$\begin{cases} \dot{x}_1 = u \\ \dot{x}_2 = x_1^3 - \varepsilon x_1 \end{cases} \qquad (2.57)$$

We have that:

$$[F, G] = \begin{pmatrix} 0 \\ \varepsilon - 3x_1^2 \end{pmatrix}$$

hence the system is locally controllable. The X–, Y–trajectories are quartic polynomials of the following types, respectively:

$$x_2 = -\frac{x_1^4}{4} + \varepsilon\,\frac{x_1^2}{2} + \alpha \qquad \alpha \in \mathbb{R}$$

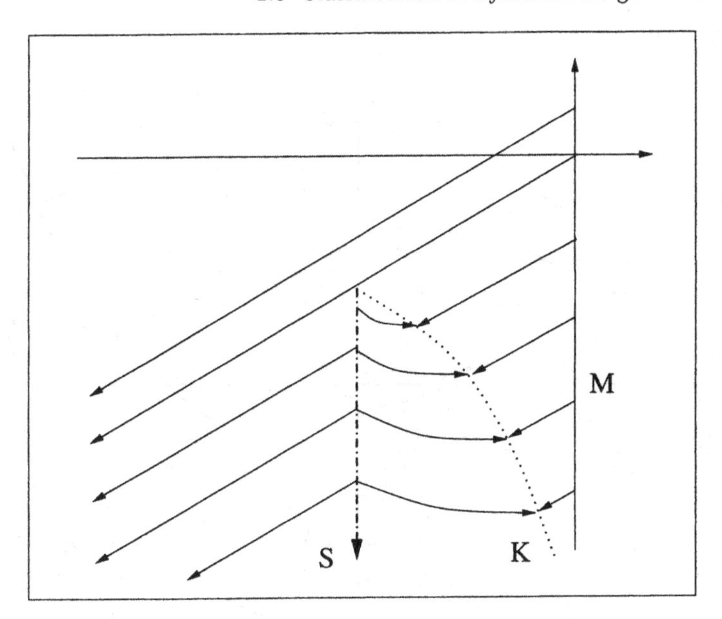

Fig. 2.22. Example 12, $(S,K)_1$ Frame Point

$$x_2 = \frac{x_1^4}{4} - \varepsilon\,\frac{x_1^2}{2} + \alpha \qquad \alpha \in \mathbb{R}.$$

The equation for turnpikes is:

$$0 = \Delta_B(x_1, x_2) = \varepsilon - 3x_1^2$$

whose set of solutions is:

$$\left\{(x_1, x_2) : x_1 = \pm\sqrt{\frac{\varepsilon}{3}}\right\}. \tag{2.58}$$

Every turnpike is subset of (2.58). The algorithm constructs the turnpikes:

$$S_1' = \left\{(x_1, x_2) : x_1 = \sqrt{\frac{\varepsilon}{3}}, x_2 \leq -\frac{5}{36}\varepsilon^2\right\} \cap \mathcal{R}(\tau)$$

$$S_2' = \left\{(x_1, x_2) : x_1 = -\sqrt{\frac{\varepsilon}{3}}, x_2 \geq \frac{5}{36}\varepsilon^2\right\} \cap \mathcal{R}(\tau).$$

The points $(\pm\sqrt{\varepsilon}, \mp\frac{\varepsilon^2}{4})$ are conjugate to the origin along $\gamma^{\pm}$. Two overlap curves K_2, K_1, respectively, originate at these points. The algorithm deletes partially the turnpikes S_1', S_2' determining two new turnpikes $S_1 \subset S_1'$, $S_2 \subset S_2'$. The new turnpikes S_1, S_2 end on K_1, K_2 respectively.
In Fig. 2.23, $\mathcal{R}(\tau)$ is represented.

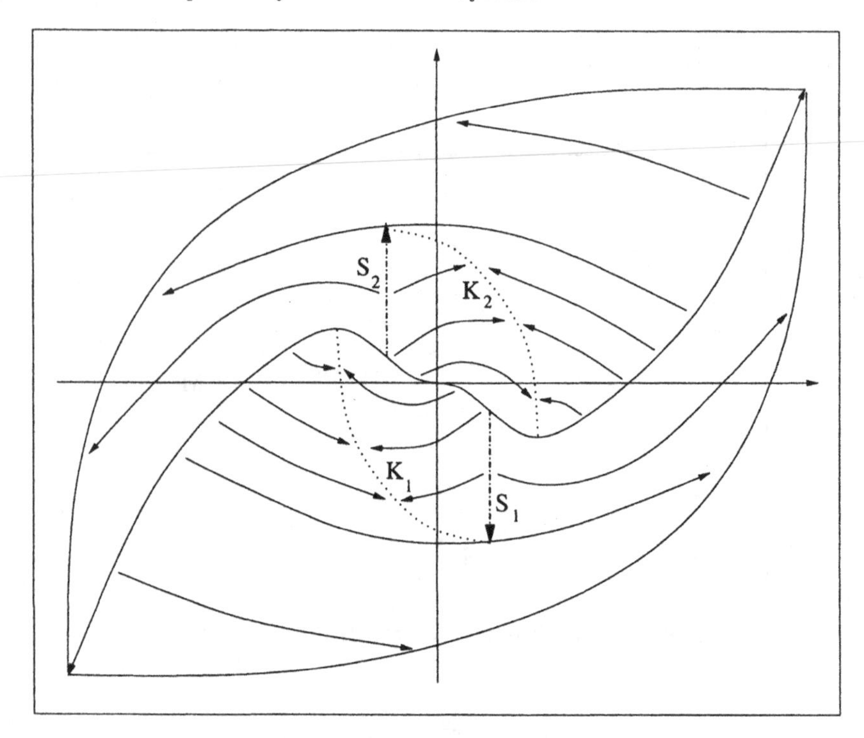

Fig. 2.23. Example 13, $(S,K)_2$, (K,K) Frame Points

Example 14: $(S,K)_3$ Frame Point.
Consider the system Σ with

$$F \equiv \begin{pmatrix} 0 \\ 1 \end{pmatrix}, \quad G(x) = \begin{pmatrix} x_1 - x_2 \\ x_1 \end{pmatrix}.$$

Then $[F,G] \equiv (-1,0)$, $\Delta_A(x) = x_2 - x_1$ and $\Delta_B(x) = x_1$. Assume that a singular trajectory γ_S runs the turnpike $S = \{x : x_1 = 0\}$ starting from a point $(0,a)$ for some $a < 0$. Let $\gamma^{y,\pm}$ be the extremal trajectory leaving the turnpike from $(0,y)$, with control ± 1, and let $\lambda^{y,\pm}$ the corresponding covector.

The trajectories $\gamma^{y,\pm}$, $y < 0$, can not switch before crossing the line $\Delta_A^{-1}(0)$ by Theorem 11. To fix the ideas let us consider first $\gamma^{y,+}$. Then the corresponding switching function $\phi^{y,+}$ satisfies:

$$\phi^{y,+}(t) = \lambda_1^{y,+}(t)(\gamma_1^{y,+}(t) - \gamma_2^{y,+}(t)) + \lambda_2^{y,+}(t)\gamma_1^{y,+}(t),$$

where $\gamma_i^{y,+}$, respectively $\lambda_i^{y,+}$, are the components of $\gamma^{y,+}$, respectively $\lambda^{y,+}$. One easily gets, for $y < 0$ and $t > 0$ sufficiently small, $\lambda_1^{y,+}(t) < 0$ and $\lambda_2^{y,+}(t) > 0$, thus $\gamma^{y,+}$ can not switch on the region over $\Delta_A^{-1}(0)$. Hence the trajectories $\gamma^{y,+}$, and similarly $\gamma^{y,-}$, never switch in a neighborhood of the origin.

Consider now a point $(0, b)$, $b > 0$. There are three trajectories reaching this point: γ_S and $\gamma^{y,\pm}$ for some $y < 0$. Along the first trajectory, the equation for the second component is $\dot{x}_2 = 1$, while along the others is $\dot{x}_2 = 1 \pm x_1$, hence we easily conclude that γ_S is not optimal after crossing the origin. A K curve starts at the origin formed by points reached optimally by $\gamma^{y,\pm}$, $y < 0$. This curve is a straight line but any second order perturbation of the system Σ renders it curved keeping the tangency with S at the origin, thus we obtain the situation of figure 2.24.

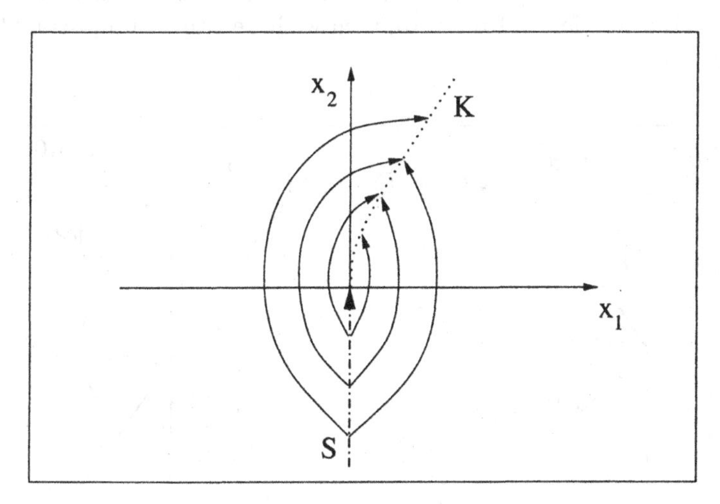

Fig. 2.24. Example 14, $(S, K)_3$ Frame Point

2.7 On the Relation Between the Switching Strategy and the Functions ϕ and θ^γ

The switching strategy of an extremal trajectory γ can be studied in two different ways: either by means of the switching function ϕ or using the function θ^γ. In this paragraph we summarize their properties. We fix an extremal trajectory γ starting from the origin, with never vanishing covector $\lambda(t)$.

The relation between the definitions of these two functions can be well understood by the following formula (cfr. Section 2.1, p. 35):

$$\phi(t) := \lambda(t) \cdot G(\gamma(t)) = \lambda(0) \cdot v^\gamma(G(\gamma(t)), t; 0).$$

The function θ^γ gives the angle of rotation of the adjoint vector v^γ with respect to its initial position $v^\gamma(G(\gamma(0)), 0; 0) = G(\gamma(0))$ (cfr. Definition 14,

p. 37). In other words, θ^γ is the angle between $G(0)$ and $G(\gamma(t))$ pulled back to the origin via the adjoint equation along γ. Notice that $\theta^\gamma(0) = 0$. We recall that switchings between bang arcs happen when ϕ changes sign, while along singular arcs we have $\phi \equiv 0$. How the switching strategy is described in terms of the function θ^γ is explained in the following and can be easily understood also by means of Figure 2.2, p. 37. Assume that the extremal trajectory γ starts from the origin with constant control $+1$ and let $\bar{t} > 0$ be the first time at which $\phi = 0$. If $G(\gamma(\bar{t})) \neq 0$, then the vectors $\lambda(0)$ and $v^\gamma(G(\gamma(\bar{t})), \bar{t}; 0)$ are orthogonal. Let $\bar{\theta} \in [-\pi/2, +\pi/2]$ be the value of θ^γ at time $\bar{t}$: $\bar{\theta} := \theta^\gamma(\bar{t})$. If $G(\gamma(t))$ is always different from zero, then we have situation illustrated in Figure 2.25:

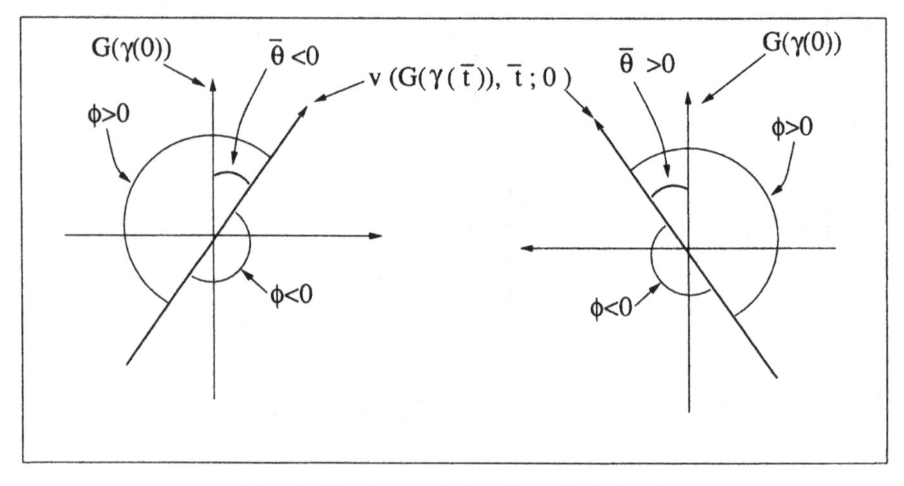

Fig. 2.25. ϕ and θ^γ for $G(\gamma(0)) = (0, 1)$.

- at time t we have $\phi(t) = 0$ if and only if $\theta^\gamma(t) = \bar{\theta} \pm n\pi$, $n \in \mathbb{N}$;
- $\phi(t) > 0$ (resp. $\phi(t) < 0$) for every t on some interval $]a, b[$ if for every $t \in]a, b[$ (modulo 2π):

$$\begin{cases} \theta^\gamma(t) \in]\bar{\theta}, \bar{\theta} + \pi[\text{ if } \bar{\theta} < 0 \\ \theta^\gamma(t) \in]\bar{\theta} - \pi, \bar{\theta}[\text{ if } \bar{\theta} > 0, \end{cases} \qquad \left(\text{resp.} \begin{cases} \theta^\gamma(t) \in]\bar{\theta} + \pi, \bar{\theta} + 2\pi[\text{ if } \bar{\theta} < 0 \\ \theta^\gamma(t) \in]\bar{\theta}, \bar{\theta} + \pi[\text{ if } \bar{\theta} > 0. \end{cases} \right).$$

The situation for a trajectory starting with constant control -1 is similar. To conclude we have the following:

- for a bang-bang trajectory switchings happen when:
  -) ϕ change sign;
  -) θ^γ crosses the values $\bar{\theta} \pm n\pi$, $n \in \mathbb{N}$. Moreover for an abnormal extremal $\bar{\theta} = 0$ (cfr. Section 4.3, p. 171, Proposition 20, p. 174).

- If γ is a singular arc on $[a, b]$, then for every t in this interval we have:
  -) $\phi(t) = \dot{\phi}(t) = 0$
  -) $\theta^\gamma(t) = \bar{\theta} \pm n\pi$, $\dot{\theta}^\gamma(t) = 0$.

Remark **29** The regularity of ϕ and θ^γ are given respectively by Lemma 6 and Lemma 8. That is ϕ is C^1, while θ^γ is C^1 where it is defined (i.e. at times when G does not vanish). Notice that in general, $\phi(\cdot)$ and θ^γ are not C^2 at switching points.

The main advantage in using the θ^γ function is the possibility of easily checking its monotonicity using the relation $sgn(\dot{\theta}^\gamma(t)) = sgn(\Delta_B(\gamma(t)))$ (see Lemma 8, p. 43). Hence, if $\dot{\theta}^\gamma$ is defined in t_0 we have $\dot{\theta}^\gamma(t_0) = 0$ iff $\gamma(t_0) \in \Delta_B^{-1}(0)$. In figure 2.26, an example of extremal trajectory with the corresponding functions ϕ and θ^γ is drawn.

2.7.1 The θ^γ Function on Singularities of the Synthesis

It is very interesting and instructive to read the singularities of the time optimal synthesis in terms of the θ^γ function.

Here, as an example, we focus on the $(C, S)_1$ singularity. We refer to Figure 2.27 case A. Let γ_s be a trajectory of the optimal synthesis entering a $(C, S)_1$ singularity, and let γ_r, γ_l be two optimal trajectories close to γ_s lying respectively to the right and to the left of γ_s. The trajectory γ_s may be prolonged after the $(C, S)_1$ singularity in an infinite number of ways: it may run on $\Delta_B^{-1}(0)$ for an arbitrary small positive time and then bifurcate to the left or right using either control $+1$ or -1.

To fix the ideas, assume $\Delta_B > 0$ (resp. $\Delta_B < 0$) on the left (resp. right) of $\Delta_B^{-1}(0)$. Figure 2.27, cases B,C,D, shows the shape of $\theta^{\gamma_s}, \theta^{\gamma_r}, \theta^{\gamma_l}$, respectively.

The function θ^{γ_r} has a maximum, since it crosses the set $\Delta_B^{-1}(0)$. Moreover it does not cross the lines $\bar{\theta}_r \pm n\pi$, $n \in \mathbb{N}$, in the (t, θ^γ) plane, on which switchings occur.

The function θ^{γ_l} is always increasing since γ_l never crosses the set $\Delta_B^{-1}(0)$. Moreover θ^{γ_l} crosses one of the lines $\bar{\theta}_l \pm n\pi$, since it switches.

The function θ^{γ_s} reaches the critical value $\bar{\theta}_s$ exactly at $\Delta_B^{-1}(0)$ that implies $\dot{\theta}^{\gamma_s} = 0$. At the $(C, S)_1$ Frame Point γ_s bifurcates to:

a. the trajectory switching to opposite control and continuing with $\dot{\theta}^\gamma > 0$ (Δ_B does not change sign).
b. the trajectory continuing with the same constant control having a maximum of the corresponding θ^γ function.
c. the trajectory entering the turnpike continuing with $\dot{\theta}^\gamma = 0$. In turn, this trajectory bifurcates to constant control $+1$ or -1 at each subsequent time.

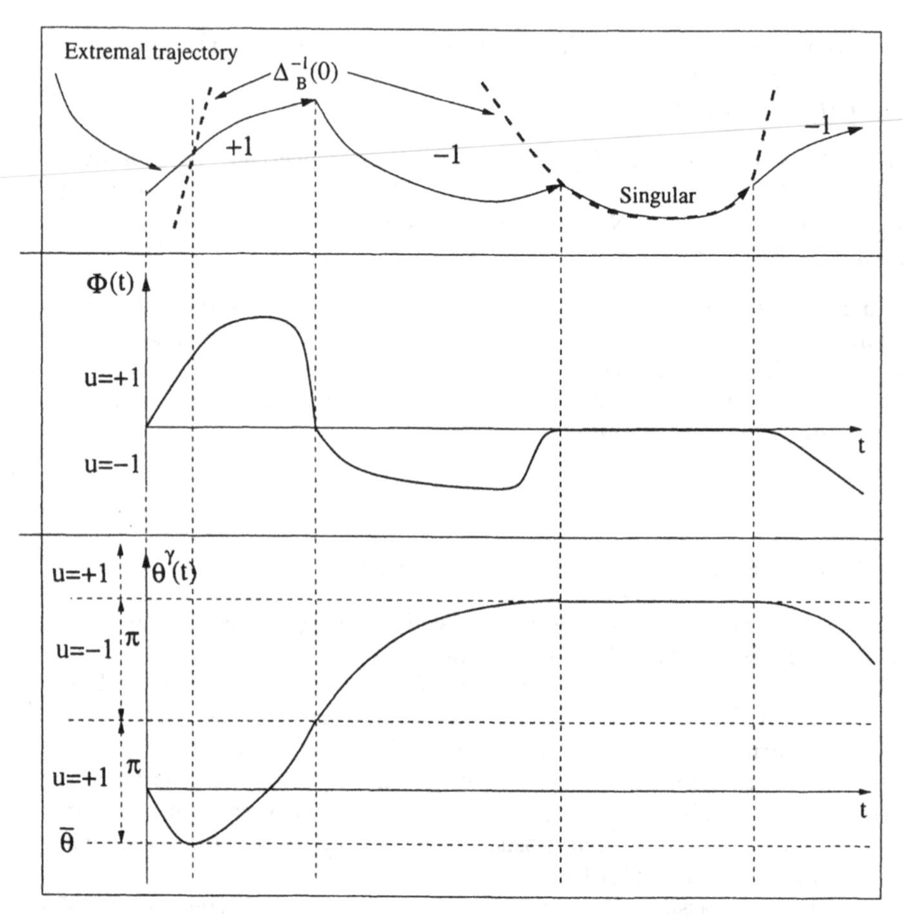

Fig. 2.26. An example of extremal trajectory with the corresponding ϕ and θ^γ functions.

2.8 Structural Stability and Classification of Optimal Feedbacks

The aim of this Section is to introduce a more detailed inductive algorithm, generating optimal synthesis, in order to ensure the structural stability. The latter property plays a key role in the classification program, guaranteeing persistence, under small perturbations, of syntheses main features. The problem of structural stability for two dimensional control systems was also studied in [58].

Remark **30** Notice that the optimal synthesis is essentially unique in the following sense. Generically there exists a finite number of embedded connected one dimensional manifolds such that, on the complement, the optimal trajec-

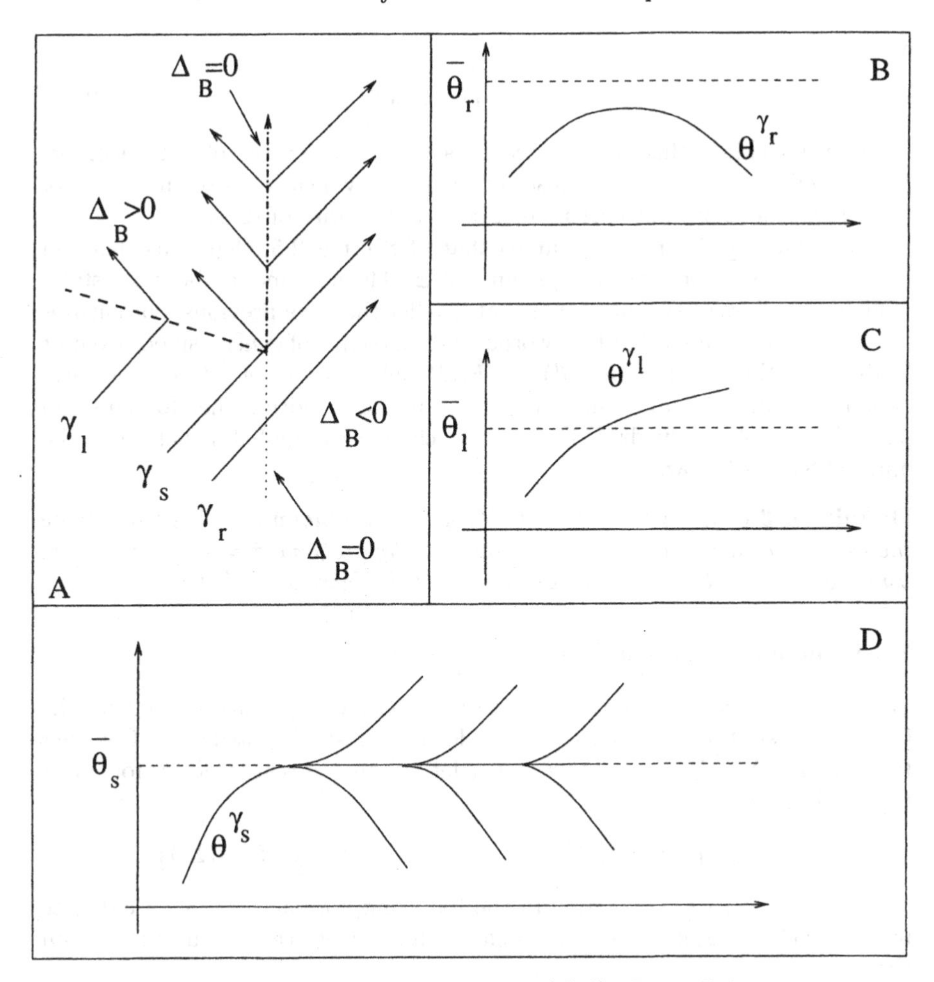

Fig. 2.27. The θ^{γ} functions on a $(C,S)_1$ Frame Point

tories are uniquely determined. This happens exactly on overlap curves and on γ_k curves. For instance, consider the point $\bar{x} = K_1 \cap S_1$ of Example 13, Section 2.6.4. Then every point on the Y–trajectory starting from $\bar{x}$ can be reached optimally in two ways: either by a $Y * X$–trajectory or by a $Y * S * Y$–trajectory.

Denote by $u_{F,G}$ the optimal feedback corresponding to $(F,G) \in \Xi$. We introduce an equivalence relation between couples of vector fields: $(F,G) \sim (F',G')$, determined by the topological equivalence of the corresponding flows:

$$\dot{x} = F(x) + u_{F,G}(x)G(x), \tag{2.59}$$

and:

$$\dot{x} = F'(x) + u_{F',G'}(x)G'(x). \tag{2.60}$$

Roughly speaking, this equivalence asks for the existence of a homeomorphism, defined on a suitable subset of the plane, which maps oriented arcs of trajectories of (2.59) onto oriented arcs of trajectories of (2.60).

The new algorithm is similar to that of Section 2.5, that constructs an optimal synthesis for a generic system $\Sigma \in \Xi$. The new one can be successfully applied to a class of systems somewhat smaller than the previous, but ensures the structural stability. In other words, given a couple of generic smooth vector fields F, G, the relation $(F', G') \sim (F, G)$ holds whenever F' is sufficiently close to F and G' is sufficiently close to G, in the C^3 norm. Therefore, a small perturbation of the fields F, G does not change the global structure of the optimal feedback flow.

Definition 35 *Given a vector field V, we use the notation $e^{tV}(\bar{x})$ to denote the value at time t of the solution to the Cauchy problem: $\dot{x} = V(x)$, $x(0) = \bar{x}$, while $(e^{tV})_*$ denotes the Jacobian matrix of the map: $x \mapsto e^{tV}(x)$.*

2.8.1 Feedback Equivalence

We now introduce an equivalence relation between systems, expressing the fact that their time-optimal feedback flows have similar structure. Consider two systems $\Sigma_1, \Sigma_2 \in \Xi^*$. For $i = 1, 2$ let $\mathcal{R}_i$ be the reachable set for Σ_i at time τ. Define

$$\mathcal{K}_i = \{x \mid x \in K \setminus \partial K,\ K \text{ is an overlap curve of } \Gamma_{\mathcal{A}}(\Sigma_i)\},$$

and set $\mathcal{R}'_i = \mathcal{R}_i \setminus \mathcal{K}_i$, $i = 1, 2$. In the following, for each $x \in \mathcal{R}_i$, we denote by $t \mapsto \gamma^i_x(t)$ a trajectory of Σ_i which reaches x from the origin in minimum time.

Definition 36 (Equivalence of Feedback Flows). *We say that the time-optimal feedback flows for Σ_1 on $\mathcal{R}_1$ and Σ_2 on $\mathcal{R}_2$ are equivalent, or simply that $\Sigma_1 \sim \Sigma_2$, if there exists a homeomorphism $\Psi : \mathcal{R}'_1 \mapsto \mathcal{R}'_2$ such that:*

(E1) Ψ maps arcs of optimal trajectories for Σ_1 onto arcs of optimal trajectories for Σ_2. More precisely, for every $x \in \mathcal{R}'_1$ one has $\{\Psi(\gamma^1_x(t)) : t \in Dom(\gamma^1_x)\} = \{\gamma^2_{\Psi(x)}(t) : t \in Dom(\gamma^2_{\Psi(x)})\}$.

(E2) Ψ induces a bijection on frame curves that are not overlap curves, i.e. for each frame curve D_1, which occurs in the construction of the optimal feedback for Σ_1 and is not a K-curve, we have that $\varphi(D_1)$ is a frame curve of the same type corresponding to Σ_2, and viceversa.

(E3) If A is an open region of $\mathcal{R}'_1$ enclosed by frame curves and entirely covered by Y- or X-trajectories, then $\Psi(A)$ is enclosed by the corresponding frame curves and is covered by Y- or X-trajectories, respectively.

Remark **31** The exclusion of overlap curves is operated to have not too small equivalence classes and hence not too many of them. Consider Example 4 of Section 2.6.4 and let γ_1 and γ_2 be the X, resp. Y, trajectory starting from the (X, S) point of the first, resp. fourth, orthant. If we include the overlap curves in the definition of equivalence, then the relative position of the endpoints of γ_1 and γ_2 on the K curve is an invariant of the equivalence class and the same for all relative positions of points that can be obtained similarly from them concatenating arcs of X and Y trajectories (see the Definition of equivalence of Frame Points at the end of this Section).

Remark **32** As mentioned in Section , optimal synthesis is the correct concept of solution to a family of optimal control problems parameterized by initial data, while discontinuous feedbacks may generate not optimal trajectories. However, for our model problem, under generic conditions, Caratheodory solutions to discontinuous optimal feedbacks are optimal, see [97], thus we refer to synthesis or feedbacks in the same way.

Definition 37 (Structural Stability) *We say that a system $\Sigma \in \Xi^*$ is structurally stable if there exists a neighborhood $\mathcal{N}$ of Σ in the space Ξ (endowed with the C^3 norm, see formula (2.6), Section 2.2), such that the feedback flows for Σ and Σ' are equivalent, for every $\Sigma' \in \mathcal{N}$.*

2.8.2 An Algorithm for the Synthesis

In the description of the inductive algorithm we make various generic assumptions. Some of these, indicated by **(GA)**, are given in the first step. Others are given at a generic step N, finally some are determined at the end of the algorithm.

In the first step, the algorithm constructs the two trajectories $\gamma^{\pm}(t) = e^{t(F \pm G)}(0)$ and marks some special points along these curves, from which additional frame curves bifurcate. At step N, the algorithm constructs precisely those trajectories which are concatenation of N bang- or singular arcs and satisfy Pontryagin Maximum Principle.

In the following, we fix a locally controllable system $\Sigma = (F, G) \in \Xi$, $\tau > 0$ and we assume that Σ verifies the generic properties **(P1)**,...,**(P7)**. By Theorem 13, these assumptions imply that, for some integer N_0, every optimal trajectory γ starting from the origin, with $Dom(\gamma) \subseteq [0, 1]$, is a concatenation of at most N_0 bang- or singular arcs. In the description of the algorithm we give additional conditions implying that at every step only a finite number of frame curves and frame points are constructed by the algorithm, and these curves and points are structural stable.

In the following, if a trajectory γ with $Dom(\gamma) = [0, b]$ is constructed as part of the synthesis, we then regard all trajectories $\gamma|_{[0,a]}$ with $a < b$ as constructed trajectories. Similarly, we regard as frame curve every connected subset of a frame curve, having frame points as endpoints. To every constructed trajectory we associate a covector field.

Remark **33** Some proofs of this section are collected in Appendix A. These proofs can be missed at first reading.

Algorithm $\mathcal{A}$, STEP 1.
We define:

$$\theta^{\pm}(t) := \theta^{\gamma^{\pm}}(t)$$
$$v^{\pm} := v^{\gamma^{\pm}}.$$

From Lemma 7, p. 42 we have: $v^{\pm}((F\pm G)(\gamma^{\pm}(t)), t; 0) = (F\pm G)(0) = \pm G(0)$. Moreover set:

$$t_f^{\pm} := \min\{t \in [0,\tau]; \quad |\theta^{\pm}(s_1) - \theta^{\pm}(s_2)| = \pi \text{ for some } s_1, s_2 \in [0,t]\}, \tag{2.61}$$

with the understanding that $t_f^{\pm} = \tau$ if $|\theta^{\pm}(s_1) - \theta^{\pm}(s_2)| < \pi$ for all $s_1, s_2 \in [0,\tau]$.

Remark **34** Notice that $t_f^{\pm}$ is the last time at which $\gamma^{\pm}$ is extremal so we have $t_{op}^{\pm} \le t_f^{\pm}$ (cfr. Proposition 3, p. 92, below). Notice that $t_{op}^{\pm}$ is known only at the end of the algorithm.

Next, we single out times $t_i^+, t_i'^+ \in [0, t_f^+]$ where the function θ^+ assumes increasingly large local maxima, and increasingly small local minima, respectively, see Figure 2.28. By induction, define:

Definition 38

$$t_0^+ = t_0'^+ = 0$$
$$t_1^+ = \inf\{t > 0 : \quad \theta^+ \textit{ has a local max. at } t, \quad \theta^+(t) > 0\}$$
$$t_i^+ = \inf\{t > t_{i-1}^+ : \quad \theta^+ \textit{ has a local max. at } t, \quad \theta^+(t) > \theta^+(t_{i-1}^+)\}$$
$$t_1'^+ = \inf\{t > 0 : \quad \theta^+ \textit{ has a local min. at } t, \quad \theta^+(t) < 0\}$$
$$t_i'^+ = \inf\{t > t_{i-1}'^+ : \quad \theta^+ \textit{ has a local min. at } t, \quad \theta^+(t) < \theta^+(t_{i-1}^+)\}$$
$$s_i^+ = \begin{cases} \max\{t \in [t_{i-1}^+, t_i^+] : \quad \theta^+(t) = \theta^+(t_{i-1}^+)\} \textit{ if } t_i^+ \textit{ is defined,} \\ \max\{t \in [t_{i-1}^+, t_f^+] : \quad \theta^+(t) = \theta^+(t_{i-1}^+)\} \textit{ otherwise} \end{cases}$$
$$s_i'^+ = \begin{cases} \max\{t \in [t_{i-1}'^+, t_i'^+] : \quad \theta^+(t) = \theta^+(t_{i-1}'^+)\} \textit{ if } t_i'^+ \textit{ is defined,} \\ \max\{t \in [t_{i-1}'^+, t_f'^+] : \quad \theta^+(t) = \theta^+(t_{i-1}'^+)\} \textit{ otherwise.} \end{cases}$$

Similarly the times t_i^-, $t_i'^-$, s_i^-, $s_i'^-$, were defined.

Remark **35** We can have three situations (cfr. Figure 2.29):

1) $|\theta^{\pm}(a) - \theta^{\pm}(b)| < \pi$ for every $a, b \in [0,\tau]$. In this case $t_f^{\pm} = \tau$ and $|\theta^{\pm}(t_f^{\pm})| < \pi$;

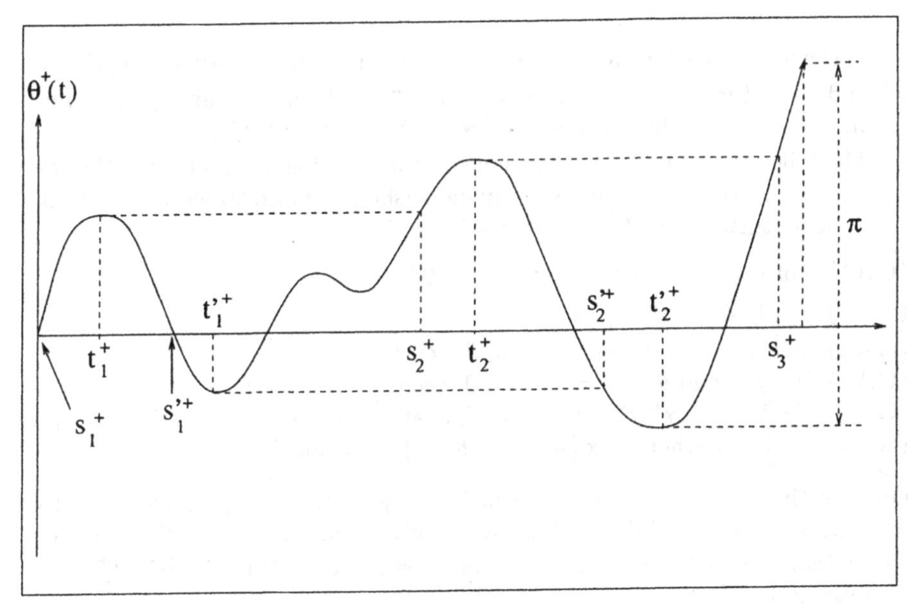

Fig. 2.28. An example of function $\theta^{+}(t)$ in the case $t_f^{+} < \tau$.

2) $|\theta^{\pm}(a) - \theta^{\pm}(b)| = \pi$ for some $a, b \in]0, \tau]$. In this case $|\theta^{\pm}(t_f^{\pm})| < \pi$, $\theta^{\pm}$ has a maximum or a minimum in $]0, t_f^{\pm}[$ and either $s_1^{\pm} \neq 0$ or $s_1'^{\pm} \neq 0$
3) $|\theta^{\pm}(t_f^{\pm})| = \pi$. In this case, $s_1^{\pm}$ and $s_1'^{\pm}$ are not defined and generically we get $t_f^{\pm} < \tau$.

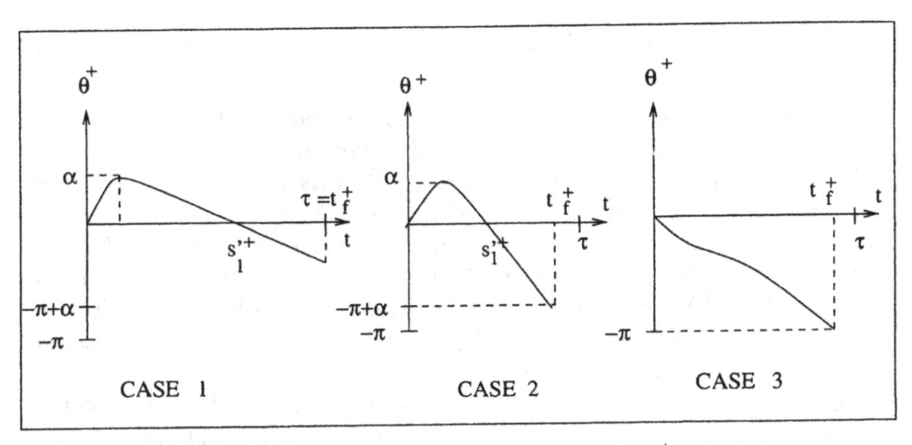

Fig. 2.29. The three possible situation for the $\theta^{\pm}$ functions

In the first step of the algorithm, we construct the trajectory $\gamma^+ : [0, t_f^+] \mapsto \mathbb{R}^2$ and regard it also as frame curve. On γ^+ we define as frame points: the origin, $\gamma^+(t_f^+)$, and all points $\gamma^+(s_i)$, $\gamma^+(t_i^+)$, $\gamma^+(s_i')$, $\gamma^+(t_i'^+)$.

The following stability assumptions on the function θ^+ imply that the sequences $\{t_i^+\}$, $\{t_i'^+\}$ are finite, strictly increasing, and also stable with respect to small perturbations of the vector fields F, G.

(GA1) For every $t \in [0, t_f^+]$, $G(\gamma^+(t)) \neq 0$.
(GA2) $\dot\theta^+(0) \neq 0, \quad \dot\theta^+(t_f^+) \neq 0$.
(GA3) If $\dot\theta^+(t) = 0$, then $\theta^+(t) \neq 0$, $\ddot\theta^+(t) \neq 0$.
(GA4) If $t \neq s$ and $\dot\theta^+(s) = \dot\theta^+(t) = 0$, then $\theta^+(s) \neq \theta^+(t)$.
(GA5) $(\nabla \Delta_B \cdot V)(\gamma^+(t)) \neq 0$, $V = X, Y$ at all points $t \in \{t_i^+,\ t_i'^+;\ \ i \geq 1\}$.
(GA6) If $t_f^+ = \tau$, then $\max\{|\theta^+(t) - \theta^+(\tau)|;\ \ t \in [0, \tau]\} < \pi$.

Observe that t_f^+ is the first time which is negatively conjugate along γ^+ to some previous time, while s_i, s_i' are positively conjugate to t_{i-1}, t_{i-1}', respectively. Moreover, θ^+ is monotonically increasing on each interval $[s_i^+,\ t_i^+]$ and decreasing on $[s_i',\ t_i']$.

We then perform exactly the same construction for the trajectory γ^-, replacing G with $-G$ and interchanging X with Y throughout. Moreover we ask similar generic conditions.

The following results (Propositions 3,4,5) motivate our construction.

Proposition 3 *Under the stability assumption* **(GA1)––(GA6)** *we have:*

(I) the trajectory $\gamma^+ : t \mapsto e^{tY}(0)$ is extremal up to time t_f^+.

(II) if t^* lies in one of the open intervals $]s_i^+, t_i^+[$ or $]s_i'^+,\ t_i'^+[$ then there exists an extremal control of the form:

$$u(t) = \begin{cases} 1 & \text{if} \quad t \in [0, t^*[, \\ -1 & \text{if} \quad t \in [t^*, t^* + \varepsilon[, \end{cases} \tag{2.62}$$

for some $\varepsilon > 0$. In the first case the trajectory bifurcates to the right of γ^+ (clockwise), in the second case to the left (counterclockwise). On the other hand, no extremal control of the form (2.62) exists if t^* is not contained in any one of the closed intervals $[s_i^+, t_i^+]$ or $[s_i'^+,\ t_i'^+]$;

(III) to the right of each point $\gamma^+(t_i^+)$ $(t_i^+ > 0)$ and to the left of each point $\gamma^+(t_i'^+)$ $(t_i'^+ > 0)$ it originates either a turnpike, when the inner products $\nabla\Delta_B \cdot X$ and $\nabla\Delta_B \cdot Y$ have opposite signs, that is when the two vector fields X, Y points to opposite sides of the set of zeros of Δ_B, or a switching curve when the signs are equal;

(IV) if $t_f^+ < \tau$ then a switching curve originates from $\gamma^+(t_f^+)$. Such curve bifurcates to the left of γ^+ when $\dot\theta^+(t_f^+) > 0$, and to the right of γ^+ when $\dot\theta^+(t_f^+) < 0$;

A similar result holds for γ^- under similar generic assumptions.

Proof. See Appendix A. ∎

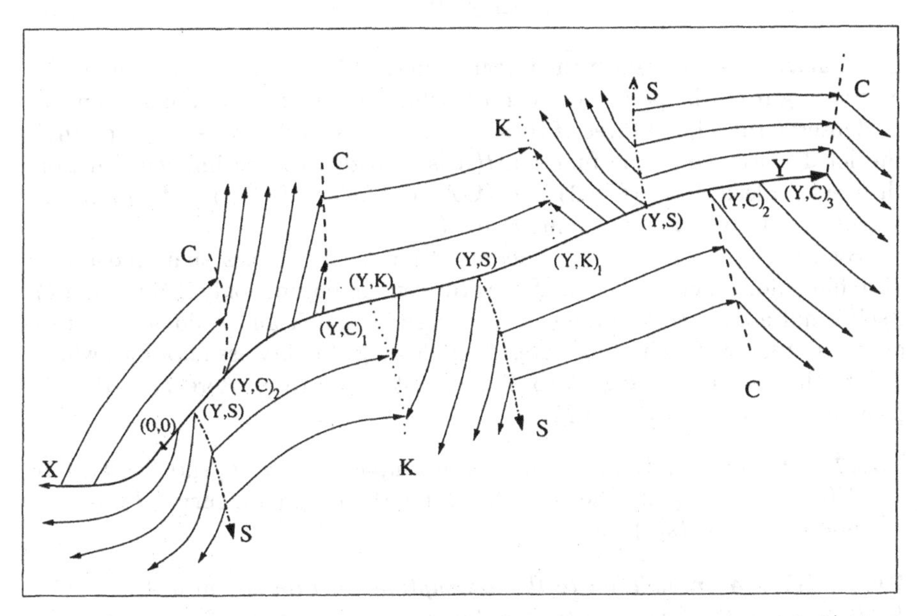

Fig. 2.30. An example of optimal synthesis corresponding to the function $\theta^+(t)$ of Figure 2.28.

In order to study the behavior of optimal trajectories in a neighborhood of the points $q_i := \gamma^+(s_i^+)$, $q_i' := \gamma^+(s_i'^+)$, and complete our description of the time optimal feedback in a neighborhood of the curve γ^+, additional stability conditions are needed. As a preliminary, observe that, for each i, a curve of conjugate points (i.e. a switching curve) starting at q_i can be defined as follows.

Let $i \geq 2$ and assume first that at $p_{i-1} := \gamma^+(t_{i-1})$ a turnpike originates. For $\varepsilon \geq 0$, define the time $t(\varepsilon)$ by requiring that the point

$$\Gamma_i(\varepsilon) := e^{t(\varepsilon)Y} e^{\varepsilon(F+\varphi G)} e^{t_{i-1}Y}(0) \tag{2.63}$$

be conjugate to $e^{\varepsilon(F+\varphi G)} e^{t_{i-1}Y}(0)$ along the integral curve of Y. On the other hand, if at p_{i-1} it originates a curve of conjugate points (i.e. a switching curve):

$$e^{\psi(\varepsilon)X} e^{(t_i^+-\varepsilon)Y}(0),$$

where ψ is defined in (A.19) of Appendix A, then define $t(\varepsilon)$ by requiring that

$$\Gamma_i(\varepsilon) := e^{t(\varepsilon)Y} e^{\psi(\varepsilon)X} e^{(t_{i-1}-\varepsilon)Y}(0) \tag{2.64}$$

is conjugate to $e^{\psi(\varepsilon)X} e^{(t_{i-1}-\varepsilon)Y}(0)$ along the integral curve of Y. Finally, if $i = 1$, $s_1 > 0$, we define $t(\varepsilon)$ by requiring that

$$\Gamma_1(\varepsilon) := e^{t(\varepsilon)Y} e^{\varepsilon X}(0) \tag{2.65}$$

is conjugate to $e^{\varepsilon X}(0)$ along the integral curve of Y. The conjugate curves Γ'_i, originating from the points $s_i'^+$, can be defined in an entirely similar manner.

Observe that, in all three of these cases, one has $t(0) = s_i^+ - t_{i-1}$, and that the local existence of the function $t(\cdot)$ is provided by the implicit function theorem. Indeed, from **(GA3)** and **(GA4)** it follows $\dot\theta^+(s_i^+) \neq 0$, and hence $\nabla \Delta_B(q_i) \neq 0$, because of Lemma 8, p. 43.

By Envelope Theory (see Theorem 9) the trajectories that undergo a switching along a curve Γ_i can afterwards remain optimal only if the curve Γ_i itself is not a trajectory of the control system, i.e., if X and Y do not point to opposite sides of Γ_i. This motivates the following stability assumptions, which ensure that X is not tangent to Γ_i at points close to q_i. Here $\dot\Gamma_i = d\Gamma_i/d\varepsilon$ provides a tangent vector to Γ_i.

(GA7) At every point $q_i := \gamma^+(s_i^+)$, the conjugate curve Γ_i satisfies $\dot\Gamma_i(0) \wedge X(q_i) \neq 0 \quad \forall i \geq 2$. The same holds for the conjugate curves Γ'_i, at the points $q'_i := \gamma^+(s_i'^+)$.

Proposition 4 *In addition to the assumptions of Proposition 3, let* **(GA7)** *hold. Then, to the right of every point $q_i := \gamma^+(s_i^+)$ with $i \geq 2$, the time optimal synthesis contains either the curve of conjugate points Γ_i defined at (2.63), (2.64), or an overlap curve, starting at q_i. The first case occurs precisely when the vector fields X, Y point to the same side of Γ_i, in a neighborhood of q_i. The analogous results hold for the points $q'_i := \gamma^+(s_i'^+)$.*

Proof. See Appendix A. ■

It remains to consider the points $q_1 = \gamma^+(s_1^+)$, $q'_1 = \gamma^+(s_1'^+)$. From the definition, we have that only one of the two numbers $s_1^+, s_1'^+$ is different from zero. Assume that $s_1^+ \neq 0$. The case $s_1'^+ \neq 0$ can be treated in an entirely similar manner. We introduce a new system of coordinates in such a way that:

$$Y \equiv \begin{pmatrix} 1 \\ 0 \end{pmatrix} \quad X(x) = \begin{pmatrix} -1 + a_1 x_1 + a_2 x_2 + O(|x|^2) \\ b_1 x_1 + b_2 x_2 + O(|x|^2) \end{pmatrix}, \quad b_1 > 0, \tag{2.66}$$

$$X(x) = \begin{pmatrix} c_0 + c_1(x_1 - s_1^+) + c_2 x_2 + O(|x - (s_1^+, 0)|^2) \\ d_1(x_1 - s_1^+) + d_2 x_2 + O(|x - (s_1^+, 0)|^2) \end{pmatrix}, \quad \begin{cases} c_0 < 1 \\ d_1 < 0. \end{cases} \tag{2.67}$$

The expression for X in formula (2.66) is useful in a neighborhood of $(0,0)$, while the expression (2.67) gives information in a neighborhood I of $(s_1^+, 0)$. The signs of b_1, d_1 and condition $c_0 < 1$ follow from the stability conditions

(GA2),**(GA3)**. Indeed, since $\Delta_B(0) = -b_1/2$, from $s_1^+ > 0$, we get $\Delta_B(0) < 0$, thus $b_1 > 0$. Also $\Delta_B(s_1^+, 0) = d_1(c_0 - 1)/4 > 0$, hence $d_1 \neq 0$ and $c_0 \neq 1$. Finally at $x_\varepsilon := (s_1^+ + \varepsilon, 0)$, $\varepsilon > 0$ there are trajectories switching from Y to X, then by Theorem 11, $f(x_\varepsilon) < 0$. Then from $\Delta_B(x_\varepsilon) > 0$ we get $0 < \Delta_A(x_\varepsilon) = -d_1/2$, finally we conclude $d_1 < 0$ and $c_0 < 1$. If $c_0 \neq 0$, then the trajectories:

$$\sigma_2 \mapsto e^{\sigma_2 X} e^{(s_1^+ + \sigma_1)Y}(0), \qquad \tau_2 \mapsto e^{\tau_2 Y} e^{\tau_1 X}(0),$$

for σ_1, τ_1 sufficiently small, cross each other near the point $(s_1^+, 0)$. An overlap curve at $(s_1^+, 0)$ is determined by a system of three equations of the form

$$\begin{cases} s_1^+ + \sigma_1 + \sigma_2 = \tau_1 + \tau_2 \\ s_1^+ + \sigma_1 + c_0\sigma_2 + O(1)(\sigma_1^2 + \sigma_1\sigma_2 + \sigma_2^2) = -\tau_1 + \tau_2 + O(1)\tau_1^2 \\ -\frac{b_1}{2}\tau_1^2 + O(1)\tau_1^3 = d_1\sigma_1\sigma_2 + \frac{c_0 d_1}{2}\sigma_2^2 + O(1)(\sigma_2^3 + \sigma_1^2\sigma_2 + \sigma_1\sigma_2^2). \end{cases} \tag{2.68}$$

Here with the Landau symbol $O(1)$ we indicate some function which is uniformly bounded, together with its derivatives. The first equation yields

$$\tau_2 = s_1^+ + \sigma_1 + \sigma_2 - \tau_1.$$

Hence:

$$\begin{cases} (c_0 - 1)\sigma_2 + 2\tau_1 = O(1)(\sigma_1^2 + \sigma_1\sigma_2 + \sigma_2^2 + \tau_1^2) \\ d_1\sigma_1\sigma_2 + \frac{c_0 d_1}{2}\sigma_2^2 + \frac{b_1}{2}\tau_1^2 = O(1)(\sigma_2^3 + \sigma_1^2\sigma_2 + \sigma_1\sigma_2^2). \end{cases}$$

A trivial branch of solutions is given by $\tau_1 = \sigma_2 = 0$. When $\sigma_1, \sigma_2, \tau_1$ are sufficiently small, one can solve the first equation for τ_1 as a function of (σ_1, σ_2). This yields:

$$\frac{\partial \tau_1}{\sigma_2}(0,0) = \frac{1 - c_0}{2}, \qquad \frac{\partial \tau_1}{\sigma_1}(0,0) = 0.$$

Then the second equation, for $\sigma_1, \sigma_2, \tau_1$ sufficiently small, is equivalent, up to higher order terms, to:

$$\left[d_1\sigma_1 + \frac{1}{2}\left(c_0 d_1 + b_1 \left(\frac{1 - c_0}{2}\right)^2 \right) \sigma_2 \right] \sigma_2 = 0. \tag{2.69}$$

We have a nontrivial branch of solutions if the following stability condition holds:

(GA8) $c_0 \neq 0, \qquad \delta \doteq c_0 d_1 - b_1 \left(\frac{1-c_0}{2}\right)^2 \neq 0.$

Proposition 5 *In addition to the assumptions of Proposition 3, assume $s_1^+ > 0$ and let* **(GA8)** *hold. Then, to the right of $q_1 := \gamma^+(s_1^+)$, the time optimal synthesis contains either the curve of conjugate points Γ_1 defined at (2.65), or an overlap curve, starting at q_1. The first case occurs precisely when $\delta < 0$. The analogous results hold for the point $q_1' := \gamma^+(s_1'^+)$.*

Proof. To determine the switching curve Γ_1 of (2.65), let $(\bar{\gamma}, \bar{\lambda})$ be an extremal trajectory of our system, corresponding to constant control -1 on the interval $[0, \tau_1[$ and to constant control $+1$ on the interval $[\tau_1, \tau_2]$. We have:

$$\bar{\gamma}(\tau_1) = \begin{pmatrix} -\tau_1 + O(\tau_1^2) \\ -\frac{1}{2}b_1\tau_1^2 + O(\tau_1^3) \end{pmatrix}, \quad \bar{\gamma}(\tau_2) = \begin{pmatrix} -\tau_1 + \tau_2 + O(\tau_1^2) \\ -\frac{1}{2}b_1\tau_1^2 + O(\tau_1^3) \end{pmatrix}.$$

Now if τ_1, τ_2 are switching times for $\bar{\gamma}$ we must have:

$$\bar{\lambda}(\tau_1) \cdot G(\bar{\gamma}(\tau_1)) = 0, \tag{2.70}$$

$$\bar{\lambda}(\tau_2) \cdot G(\bar{\gamma}(\tau_2)) = 0. \tag{2.71}$$

Moreover, from equation **i)** of the PMP, it follows $\bar{\lambda}(\tau_1) = \bar{\lambda}(\tau_2) =: \bar{\lambda}$. Set $\bar{\lambda} = (\bar{\lambda}_1, \bar{\lambda}_2)$ and normalize $\bar{\lambda}$ in such a way that $\bar{\lambda}_2 = \sqrt{1 - \bar{\lambda}_1}$. Equations (2.70), (2.71) become:

$$f_1(\lambda_1, \tau_1, \tau_2) := \bar{\lambda}_1(1 + a_1\tau_1) + \sqrt{1 - \bar{\lambda}_1}(\frac{1}{2}b_1\tau_1) + O(\tau_1^2) = 0,$$

$$f_2(\lambda_1, \tau_1, \tau_2) := \frac{1}{2}\bar{\lambda}_1((1 - c_0) - c_1(-\tau_1 + \tau_2 - s_1^+)) + \sqrt{1 - \bar{\lambda}_1}(-\frac{1}{2}d_1(-\tau_1 + \tau_2 - s_1^+)) + O(\tau_1^2) = 0.$$

These are two equations for the variable $(\bar{\lambda}_1, \tau_1, \tau_2)$ and $(0, 0, s_1^+)$ is a solution. The 2×2 Jacobian matrix of partial derivatives of f_1 and f_2 with respect to $(\bar{\lambda}_1, \tau_2)$ has determinant, at the point $(0, 0, s_1^+)$, equal to $-d_1/2$. Since $d_1 \neq 0$, we can solve the system in a neighborhood of $(0, 0, s_1^+)$ expressing $(\bar{\lambda}_1, \tau_2)$ as a function of τ_1. This yields:

$$\left.\frac{\partial \tau_2}{\partial \tau_1}\right|_{(0,0,s_1^+)} = -\frac{(b_1(1 - c_0) - 2d_1)}{2d_1} =: m, \tag{2.72}$$

hence $\tau_2 = s_1^+ + m\tau_1 + O(\tau_1^2)$ and the parametric expression for the switching curve starting at $(s_1^+, 0)$ is:

$$x_1(\tau_1) = s_1^+ + (m - 1)\tau_1 + O(\tau_1^2),$$

$$x_2(\tau_1) = -\frac{1}{2}b_1\tau_1^2 + O(\tau_1^3).$$

From $b_1 > 0$, $d_1 < 0$ and $c_0 < 1$ we get $m \neq 1$. Thus we can express x_2 as a function of x_1:

$$x_2(x_1) = -\frac{2d_1^2}{b_1(1 - c_0)^2}(x_1 - s_1^+)^2 + O((x_1 - s_1^+)^3). \tag{2.73}$$

From (2.69), being $\delta \neq 0$, we can express σ_1^+ as a function of σ_2.

After straightforward calculations one obtains for the overlap curve K:

$$x_2(x_1) = -\frac{4\,b_1\,(-1+c_0)^2\,d_1{}^2}{\left(b_1\,(-1+c_0)^2 - 4\,c_0\,d_1\right)^2}(x_1 - s_1^+)^2 + O(|x_1 - s_1^+|^3).$$

On the other hand, the support of the X−trajectory γ through $(s_1^+, 0)$ is given by:

$$x_2(x_1) = \frac{d_1}{2c_0}(x_1 - s_1^+)^2 + o(|x_1 - s_1^+|^2).$$

Therefore Γ_1, K and γ have a first order tangency at $(s_1^+, 0)$. Notice that the condition $\delta \neq 0$ means precisely that γ does not have a second order tangency with Γ_1. Now, if $\delta < 0$ then $c_0 > 0$, σ_1^+ and σ_2 have opposite signs and the system (2.68) does not describe an admissible overlap curve. On the other side the curve γ lies between γ^+ and Γ_1. Therefore the trajectories $t_2 \to e^{t_2 Y} e^{t_1 X}(0)$ switch along Γ_1 before crossing γ. By a sufficiency argument, see [111], these trajectories are optimal. If $\delta > 0$ then σ_1^+, σ_2^+ have the same sign, hence the trajectories:

$$\sigma_2 \mapsto e^{\sigma_2 X} e^{(s_1^+ + \sigma_1)Y}(0)$$

that reach K, switch after s_1^+ and thus are extremal. The curve Γ_1 lies between γ and γ^+. We can conclude again by a sufficiency argument. ■

Remark **36** Therefore the switching curve starting at the first switching time is tangent to *Supp*(γ). Notice that the curve bifurcating from $(0, s_1^+)$ is an abnormal extremal whose expression can be obtained from (2.66):

$$\gamma_A(t) = \begin{pmatrix} x_1(t) \\ x_2(t) \end{pmatrix} = \begin{pmatrix} s_1^+ + c_0 t + O(t^2) \\ d_1(\frac{1}{2}c_0 t^2) + O(t^3) \end{pmatrix}. \tag{2.74}$$

Of course, entirely similar results hold for the curve γ^-. This completes the analysis of STEP 1 of the algorithm. We now define by induction the step N, where $N > 1$.

STEP N. At step N the algorithm prolongs the trajectories constructed at step N-1, by joining an arc that is an X, Y or Z–trajectory. In such a way it constructs all extremal trajectories consisting of ≤N arcs.

Consider all frame curves generated by the previous step which are not overlap curves. Let D be such a frame curve. Consider the constructed trajectories γ, $\gamma : [0, t_0] \mapsto \mathbb{R}^2$, $\gamma(0) = 0$, such that $\gamma(t_0) = x \in D$. For each one of these trajectories, calling u the corresponding control, we construct two new trajectories $\gamma_1, \gamma_2 : [0, 1] \mapsto \mathbb{R}^2$, $\gamma_1(0) = \gamma_2(0) = 0$, corresponding, respectively, to the following controls u_1 and u_2:

$$u_1(t) = u_2(t) = u(t) \qquad \forall t \in [0, t_0],$$

$$u_1(t) = +1 \quad u_2(t) = -1 \qquad \forall t \in [t_0, \tau].$$

If there exists a conjugate point $\gamma_1(t_1)$ ($t_1 > t_0$) to $\gamma_1(t_0)$ along (u_1, γ_1) then we construct only $\gamma_1|_{[0,t_1]}$, and do the same for γ_2. Indicate with $\bar{\lambda} : [0, t_0] \mapsto (\mathbb{R}^2)_*$ the covector field associated by induction to $\gamma_1|_{[0,t_0]}$. Then γ_1 is constructed only if there exists a covector field λ along (u_1, γ_1) such that:

$$\lambda|_{[0,t_0]} = \bar{\lambda} \qquad\qquad \lambda(t_0) \cdot G(\gamma_1(t_0)) = 0 \tag{2.75}$$

and, if $\gamma_1(t_1)$ is a conjugate point to $\gamma_1(t_0)$ ($t_1 > t_0$) along (u_1, γ_1), then:

$$\lambda(t) \cdot G(\gamma_1(t)) > 0 \quad \forall t \in]t_0, t_1[\qquad\qquad \lambda(t_1) \cdot G(\gamma_1(t_1) = 0 \tag{2.76}$$

otherwise:

$$\lambda(t) \cdot G(\gamma_1(t)) > 0 \qquad \forall t \in]t_0, \tau]. \tag{2.77}$$

If there exists such a covector field then associate it to γ_1. Proceed similarly for the other case defined above (γ_2) changing signs in (2.75), (2.76), (2.77). Notice that, under generic conditions, we construct all extremal trajectories with N arcs.

Consider the connected components of $\{x : \Delta_B(x) = 0, |\varphi(x)| \leq 1\}$ (φ is defined in (2.13)) that intersect some constructed trajectories. Assume that S is such a component, γ is a constructed trajectory with associated covector field λ and t_0 is the first time of intersection of γ with S. If $\lambda(t_0) \cdot G(\gamma(t_0)) = 0$, then we consider the maximal trajectory γ_1 that satisfies:

$$\gamma_1|_{[0,t_0]} \equiv \gamma, \qquad \gamma_1(t) \in S \quad \forall t > t_0.$$

We construct the trajectory $\gamma_1|([0, \tau] \cap Dom(\gamma_1))$. If u is the control corresponding to γ, we have that γ_1 corresponds to the control u_1 given by:

$$u_1|_{[0,t_0[} \equiv u, \qquad u_1(t) = \varphi(\gamma_1(t)) \ \forall t \geq t_0.$$

We associate to γ_1 the covector field λ_1 that verifies $\lambda_1 \equiv \lambda$ on $[0, t_0]$.

Now, it may happen that a point $x \in \mathbb{R}^2$ is reached by more than one extremal trajectory. In this case, trajectories that are not globally optimal are removed from the synthesis constructed up to step N. When certain trajectories are deleted, each corresponding frame point and frame curve (or part of it) are also deleted. More precisely, let $x \in \mathbb{R}^2$ be a point reached by some trajectories constructed in this step. There are a finite number of constructed trajectories (not necessarily constructed in this step) $\gamma_1, \ldots, \gamma_n$ that reach x at certain times $t_i^+ \in Dom(\gamma_i)$, $i = 1, \ldots, n$. Let $\bar{t} = \min_i t_i^+$. If $t_i^+ > \bar{t}$ then we cut the trajectory γ_i after the time t_i^+, i.e. consider only $\gamma_i|_{[0,t_i^+]}$. It can happen that $\bar{t} = t_i^+$ for more than one i (generically at most two), in this case the algorithm constructs an overlap curve.

Define the following new frame curves:

a) Maximal regular turnpikes S for which there exists a trajectory γ, constructed in this step, that verifies $\gamma(I) = S$ for some $I \subset Dom(\gamma)$,
b) Overlap curves constructed in this step,
c) Conjugate curves to frame curves, constructed in the previous step, along $X-$ or $Y-$trajectories.

We give generic stability conditions for frame curves.
Consider a turnpike S, and γ, $I = [t_0, t_1]$ as in a). Let φ be the control defined in (2.13). As stability condition we then assume:

$$| \varphi(\gamma(t)) | < 1 \quad \forall t \in [t_0, t_1[. \tag{2.78}$$

Consider an overlap curve K. In general it is defined by a system of equations of the following type:

$$\begin{cases} (exp\ tX)\ x_1(s) = (exp\ t'Y)\ x_2(s') \\ s + t \ = \ s' + t' \end{cases} \tag{2.79}$$

where $x_1(s)$ and $x_2(s')$, are parameterizations of two frame curves defined in the previous steps. As stability condition, we assume that the system (2.79) has rank 3 on K, except for the possible intersections with $\gamma^{\pm}$. It is easy to verify that (2.79) may have rank 2 on $K \cap \gamma^{\pm}$, see Example 1 of Section 2.6.4, p. 62.

Next consider stability conditions for a conjugate curve to D. Let D be a frame curve, D' the conjugate curve along the $Y-$trajectories. In other words, for each $x \in D$ there exist a trajectory $\gamma_x : [0, \sigma_x] \mapsto \mathbb{R}^2$ and $t_x \in]0, \sigma_x[$ such that $\gamma_x(t_x) = x$, $\gamma_x|_{[t_x, \sigma_x]}$ is a $Y-$trajectory and $\gamma_x(\sigma_x) \in D'$ ($\gamma_x(\sigma_x)$ is conjugate to x). Denote by u_x the control corresponding to γ_x and by v_x the vector fields along (u_x, γ_x), i.e. the solutions to (2.4) for $\gamma = \gamma_x$. The stability conditions are the following:

$$det \left[v_x\Big(G(x), t_x; t\Big), G\Big(\gamma_x(t)\Big) \right] \neq 0 \qquad \forall x \in D \quad \forall t \in]t_x, \sigma_x[\tag{2.80}$$

$$\frac{\partial}{\partial t} det \left[v_x\Big(G(x), t_x; t\Big), G\Big(\gamma_x(t)\Big) \right] \Big|_{t=\sigma_x} \neq 0 \quad \forall x \in D. \tag{2.81}$$

Otherwise stated, these conditions read: $\theta^{\gamma_x}(t) \neq \theta^{\gamma_x}(t_x)$ for $t \in]t_x, \sigma_x[$, $(d/dt)\theta^{\gamma_x}(\sigma_x) \neq 0$. We use the same stability conditions for a conjugate curve along the $X-$trajectories.

By definition, the intersections between frame curves are frame points. Now, we give stability conditions for frame points. We assume that all frame points are generic, hence we have to consider only the cases listed in Section 2.6.2, p. 60. Stability conditions for frame points that belong to $\gamma^{\pm}$ are given in the first step. Those on the frontier of the reachable set are considered at the end of the algorithm, thus we remain with ten cases.

Let x be a frame point, $\gamma : [0, b] \to \mathbb{R}^2$ be one constructed trajectory that satisfies $\gamma(b) = x$ and λ the associated covector field. Let t_0 be the last switching time of γ before b and let D be the frame curve to which $\gamma(t_0)$ belongs. We assume that $\gamma(t_0) \in D \setminus \partial D$. For each $y \in D$ there exists a constructed trajectory γ_y, with associated covector field λ_y, that switches at y.

Consider first the case of a frame point of type $(C, C)_1$. We have $\Delta_B(x) = 0$ (see Appendix A) and $\nabla\Delta_B(x) \neq 0$ (from condition **(P2)** of Section 2.4). From the stability condition (2.78) we have:

$$X(\gamma(b)) \cdot \nabla\Delta_B(\gamma(b)) \neq 0, \qquad Y(\gamma(b)) \cdot \nabla\Delta_B(\gamma(b)) \neq 0,$$

then by the implicit function theorem, for each y in a neighborhood of $\gamma(t_0)$ in D, there exists a time σ_y such that $\Delta_B(\gamma_y(\sigma_y)) = 0$. Let $y(s)$ be a parametrization of a neighborhood in D of $\gamma(t_0) = y(s_0)$. Define:

$$\psi(y(s)) \doteq \lambda_y(\sigma_y) \cdot G(\gamma_y(\sigma_y)), \tag{2.82}$$

we assume the stability condition:

$$\left.\frac{\partial\psi}{\partial s}\right|_{s=s_0} \neq 0. \tag{2.83}$$

Consider the switching curve C_2 as in the description of the $(C, C)_1$ frame points given in Example 6 of Section 2.6.4, p. 62. Let $C_2(x)$ be a tangent vector to C_2 at x. The last stability condition is:

$$Y(x) \wedge C_2(x) \neq 0, \qquad X(x) \wedge C_2(x) \neq 0. \tag{2.84}$$

If x is of type $(C, S)_1$, resp. $(S, K)_1$, then we have the same stability conditions, replacing in (2.84) the vector $C_2(x)$ with $C(x)$, resp. $K(x)$, tangent to C, resp. to K, at x.

If x is of type $(C, K)_2$ and $\Delta_B(x) = 0$ then we assume that C and K are not tangent at x and we have the same stability conditions, replacing the vector $C_2(x)$ with $K(x)$ in (2.84).

Notice that for $(C, C)_1, (C, S)_1, (S, K)_1$ and $(C, K)_2$ frame points, the condition (2.84) plays the role of condition **(GA7)** for the points q_i.

If x is of type $(C, C)_2$, $(S, K)_3$ or (K, K) then no stability condition is needed, since stability is guaranteed by other assumptions.

If x is of type $(C,S)_2$ let $\varphi(y)$ be the control defined in (2.13) for every $y \in S$, and let $y(s)$ be a parametrization of a neighborhood of $x = y(s_0)$ in S. We assume:

$$\left.\frac{\partial \varphi}{\partial s}\right|_{s=s_0} \neq 0. \tag{2.85}$$

If x is of $(C,K)_1$ type then, as above, let $C(x)$, resp. $K(x)$, be the vector tangent to C, resp. K, at x. The stability condition is:

$$C(x) \wedge K(x) \neq 0. \tag{2.86}$$

If x is of type $(C,K)_2$ and $\Delta_B(x) \neq 0$, let $C(y)$ be the vector tangent to C at y, where y ranges in a neighborhood of x in C. The stability condition is:

$$\left.\frac{\partial}{\partial y}(Y(y) \wedge C(y))\right|_{y=x} \neq 0, \qquad \left.\frac{\partial}{\partial y}(X(y) \wedge C(y))\right|_{y=x} \neq 0. \tag{2.87}$$

Finally let x be of type $(S,K)_2$. If φ is the control of (2.13) then the stability condition is:

$$|\varphi(x)| < 1. \tag{2.88}$$

End of step N.

If at step N the algorithm $\mathcal{A}$ does not construct any new frame curve, then at step $N+1$ the algorithm $\mathcal{A}$ does no operation. Therefore, in this case, we say that the algorithm $\mathcal{A}$ stops at step N.

The conclusions of Section 2.4, p. 48 ensure the existence of $N \in \mathbb{N}$ such that $\mathcal{A}$ stops at step N. Consider the set $\mathcal{R}_\mathcal{A}$ of points reached by constructed trajectories of $\mathcal{A}$. By definition $Fr(\mathcal{R}_\mathcal{A})$ is a frame curve and its intersections with other frame curves are frame points. As first stability condition we assume:

($\mathcal{F}1$) The inequality (2.77), or the corresponding condition with $>$ replaced by $<$, holds for every trajectory reaching a point of $Fr(\mathcal{R}_\mathcal{A})$, except the intersections with turnpikes and switching curves.

We now give the stability conditions for these new frame points. We have to exclude that for every $\varepsilon > 0$ the synthesis constructed for time $\tau + \varepsilon$ has a new frame point. Indeed, in this case the synthesis can not be structurally stable. Hence, we give conditions to exclude that a frame point on $Fr(\mathcal{R}_\mathcal{A})$ can evolve in a frame point of different type immediately after time τ. For example we have to exclude that $x = \gamma^+ \cap Fr(\mathcal{R}_\mathcal{A})$ verifies $x = \gamma^+(t_i^+)$ for some i (see Definition 38, p. 90). Otherwise, x evolves in a (X,S) or (X,C) frame point after time τ.

($\mathcal{F}2$) Let x be a frame point of $Fr(\mathcal{R}_{\mathcal{A}})$. We assume $\Delta_A(x) \neq 0$ and if x is not of type (F, S) then we also assume that $\Delta_B(x) \neq 0$. If x is of type $(X, F)_{1,2}$ then $x = \gamma^{\pm}(\tau)$ and we assume that $\tau \notin \{t_i^+, t_i'^+, s_i^+, s_i'^+\}$ (see Definition 38, p. 90). If x is of type (F, C) then we assume that $X(x), Y(x)$ are not tangent to C at x and x is not reached by two different optimal trajectories. If x is of type (F, S) then we assume $|\varphi(x)| < 1$, where φ is defined in (2.13). Finally, if x is of type (F, K) then we assume that x is not a switching point for the constructed trajectories arriving at x (this condition is yet ensured by ($\mathcal{F}1$)).

We define an equivalence relation $\sim$ on the set of frame points.

Definition 39 (Equivalence of Frame Points). *If x_1, x_2 are two frame points, that are not of type $(C, C)_2, (K, K)$, we let $x_1 \sim x_2$ if and only if there exist some points $y_0 = x_1, y_2, \ldots, y_n = x_2$ such that the following holds. Each y_i belongs to a frame curve D_i. If y_i is a frame point then it is not of (K, K) type; if y_i is not a frame point then D_i is not of K type. For every y_i, $i = 1, \ldots, n-1$, there exist a constructed trajectory γ_i and $a_i, b_i \in Dom(\gamma_i)$ verifying $\gamma_i(a_i) = y_i, \gamma_i(b_i) = y_{i+1}$, $\gamma_i|_{[a_i, b_i]}$ is an X or Y trajectory and $\gamma_i(]a_i, b_i[) \cap \gamma^{\pm}([0, t_f^{\pm}]) = \emptyset$. That is there exists a curve, connecting x_1 with x_2, formed by X and Y arcs of constructed trajectories and that does not intersect the frame curves $\gamma^{\pm}$, the relative interior of overlap curves and (K,K) frame points.*

Remark **37** Notice that if we let γ_i intersect $\gamma^{\pm}$, then the origin is obviously in relation with every frame point reached by a bang-bang constructed trajectory. We have to exclude the points $(C, C)_2, (K, K)$ in the definition of $\sim$. In fact, it is not a generic situation for two frame points to be in relation unless one of them is of types $(C, C)_2$ or (K, K), which are constructed by $\mathcal{A}$ exactly because they are in relation with other frame points. But locally, at these points, the synthesis generated by $\mathcal{A}$ does not present a singularity.

If $\mathcal{A}$ stops at step N and:

($\mathcal{A}1$) All frame curves and points satisfy the stability conditions,
($\mathcal{A}2$) If x_1, x_2 are two frame points and $x_1 \sim x_2$ then $x_1 = x_2$,

then we say that the algorithm $\mathcal{A}$ succeeds for Σ.

Proceeding as in Section 2.5, p. 56, we construct the synthesis $\Gamma_{\mathcal{A}}(\Sigma)$ generated by the algorithm $\mathcal{A}$ at time τ. It is easy to see that conditions $(\mathcal{A}_1), (\mathcal{A}_2)$ are generic. Thus by Corollary 1, p. 50, $\mathcal{A}$ succeeds for a generic set of systems. Finally, by Theorem 14, $\Gamma_{\mathcal{A}}(\Sigma)$ is an optimal synthesis. Therefore we have the following:

Theorem 17 *Given $\tau > 0$, there exists a generic set $\Pi_\tau \subset \Xi$ such that for every $\Sigma \in \Pi$ the algorithm $\mathcal{A}$ succeeds for Σ at time τ and $\Gamma_{\mathcal{A}}(\Sigma)$ is an optimal synthesis.*

2.8.3 Structural Stability

In this section we prove the structural stability of systems for which $\mathcal{A}$ succeeds. It follows that these systems are observable, that is a small perturbation does not change the topological structure of the optimal synthesis.

Theorem 18 *If $\mathcal{A}$ succeeds for Σ then Σ is structurally stable.*

Proof. We recall that Σ is locally controllable and satisfies **(P1)**,...,**(P7)** of Section 2.4, p. 48. Let Σ' be a system in a small neighborhood of Σ. More precisely, assume that:

$$\|F - F'\|_{C^3} + \|G - G'\|_{C^3} < \varepsilon$$

where $\Sigma' = (F', G')$. From Lemma 3, p. 39 Σ' is locally controllable if:

$$F'(0) \wedge [F', G'](0) \neq 0.$$

From **(P1)** (of Section 2.4, p. 48):

$$F(0) \wedge [F, G](0) \neq 0$$

and:

$$|F'(0) \wedge [F', G'](0) - F(0) \wedge [F, G](0)| \leq C\varepsilon(\|F\|_{C^3} + \|G\|_{C^3}),$$

for some $C > 0$, hence for ε sufficiently small Σ' is locally controllable.

The conditions **(P1)**,...**(P7)** involve the components of the vector fields F, G and their derivatives, hence they can be established for Σ', if ε is sufficiently small, in the same way. Now, we can apply the algorithm $\mathcal{A}$ to Σ'. The conditions **(GA1)**–**(GA8)** and the stability conditions for frame points and curves are generic and hold for Σ, therefore they hold also for Σ' for ε small. Moreover, an iterative application of the inverse function theorem guarantees that, for ε sufficiently small, $\mathcal{A}$ produces the same frame curves, except for K-curves, and the same frame points, except (K, K) points, for Σ' and these frame curves and points satisfy the stability conditions as well. To prove this we proceed by induction as in the description of the algorithm $\mathcal{A}$. Firstly, we choose ε in such a way that the Y–trajectory through the origin $\tilde{\gamma}^+$ of Σ' is near in the $\mathcal{C}^0$ norm to γ^+ (of Σ) and it has the same sequence of frame points. Using Propositions 3, p. 92, 4, p. 94 and 5, p. 95, we obtain that the synthesis near $\tilde{\gamma}^+$ is the same as the synthesis near γ^+. Then, it is possible to continue following the same procedure of the algorithm $\mathcal{A}$.
As an example, let us check that if $\Gamma_{\mathcal{A}}(\Sigma)$ has a stable $(C, S)_1$ frame point x then the same holds for Σ'. From the condition **(P2)** (of Section 2.4, p. 48)), we know that zero is a regular value for the functions Δ_A, Δ_B of Σ. This is a structurally stable condition, hence for ε sufficiently small the same holds for Σ'. Moreover, the sets $\{x : \Delta_B(x) = 0\}$ for Σ and Σ' are close to each other.

From condition (2.78), it follows that there exists a regular turnpike for Σ'. Recall now the terminology used in the definition of stability conditions for $(C,S)_1$ points. We can construct the function ψ for Σ' as in (2.82). Moreover, by (2.83) we have that ψ has a unique zero near x. Therefore there is a point y' near $y(s_0)$ such that the trajectory $\gamma_{y'}$ that intersect S at time $\sigma_{y'}$ satisfies $\lambda_{y'}(\sigma_{y'}) \cdot G(\gamma_{y'}(\sigma_{y'})) = 0$. The algorithm $\mathcal{A}$ constructs a turnpike as frame curve also for Σ', starting from a point x' near x. Finally, the condition (2.84) ensures that the trajectories γ_y, with $y \in D$ on one side of y', switch before crossing the $X-$ or the $Y-$ trajectory starting from x'. Therefore x' is a $(C,S)_1$ point (the other possible case is a (S,K) point). At the end, the conditions $(\mathcal{F}1)$, $(\mathcal{F}2)$ are fulfilled for Σ' as well.

For each frame curve $D \in \Gamma_{\mathcal{A}}(\Sigma)$ of type C or S, we have an ordered sequence of points $x_1, \ldots, x_n \in D$ that are in relation with some frame point not of type (K,K). These are exactly the points considered in the construction of lines of the graph $\mathcal{G}$ associated to Σ, (see the next section) . From $(\mathcal{A}2)$, we have that for ε sufficiently small the corresponding frame curve D' of Σ' has the same number of distinct points y_i with the same properties of x_i. Moreover, the points y_i are in the same order as the x_i. Hence we have that $(\mathcal{A}1)$,$(\mathcal{A}2)$ are fulfilled and $\mathcal{A}$ succeeds for Σ'.

Recall Definition 36, p. 88. It is possible to construct the homeomorphism Ψ with an inductive procedure. Let $\Psi(\gamma^+) = \tilde{\gamma}^+$ and let the image of every frame point of $\Gamma_{\mathcal{A}}(\Sigma)$ be the corresponding frame point of $\Gamma_{\mathcal{A}}(\Sigma')$. Then we map every X–trajectory (Z–trajectory) arising from a point x of γ^+ onto the corresponding X–trajectory (Z–trajectory) arising from $\Psi(x)$. The same can be done for $\gamma^-, \tilde{\gamma}^-$. In this way we have defined Ψ on the trajectories formed by two arcs. By induction, if Ψ is constructed on trajectories with n arcs we consider the frame curves D formed by terminal points of these trajectories. Let D' and the points $x_i \in D$, $y_i \in D'$ be as above. By construction we have $\Psi(D) = D'$ and $\Psi(x_i) = y_i$. As for γ^+ and $\tilde{\gamma}^+$, we can now define Ψ on the X–,Y– and Z–trajectories originating from D and D'. In a finite number of steps the definition of Ψ is completed. Every trajectory γ of $\Gamma_{\mathcal{A}}(\Sigma)$ is formed by arcs of the same type of the arcs forming $\Psi(\gamma)$ and meets the same frame curves in the same order as $\Psi(\gamma)$. This ensures the conditions (E1)–(E3) for Ψ and then the equivalence between Σ and Σ'. ■

2.8.4 Graphs

In this section we introduce the definition of graph and describe a procedure to associate a graph to every system for which the algorithm $\mathcal{A}$ succeeds. The points and edges of this topological graph correspond to frame points and curves of the system. Moreover, some additional lines must be included in the definition of graph, to describe the history of all trajectories that form the optimal synthesis. These lines are precisely the trajectories "transporting" special information mentioned at the end of Section 2.6.1. In Remark 38, p. 106 we give some examples to motivate the definition of graph.

From now on, we consider only the systems of Ξ for which the algorithm $\mathcal{A}$ succeeds at time τ.

Definition 40 (Graph). *A graph $\mathcal{G}$ is a finite set of points of $\mathbb{R}^2$ and smooth connected embedded one dimensional manifolds with boundary connecting the points, called edges. Moreover, inside each region enclosed by edges there are possibly some other smooth manifolds, called lines, connecting points and edges. We assume that edges and lines do not cross each other.*

Every edge can be of one of the following types: X, Y, F, S, C, K; corresponding to the types of frame curves. An edge of type X, Y or S has an orientation and hence an initial and a terminal point. The edges of type C have a positive side, corresponding to the fact that constructed trajectories cross a frame curve of type C passing from one side to another.

Every region enclosed by edges, that are not all of F type, has a sign $+$ or $-$. This corresponds to the fact that a region of the reachable set, that contains no frame curve, is covered by X- or by Y-trajectories. On each region we can have some curves connecting points and edges. These correspond to constructed trajectories that pass through frame points. See Remark 38, p. 106 below.

We say that two edges E_1, E_2 are related and we write $E_1 \sim E_2$ if they have in common a point of the graph.

We now describe a canonical way of associating a graph to a system. Given a system Σ (for which $\mathcal{A}$ succeeds at time τ) we associate a graph $\mathcal{G}$ to Σ in the following way. For every frame point we construct a point of $\mathcal{G}$ having the same coordinates in $\mathbb{R}^2$. For every frame curve D, with no frame point in $D \setminus \partial D$, $\partial D = \{x_1, x_2\}$, we construct an edge E of $\mathcal{G}$ of the same type connecting the points of $\mathcal{G}$ corresponding to x_1, x_2. If D is an X, Y or S-curve then D has the orientation of increasing time and we endow E with the corresponding orientation. If D is of type C, then some constructed trajectories enter one side of D. We define the corresponding side of E to be positive.
For every region $A \subset \mathcal{R}$ enclosed by frame curves there is a region A', in the plane of the graph, enclosed by the corresponding edges. If A is covered by Y-trajectories, we assign to A' the positive sign, otherwise we assign to A' the negative sign.
We pass now to the construction of lines. These lines are necessary to describe the behavior of every optimal trajectory of the synthesis, see Remark 38, p. 106. Consider a frame point x of $Clos(A)$, which is not of (K, K) type (recall the terminology of Section 2.6.2, p. 60), and the constructed trajectory γ_x verifying $\gamma_x(t_x) = x$ for some t_x. Assume that $\gamma_x(I) \subset A$, for some $I = [a, b] \subset Dom(\gamma_x)$, $t_x \in I$. Notice that it can happen $a \neq t_x \neq b$, e.g. if x is of type $(X, K)_3$. If $t_x \neq a$ and $\gamma_x(a) \in D$ frame curve, then we construct a line in A' going from a point y of the edge E, corresponding to D, to the point x' of $\mathcal{G}$ corresponding to x. If $\gamma_x(a)$ is a frame point then we choose y to be the corresponding point of E, otherwise we choose y in $E \setminus \partial E$. If D is of C type, and $\gamma_x(a) \in D \setminus \partial D$, then we consider the last switching point z of γ_x before

$\gamma_x(a)$. If D is of S type then there exists a constructed trajectory γ_1 that switches at $\gamma_x(a)$ and enters the region on the opposite side, with respect to D, of the region entered by γ_x. Indeed, a Y and an X constructed trajectory originate from every point of a turnpike. We let, in this case, z to be the first switching point of γ_1 after $\gamma_x(a)$. If z belongs to a frame curve D_1 then we construct a line going from a point z' of the edge E_1, corresponding to D_1, to the point y. Again if z is a frame point we let z' be the corresponding point of $\mathcal{G}$. If D_1 is a C or S frame curve then we proceed in the same way. We continue until we reach a frame curve not of C or S type. We do the same if $t_x \neq b$.
We can construct these lines in such a way that they do not cross each other. If $\mathcal{G}$ is associated to Σ in this way then we say that $\mathcal{G}$ is canonically associated to Σ.

Remark **38** Consider the system

$$\begin{cases} \dot{x}_1 = 3x_1 + u \\ \dot{x}_2 = x_1^2 + x_1 \end{cases}$$

For every time $\tau > ln(4)/3$ the reachable set in time τ contains two switching curves starting from γ^-. There are two frame points of type (X, C) that are not topologically equivalent. See Example 3 of Section 2.6.4, p. 62 for an accurate description of this system and for the classification of (X, C) frame points. In Fig. 2.31 it is portrayed the reachable set of this example and in Fig. 2.32 its associated graph $\mathcal{G}_3$. If we do not specify a sign for every region of $\mathcal{G}_3$ then the two (X, C) frame points are not distinguishable. Hence, for some system Σ with a frame point of type $(X, C)_1$ or $(X, C)_2$, we can construct a system with the same graph, except the signs of the regions, but not equivalent to Σ. This show the necessity of specifying a sign for every region.

Consider the system Σ_4 of Example 4 of Section 2.6.4, p. 62. There is a region A that is a connected component of the complement of the reachable set and is bounded. In the corresponding graph, we cannot give a sign to the region corresponding to A. Otherwise, we would have equivalent systems corresponding to different graphs. The regions enclosed by edges all of F type correspond exactly to the holes of the reachable set.

Consider now the frame point x of $(C, S)_2$ type of Example 8 of Section 2.6.4, p. 62. If we do not specify, in the corresponding graph, a positive side for the edge corresponding to the switching curve then we do not know, from the graph, if the Y or the X trajectories enter the switching curve. Again there would exist two not equivalent systems corresponding to the same graph.
The lines divide the graphs into subregions in such a way that the trajectories, contained in the same subregion, have the same history, i.e. cross the same frame curves in the same order and are composed by the same sequence of elementary arcs. For example $XYZX$..., where X,Y and Z denote respectively

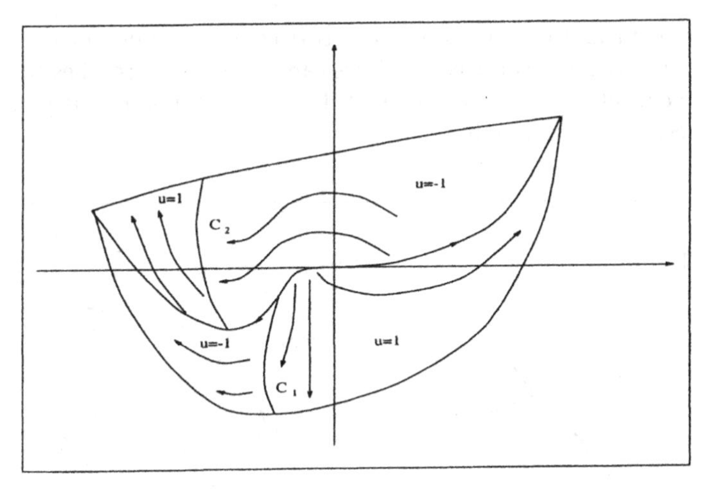

Fig. 2.31. Reachable set of Example 3 of Section 2.6.4, p. 62

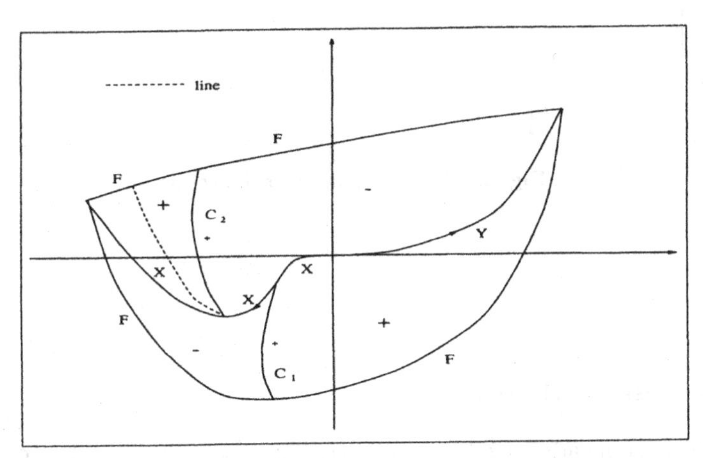

Fig. 2.32. Graph corresponding to Example 3 of Section 2.6.4, p. 62

X,Y arcs and trajectories running a turnpike. If the lines are not constructed then, in some cases, we can not decide the story of every trajectory and then we can not recognize equivalent systems. To appreciate this point, consider the following examples. Firstly, let the syntheses of Fig. 2.33 and Fig. 2.34 correspond to some system. The associated graphs contain the same points and edges. However, the syntheses are not equivalent. Indeed, the homeomorphism Ψ should map the trajectory through the (X, S) point onto the corresponding one to satisfy (E1), but obviously in this case (E2) cannot be satisfied. To have an explicit example, consider now the system Σ_4 of the fourth example of Section 2.6.4, p. 62. Let γ be the constructed trajectory that pass through the (Y, S) point and then goes on as X trajectory. If we do not consider the lines, from the graph associated to Σ_4 we cannot know if γ reaches the overlap curve or the frontier of the reachable set. Hence we cannot uniquely determine the synthesis from the graph.

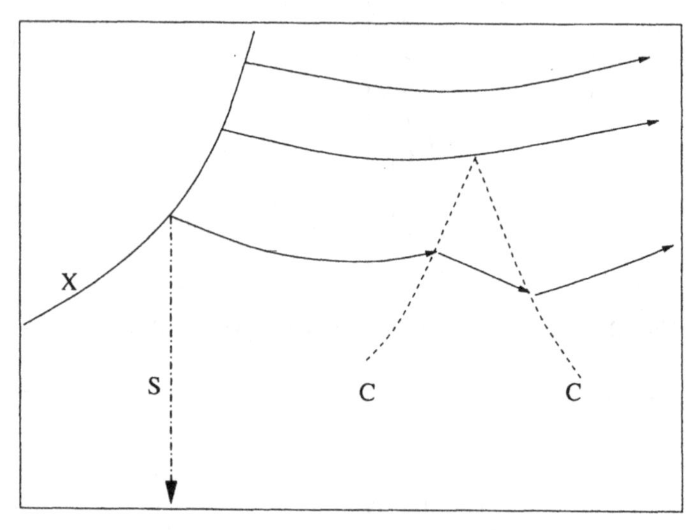

Fig. 2.33. An Example of Synthesis

2.8.5 Admissible Graphs

We now give some admissibility conditions that characterize a class of graphs. This class is be proved to be the class of graphs that correspond to systems canonically.

To every system of the Examples of Section 2.6.4, p. 62 we can associate a topological graph in the canonical way. We consider these examples restricted to a neighborhood of a frame point, then we obtain a set of graphs $\mathcal{E}$, whose elements are defined locally and each one corresponds to a type of frame point. A point x' of a graph $\mathcal{G}$ is said to be admissible if there exist a graph $\mathcal{G}' \in \mathcal{E}$ such that $\mathcal{G}$ contains a copy of $\mathcal{G}'$ to which x' belongs. We use the same terminology for the points of $\mathcal{G}$, e.g. (X, Y) point. The first condition is:

($\mathcal{G}1$) All points of $\mathcal{G}$ are admissible.

We consider graphs that contain exactly one point of the type (X, Y) and we call this point the origin of the graph. Assume that ($\mathcal{G}1$) holds. Let

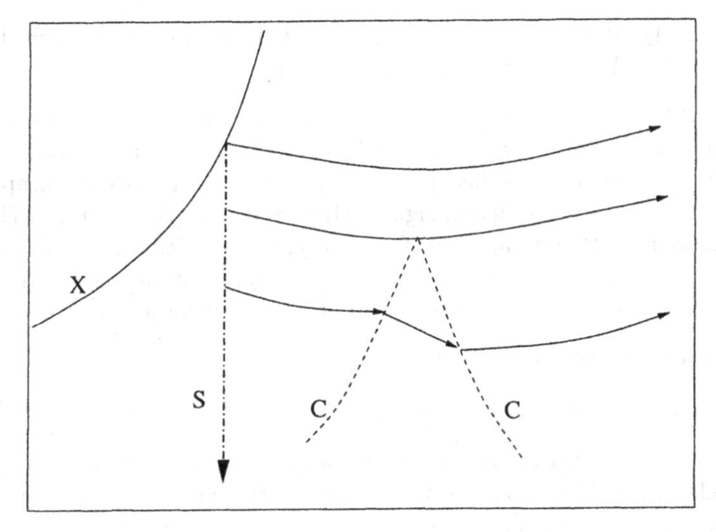

Fig. 2.34. An Example of Synthesis

E be a Y-edge and let x be the initial point of E. If x is not the origin then there exists a Y-edge E_1 for which x is the terminal point. We consider the initial point x_1 of E_1 and do the same considerations. Since $\mathcal{G}$ is finite, proceeding by induction, we find a finite collection $E_1, \ldots, E_n$ of Y-edges such that $E_i \sim E_{i+1}$, $i = 1, \ldots, n-1$, and the initial point of E_n is the origin. Then, since there is only one origin, the Y-edges form a set $\{E_1, \ldots, E_m\}$ such that the initial point of E_1 is the origin and for each $i = 1, \ldots, m-1$ the terminal point of E_i is the initial point of E_{i+1}. We call η^+ the union of these edges. Analogously we define η^- for the X-edges. In the first step of $\mathcal{A}$ we have described all the possibilities for the sequence of frame points on a curve γ^+ of a system Σ. We say that η^+ is <u>admissible</u> if there exists a system Σ such that the curve γ^+ correspond to η^+ canonically. That is there is a correspondence defined for points, edges of η^+, for lines intersecting η^+ and for the regions to which η^+ belongs, that follows the rules of canonical correspondence. This happens exactly when η^+ and γ^+ have an ordered sequence of corresponding points. The second condition is:

($\mathcal{G}2$) $\mathcal{G}$ has exactly one (X, Y) point, called the origin. The collections of edges $\eta^{\pm}$ are admissible.

Let E be a C-edge, x_1', x_2' be the endpoints of E and A' a region on one side of E. There exist two frame points x_1, x_2 corresponding to x_1', x_2'. Consider the correspondence between x_1' and x_1. Let D be the frame curve that correspond to E and $\tilde{A}_1$ the region corresponding to A'. We define A_1 to be the connected component of $\tilde{A}_1 \setminus \{x : \Delta_A(x)\Delta_B(x) = 0\}$ that contains D. Similarly we define the region A_2. We say that E is admissible if there exist x_1, x_2 such that the functions Δ_A, Δ_B have the same sign on A_1 and A_2.

Remark **39** If, for example, E connect two points of (C,S) type then A_1, A_2 are both covered by Y-trajectories or both covered by X-trajectories. In this case, since the two vector fields must point to opposite side of turnpikes, it follows that Δ_A have a different sign on the two regions A_1, A_2. From Theorem 11, p. 44 we have that along C the function f, see Definition 20, p. 44, does not change sign. Hence there exists at least one curve, intersecting C, on which $\Delta_A = 0$ and $\Delta_B = 0$. This is clearly a not generic situation.

Another admissibility condition is:

($\mathcal{G}3$) Every C-edge is admissible.

The relation $\sim$ partition the set of F-edges into a finite number of equivalence classes. If ($\mathcal{G}1$) holds then the union of the elements of an equivalence class form a closed curve.

($\mathcal{G}4$) Only one closed curve, that is union of the elements of an equivalence class of F-edges, encloses a region on which there are points and edges. Moreover, there are no frame curves and points outside this region.

Notice that it can happen to have more than one equivalence class of F-edges, e.g. the system in Example 4 of Section 2.6.4, p. 62, where $\mathcal{R}(\tau)$ has one hole.

Definition 41 (Entrance, Exit, Side) *Consider now a region A' enclosed by edges of $\mathcal{G}$. If one edge E is: of X type if A' is positive, of Y type if A' is negative, of C type with the negative side on A' or of S type then we say that E is an entrance. If E is of K, F or C type with positive side on A' then we say that E is an exit. Otherwise, we say that E is a side, i.e. if it is of Y type and A' is positive or of X type and A' is negative. The definitions are motivated by the fact that if D is a frame curve corresponding to E canonically, then through each point of D there passes a constructed trajectory that enters, resp. exits, the region corresponding to A' if and only if E is an entrance, resp. exit.*

We say that the set of lines of $\mathcal{G}$ is admissible if they do not cross each other and the following holds. Every line connects an entrance to an exit. If a point x' belongs to two entrances, resp. exits, then there is a line connecting x' with an exit, resp. entrance. Let x' be a point of one of the types $(X,C)_3, (X,K)_3, (C,C)_1, (C,S)_2, (C,K)_1, (S,K)$, and let A', B' be the two regions such that $x' \in Cl(A') \cap Cl(B')$. There are two lines l_1, l_2, both contained in A' or both in B', passing through x'; l_1 connects x' to an entrance and l_2 connect x' to an exit.
If x' is of type $(C,C)_2$ then there are two lines arriving at x' from different regions and at least one of them reaches another frame point. These are the only lines that connect two frame points.

If $x' \in E \setminus \partial E$, E is of C or S type and there is a line l arriving at x' from a region A' then there is a line arriving at x' from the other region B' such that $x' \in Cl(B')$. There are no other lines.

Remark **40** The conditions given for the set of lines follow directly from the canonical way of associating a graph to a system and from the description of frame points given in Section 2.6.2.

Consider the closed curve $\tilde{F}$, union of F-edges, described in ($\mathcal{G}4$). Let U be the connected component of the complement of the union of F-edges, that is enclosed by $\tilde{F}$ and verifies $\tilde{F} \subset Cl(U)$. If A' is a region contained in U we define $L(A')$ to be the set of lines contained in A'. The last condition is:

($\mathcal{G}5$) The set of lines of $\mathcal{G}$ is admissible. If $A' \subset U$ and A'_1 is a connected component of $A' \setminus L(A')$ then $Cl(A'_1)$ contains exactly one entrance and one exit.

The conditions in ($\mathcal{G}_5$) is given only for regions $A' \subset U$, because if the opposite happens then A' corresponds to an hole of $\mathcal{R}$, $L(A) = \emptyset$ and $Cl(A')$ contains only exits.

Definition 42 *If a graph $\mathcal{G}$ satisfies conditions ($\mathcal{G}1$),...,($\mathcal{G}5$) then we say that $\mathcal{G}$ is admissible.*

It is easy to check that if $\mathcal{G}$ corresponds to a system Σ then $\mathcal{G}$ is admissible. In the following we prove the converse.

2.8.6 Classification

To ensure that the canonical way of associating a graph to a system is well defined we have to prove that two systems are equivalent if and only if the associated graphs are equivalent.

Since we have defined the equivalence between systems in weak form, excluding overlap curves, equivalent systems may correspond to graphs having a different number of K edges. Hence we have to define an equivalence relation between graphs excluding K edges.

Given two admissible graphs $\mathcal{G}_1, \mathcal{G}_2$, we say that they are equivalent and we write $\mathcal{G}_1 \sim \mathcal{G}_2$ if there is a correspondence ψ between edges and lines of $\mathcal{G}_1$ and $\mathcal{G}_2$ such that the following holds. We let ψ be multivalued and not injective on the set of K-edges, but it has to be a bijective function restricted to the edges not of K type. Moreover, ψ is a bijective function restricted to the set of lines. Finally the following holds:

(H1) For every edge E, not of K type, $\psi(E)$ is an edge of the same type; $E_1 \sim E_2$ if and only if $\psi(E_1) \sim \psi(E_2)$, when E_1, E_2 are not both K-edges; ψ preserves orientations and positive sides.

(H2) If l is a line that connects E_1 with E_2 then $\psi(l)$ connects $\psi(E_1)$ with $\psi(E_2)$. The same holds for line connecting points. If l_1, l_2 arrive to the same point than the same happens for $\psi(l_1), \psi(l_2)$.
(H3) If A' is a region enclosed by edges $E_1, \dots, E_n$ then the region enclosed by $\psi(E_1), \dots, \psi(E_n)$ has the same sign.
(H4) If $\mathcal{K}_1, \mathcal{K}_2$ are the set of equivalence classes of K-edges (for the relation $\sim$) of $\mathcal{G}_1, \mathcal{G}_2$, then ψ induces a bijective correspondence between $\mathcal{K}_1, \mathcal{K}_2$.

We have the following:

Theorem 19 *If Σ_1, Σ_2 are two systems and $\mathcal{G}_1, \mathcal{G}_2$ the corresponding graphs then $\Sigma_1 \sim \Sigma_2$ if and only if $\mathcal{G}_1 \sim \mathcal{G}_2$.*

Proof. Assume first that $\Sigma_1 \sim \Sigma_2$ and let Ψ be as in Definition 36, p. 88. For simplicity we use the symbols Γ_1, Γ_2 for $\Gamma_A(\Sigma_1), \Gamma_A(\Sigma_2)$ respectively.
Given a frame curve D of Γ_1 that is not a K-curve let E_1, E_2 be the edges corresponding respectively to D and $\Psi(D)$. We define $\psi(E_1) = E_2$. We can proceed in the same way to define ψ on the set of lines. From (E1),(E2) it follows that (H1) and (H2) hold, and from (E3) it follows that (H3) holds.
Now, if K_1, K_2 are two K frame curves (of K type) of Γ_1, or of Γ_2, then we set $K_1 \sim K_2$ if they have a point in common. The union of the elements of an equivalence class of Γ_1 is a connected curve K. If we extend by continuity, Ψ then $\Psi(K)$ is the union of elements of an equivalence class of Γ_2. Therefore we can define ψ on K-edges in such a way that (H4) holds.

Assume now that $\mathcal{G}_1 \sim \mathcal{G}_2$. Let E_1 be an X, Y or S-edge of $\mathcal{G}_1$, $E_2 = \psi(E_1)$ and D_1, D_2 the frame curves corresponding to E_1, E_2 respectively. From (H1) we have that D_1, D_2 are of the same type. Assume that $x_1, \dots, x_n$ are the points of $D_1 \backslash \partial D_1$, ordered for increasing time, that are in relation with a frame point, not of (K, K) type, for the definition given in the previous Sections. There are exactly n lines if D_1 is of X or Y type and $2n$ lines if D_1 is of S type, starting at x_i. From (H2) it follows that there exist $y_1, \dots, y_n \in D_2 \setminus \partial D_2$, ordered for increasing time, from which some lines of $\mathcal{G}_2$ start. Observe that, if l is a line passing through x_i and $\psi(l)$ passes through y_j, then $i = j$. Indeed, if $i \neq j$ there must be a crossing between lines, but this is not allowed by the definition of a graph. We define Ψ on D_1 in such a way that Ψ is an homeomorphism, $\Psi(D_1) = D_2$ and $\Psi(x_i) = y_i$, $i = 1, \dots, n$.

For every $y \in D_1$ consider the constructed trajectories $\gamma_y \in \Gamma_1$ for which $y = \gamma_y(b_y)$ is a switching point. If D_1 is of X or Y type there is at most one such trajectory, if D_1 is of S type then there are two such trajectories. If D_1 is of type X or Y and there exists γ_y then from (H3) there exists a trajectory $\gamma_{\Psi(y)} \in \Gamma_2$ having the same property. Let $c_y > b_y$ be the first time in which γ_y reaches another frame curve and define $b_{\Psi(y)}, c_{\Psi(y)}$ similarly. We define:

$$\Psi\left(\gamma_y(t)\right) \doteq \gamma_{\Psi(y)} \left(b_{\Psi(y)} + \frac{c_{\Psi(y)} - b_{\Psi(y)}}{c_y - b_y} (t - b_y) \right) \quad \forall t \in [b_y, c_y].$$

In this way we have defined Ψ also on the frame curves that are reached by the trajectories γ_y. We proceed in the same way, defining Ψ on the images of the constructed trajectories that switch at the point of these new frame curves. After a finite number of steps we define Ψ on the whole reachable set $\mathcal{R}_1$ of the system Σ_1. Notice that we can have two different definitions of Ψ on the K frame curves, but Ψ restricted to $\mathcal{R}_1'$ (see the Definition 36, p. 88) is well defined. The condition (E1) follows by construction. The conditions (H1),(H2) ensure that corresponding trajectories have the same history, i.e. they cross the same type of frame curves in the same order and are composed by the same elementary arcs. Finally from conditions (H1)-(H4), we have that Ψ satisfies (E1)-(E3). ■

Assume now that $\mathcal{G}$ is an admissible graph. We want to find a system Σ such that $\mathcal{G}$ is associated to Σ in the canonical way. This and Theorem 19, p. 112 show that the correspondence $\Sigma \leftrightarrow \mathcal{G}$ is a bijection between the set of equivalence classes of systems for which $\mathcal{A}$ succeeds, and the set of equivalence classes of admissible graphs.

Theorem 20 *If $\mathcal{G}$ is an admissible graph then there exists a system Σ to which $\mathcal{G}$ is canonically associated.*

Proof. See Appendix A. ■

2.9 Systems on Two Dimensional Manifolds

All the geometric techniques used in this Chapter are local, thus it is possible to establish analogous results for a control system defined on a general smooth two dimensional manifold. Since $\mathcal{R}(\tau)$ is compact, we restrict, for simplicity, to an orientable compact smooth 2–D manifold. Recall that orientable compact smooth 2–D manifolds are described by a genus $g \geq 0$ which counts the number of 'handles' glued to the sphere S^2. The classification program is now performed using a new definition of graph. The key point is the introduction of a rotation system that, at each vertex, individuates a cyclic order of the incident edges and permits to define the faces of a graph. Thanks to a theorem of Heffter, a graph with a rotation system can be embedded uniquely on a 2-D orientable manifold, preserving the rotation system, if the faces are mapped to regions homeomorphic to a disk. Thus we can classify, via admissible graphs (with rotation systems), couples formed by a system and the 2-D manifold of minimal genus on which the system can live. The problem of embedding graphs into manifolds is subject of investigations for Topological Graph Theory, see [68] for an introduction. For differential and algebraic topology we refer the reader to [71, 99].

To have a more elegant representation, we let the lines, introduced above, be frame curves (hence edges of the associated graph). These new frame curves are also trajectories and transport special information about the history of optimal trajectories. We have to distinguish three kinds of curves:

- curves of kind γ_A that are abnormal extremals,
- curves of kind γ_k that are curves passing through an endpoint of an overlap (i.e. FPs of kind $(Y,K)_3$ $(S,K)_2$, $(C,K)_1$)
- curves of kind γ_0 that are the other arcs of optimal trajectories passing through FPs.

Remark **41** This refined definition of frame curves is essential for studying properties of the minimum time function. They are naturally constructed by a new algorithm, given in Chapter 3, used to provide additional information about the level curves of minimum time.

2.9.1 Graphs, Cellular Embeddings and the Heffter Theorem

We start introducing a new definition of graph.

Definition 43 *A graph is a topological Hausdorff space $\mathcal{G}$ with a closed discrete subspace $\mathcal{G}^0$, whose points are called the vertices of $\mathcal{G}$, such that the following holds. The complementary set $\mathcal{G}\backslash\mathcal{G}^0$ is a disjoint union of open subsets e_i; every e_i is homeomorphic to an open interval $I \subset \mathbb{R}$ and is called an edge of $\mathcal{G}$. For each edge e_i its boundary ∂e_i is a subset of $\mathcal{G}^0$ consisting either of one or two points; in case ∂e_i consists of two points, the set $\overline{e}_i$ is homeomorphic to a closed interval $\overline{I} = [0,1] \subset \mathbb{R}$; in case ∂e_i consists of one point, the set $\overline{e}_i$ is homeomorphic to the unit circle S^1. We indicate by $V_{\mathcal{G}}$ the set of vertices and by $E_{\mathcal{G}}$ the set of edges.*

Remark **42** Graph theory is usually developed in the framework of CW–complexes. For simplicity we stated directly the corresponding definition of graph. Notice that graphs corresponding to optimal synthesis are connected, have a finite number of vertices and edges, moreover all edges boundaries consist of two points.

Definition 44 *A local rotation of a vertex v is an oriented cyclic order (defined up to the cyclic permutations) of all edges incident to v. A rotation system R of a graph $\mathcal{G}$ is a union of all local rotations over all vertices of $\mathcal{G}$. An oriented graph, briefly orgraph, is a pair $(\mathcal{G},R)$ formed by a graph and a rotation system.*

Let $\mathcal{G}$ and $\mathcal{G}'$ be two finite graphs. A graph map $f : \mathcal{G} \to \mathcal{G}'$ consists of a vertex function $f_V : V_{\mathcal{G}} \to V_{\mathcal{G}'}$ and an edge function $f_E : E_{\mathcal{G}} \to E_{\mathcal{G}'}$ such that the incidence structure is preserved. If $(\mathcal{G},R)$ is an orgraph, we demand also that f preserves orientation at every edge. We consider also partially oriented graphs, that are graphs containing some oriented and some non oriented edges. In this case f should send oriented edges to oriented edges, preserving orientation, and non oriented edges to non oriented edges. A graph map $f : \mathcal{G} \to \mathcal{G}'$ between two graphs $\mathcal{G}$ and $\mathcal{G}'$ is called an isomorphism if both its vertex function f_V and edge function f_E are one-to-one and onto (surjective). Two graphs $\mathcal{G}$ and $\mathcal{G}'$ are called isomorphic if there exists an isomorphism $f : \mathcal{G} \to \mathcal{G}'$.

We call surface any 2-dimensional orientable compact manifold. An em--bedding $i : \mathcal{G} \to M$ of a graph $\mathcal{G}$ into a surface M is a 1-1 continuous map of the topological space $\mathcal{G}$ into the topological space M. Two embeddings i_1 and i_2 of $\mathcal{G}$ in a surface M are equivalent if there exists a homeomorphism $h : M \to M$ such that $h \circ i_1 = i_2$ (in other words, h brings the image $i_1(\mathcal{G})$ to the image $i_2(\mathcal{G})$).

If one takes an embedding $i : \mathcal{G} \to M$ of a connected graph $\mathcal{G}$ in M, then the set $M \backslash i(\mathcal{G})$ is a union of open regions A_m. Clearly, gluing up handles to each A_m, it is possible to obtain embeddings of $\mathcal{G}$ into surfaces of an arbitrary high genus. An embedding $i : \mathcal{G} \to M$ is called 2-cell (or cellular), if all open regions A_m are homeomorphic to an open disk.

Denote by $i : \mathcal{G} \to M$ a 2-cell embedding of an orgraph $\mathcal{G}$ into a surface M. Let $M \backslash i(\mathcal{G}) = A_1 \cup ... \cup A_m$ be a union of open disk regions in M. A dual graph $\mathcal{G}^{\#}$, associated to i, is a graph with the vertex set $V_{\mathcal{G}^{\#}} = \{A_1, ..., A_m\}$. An edge $e^{\#} \in E_{\mathcal{G}^{\#}}$ between A_i and A_j should be drawn (case $i = j$ is not excluded), if and only if there is an edge $e \in E_{\mathcal{G}}$ between A_i and A_j, i.e. $e \subseteq \overline{A}_i \cap \overline{A}_j$.

Rotation systems give rise to a certain system of faces on $\mathcal{G}$ and the following face tracing algorithm allows to determine all faces of a graph $\mathcal{G}$ corresponding to a rotation system R. Take an arbitrary vertex $v_1 \in V(\mathcal{G})$ and an edge a_{v_1}, incident to v_1. Let v_2 be the vertex of $\mathcal{G}$, connected with v_1 by the edge a_{v_1} and let a_{v_2} be the edge of the vertex v_2, which lies to the right of a_{v_1} in the cyclic order at v_2. Moving along the edge a_{v_2} to a vertex v_3, we define an edge a_{v_3}, which lies to the right of a_{v_2}. Proceeding inductively, we stop the process at an edge a_{v_n} if the two forthcoming edges are again a_{v_1} and a_{v_2}. Hereby a cycle $a_{v_1}, a_{v_2}, ..., a_{v_n}$ of a length n, which defines a face A_1 on $\mathcal{G}$, is traced. For tracing a next face A_2 one should start with an edge which lies to the right of any edge of the face A_1 and such, that a corner between them did not occur in A_1 - and apply the above construction. All faces $A_1, A_2, ..., A_m$ on $\mathcal{G}$ are so traced, when it remains no unused corner.

We can now state Heffter's Theorem, see [70]. This result, dating back to late 19th century, was recently rediscovered, see [104, 105], and used to provide a more neat topological classification of two dimensional dynamical systems.

Theorem 21 (Heffter's Theorem) *Let $\mathcal{G}$ be a finite graph endowed with a rotation system R. Then there exists a 2-cell embedding of $\mathcal{G}$ into an orientable surface M such that one of two rotations, induced by this embedding, coincides with R. Moreover, two embeddings are equivalent if and only if they have equivalent rotation systems.*

Each embedding $i : \mathcal{G} \to M$ induces a pair of rotation systems R and R^*, where R^* is a mirror image of R (i.e can be obtained from R by reversing of the cyclic order of all local rotations). The corresponding embeddings $i(\mathcal{G})$

and $i^*(\mathcal{G})$ are conjugate by a homeomorphism $h : M \to M$, which is not close to id_M.

2.9.2 Admissible Graphs and Syntheses Classification

We consider connected finite graphs, with oriented and non oriented edges, endowed by a rotation system R. As done in previous sections, on the orgraph $(\mathcal{G}, R)$, we put additional structure. Thus $\mathcal{G}$ presents edges of nine types: $X, Y, C, S, K, F, \gamma_A, \gamma_k$ and γ_0. The edges of type $X, Y, S, \gamma_A, \gamma_k$ and γ_0 have an orientation. As explained above we can associate to $(\mathcal{G}, R)$ a finite number of faces $A_1, \ldots, A_m$. We let the faces A_i, that are not enclosed by edges of type F, have a sign $\pm$ (corresponding to the fact that the optimal feedback is equal to ± 1). In other words we assign a sign ± 1 to the vertices of the dual graph $\mathcal{G}^{\#}$.

Remark **43** Notice that now we do not need to let edges of type C have a sign on one side, as done for the previous classification. This because the orientation of the new edges $\gamma_{A,k,0}$ permits to individuate uniquely the side optimal trajectories are entering.

We say that two orgraphs $(\mathcal{G}_i, R_i)$, $i = 1, 2$, are isomorphic if there exists a graph isomorphism f that is compatible with R_i, preserves edges types and the sign of regions.

In a way entirely similar to the one described in Section 2.8.4, we associate an orgraph $(\mathcal{G}, R)$ to a system Σ defined on a surface M. Thanks to the finer partition of $\mathcal{R}(\tau)$ provided by the new frame curves, we have that each orgraph $(\mathcal{G}, R)$ associated to a system can present only a finite number of cell types. In the sense that, if A_i is a region of the orgraph $(\mathcal{G}, R)$ and $\partial A_i = \{e_1, \ldots, e_n\}$, where e_j are the edges surrounding A_i, then necessarily there are a finite number of possibilities for ∂A_i. More precisely we have the following.

Definition 45 *Let $(\mathcal{G}, R)$ be an orgraph, A_i a face and $\partial A_i = \{e_1, \ldots, e_n\}$. An edge e is called a* side *if e is of type γ_A, γ_K, γ_0 and of type X if the region has sign $-$ and of type Y if the region has sign $+$. The face A_i is called* admissible *if:*

i) $n = 3$ or 4;
ii) if $n = 3$ there is only one side, otherwise there are two not incident sides and there is no orientation of ∂A_i compatible with both orientations of the sides;
iii) the edge containing the initial point(s) of the side(s) is called entrance *and is of type X (if the sign is $+$), Y (if the sign is $-$), S or C;*
iv) the edge containing the terminal point(s) of the side(s) is called exit *and is of type C, K, or F.*

In Figure 2.35 we represent all admissible faces.

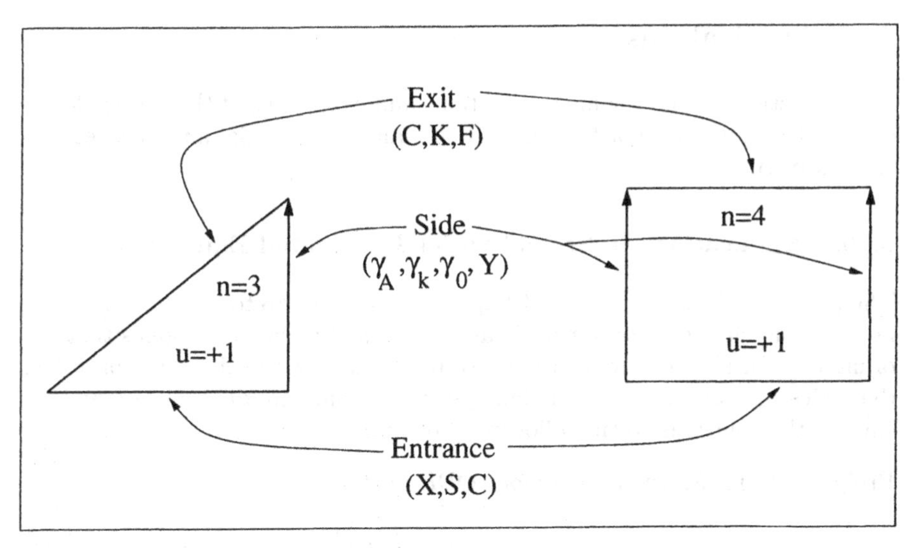

Fig. 2.35. Admissible faces in the case $u = +1$.

Proposition 6 *If* $(\mathcal{G}, R)$ *is an orgraph associated to a system, then each face* A_i *is admissible.*

Again, a set of admissibility conditions are given (beside that of admissible faces). These can be easily obtained by those of Section 2.8.5 and the description of admissible faces.

Finally, by the same methods of Section 2.8, we get

Theorem 22 *Let* Σ_1 *and* Σ_2 *be two control systems on a compact orientable 2-dimensional manifold* M. *Then* Σ_1 *is equivalent to* Σ_2 *if and only if the associated orgraphs* $(\mathcal{G}_1, R_1)$ *and* $(\mathcal{G}_2, R_2)$ *are isomorphic.*

Theorem 23 *Let* $(\mathcal{G}, R)$ *be an arbitrary admissible orgraph. Then there exists a compact 2-dimensional orientable manifold* M *and a system* Σ *on* M, *whose associated orgraph is isomorphic to* $(\mathcal{G}, R)$.

Remark **44** Notice that the manifold M of Theorem 23 is unique up to diffeomorphism because of the cellular embedding. The same system Σ can be put on a manifold with higher genus adding an arbitrary number of handles.

2.10 Applications

In this section we show some applications and extensions of the theory developed in this Chapter. For the case of two dimensional initial and final sets see Appendix B.

2.10.1 Applications to Second Order Differential Equations

The program, developed in this Chapter, can be used to to provide a particularly simple classification for optimal syntheses of a class of systems that are of interest for physical applications. More precisely, we study some controlled dynamics appearing in Lagrangian systems of mathematical physics and we classify the solutions to the following problem:

Problem: *Consider the autonomous ODE in* $\mathbb{R}$*:*

$$\ddot{y} = f(y, \dot{y}), \tag{2.89}$$
$$f \in C^\infty(\mathbb{R}^2), \quad f(0,0) = 0 \tag{2.90}$$

that describes the motion of a point under the action of a force depending on the position and velocity of the point (for instance due to a magnetic field or the viscosity of a fluid). Let us apply an external force, that we suppose bounded (e.g. $|u| \leq 1$*):*

$$\ddot{y} = f(y, \dot{y}) + u. \tag{2.91}$$

Controlling the external force we want to reach in minimum time a point (y_0, v_0)*, of the configuration space, from the rest state* $(0,0)$*.*

First of all observe that if we set $x_1 = y$, $x_2 = \dot{y}$, (2.91) becomes:

$$\dot{x}_1 = x_2 \tag{2.92}$$
$$\dot{x}_2 = f(x_1, x_2) + u, \tag{2.93}$$

that can be written in our standard form $\dot{x} = F(x) + uG(x)$, $x \in \mathbb{R}^2$ by setting $x = (x_1, x_2)$, $F(x) = (x_2, f(x))$, $G(x) \equiv (0,1)$.
In this case we have:

$$\Delta_A(x) = \det(F(x), G(x)) = x_2 \tag{2.94}$$
$$\Delta_B(x) = \det(G(x), [F(x), G(x)]) = 1. \tag{2.95}$$

From these it follows:

$$\Delta_A^{-1}(0) = \{x \in \mathbb{R}^2 : x_2 = 0\} \tag{2.96}$$
$$\Delta_B^{-1}(0) = \emptyset. \tag{2.97}$$

By direct computations it is easy to see that the generic conditions **(P1)**,...,**(P7)** (see Section 2.4, p. 48), are satisfied under the condition:

$$f(x_1,0) = \pm 1 \;\Rightarrow\; \partial_1 f(x_1,0) \neq 0 \tag{2.98}$$

that obviously implies $f(x_1,0) = 1$ or $f(x_1,0) = -1$ only in a finite number of points.
The reader can easily prove that for our problem (2.92), (2.93), with the condition (2.98), the "shape" of the optimal synthesis is that shown in Figure 2.36.

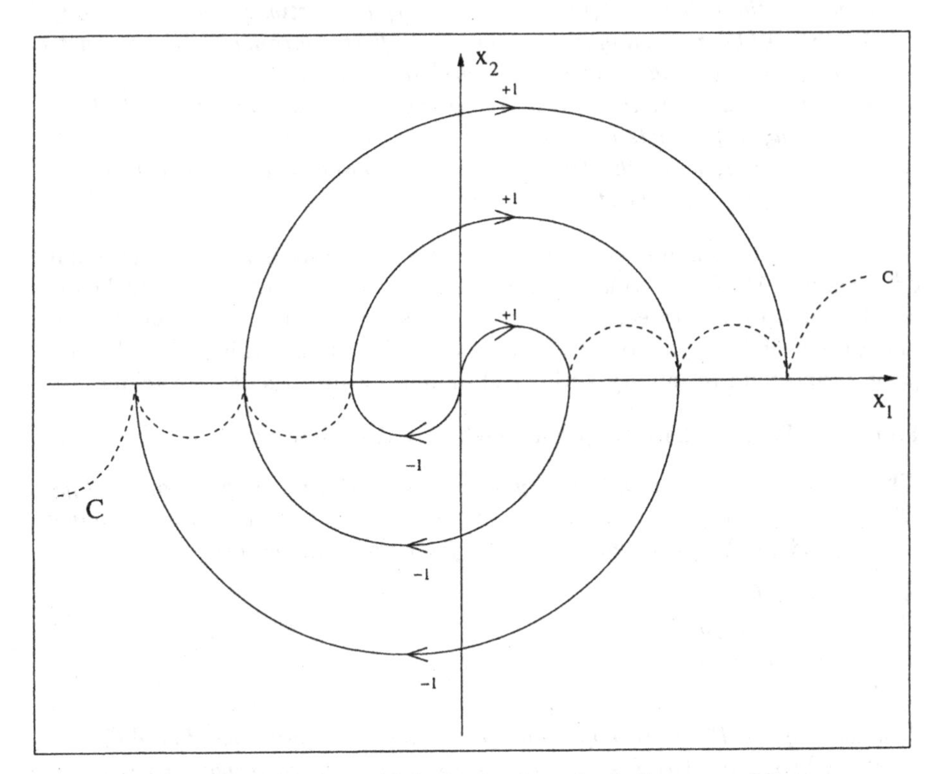

Fig. 2.36. The shape of the optimal synthesis for our problem.

In particular the partition of the reachable set is described by the following:

Theorem 24 *The optimal synthesis of the control problem (2.92) (2.93) with the condition (2.98), satisfies the following:*

1. *there are no turnpikes;*
2. *the trajectory* $\gamma^{\pm}$ *(starting from the origin and corresponding to constant control* ± 1*) exits the origin with tangent vector* $(0, \pm 1)$ *and, for an in-*

terval of time of positive measure, lies in the set $\{(x_1,x_2): x_1,x_2 \geq 0\}$ *respectively* $\{(x_1,x_2): x_1,x_2 \leq 0\}$*;*

3. $\gamma^{\pm}$ *is optimal up to the first intersection (if it exists) with the* x_1*-axis. At the point in which* γ^{+} *intersects the* x_1*-axis it generates a switching curve that lies in the half plane* $\{(x_1,x_2): x_2 \geq 0\}$ *and ends at the next intersection with the* x_1*-axis (if it exists). At that point another switching curve generates. The same happens for* γ^{-} *and the half plane* $\{(x_1,x_2): x_2 \leq 0\}$*;*
4. *let* y_i*, for* $i = 1,...,n$ *(*n *possibly* $+\infty$*) (respectively* z_i*, for* $i = 1,...m$ *(*m *possibly* $+\infty$*)) be the set of boundary points of the switching curves contained in the half plane* $\{(x_1,x_2): x_2 \geq 0\}$ *(respectively* $\{(x_1,x_2): x_2 \leq 0\}$*) ordered by increasing (resp. decreasing) first components. Under generic assumptions,* y_i *and* z_i *do not accumulate. Moreover:*
   - *For* $i = 2,...,n$*, the trajectory corresponding to constant control +1 ending at* y_i *starts at* z_{i-1}*;*
   - *For* $i = 2,...,m$*, the trajectory corresponding to constant control* -1 *ending at* z_i *starts at* y_{i-1}*.*

Remark **45** The union of $\gamma^{\pm}$ with the switching curves is a one dimensional C^0 manifold M. Above this manifold the optimal control is $+1$ and below is -1. The optimal trajectories turn clockwise around the origin and switch along the switching part of M . If $n, m < \infty$, they stop turning after the last y_i or z_i and tend to infinity with $x_1(t)$ monotone after the last switching.

From **4** of Theorem 24 it follows immediately the following:

Theorem 25 *To every optimal synthesis for a control problem of the type (2.92) (2.93) with the condition (2.98), it is possible to associate a couple* $(n,m) \in (\mathbb{N} \cup \infty)^2$ *such that one of the following cases occurs:*

A. $n = m$*,* n *finite;*
B. $n = m + 1$*,* n *finite;*
C. $n = m - 1$*,* n *finite;*
D. $n = \infty$*,* $m = \infty$*.*

Moreover, if Γ_1*,* Γ_2 *are two optimal syntheses for two problems of kind (2.92), (2.93), (2.98), and* (n_1,m_1) *(resp.* (n_2,m_2)*) are the corresponding couples, then* Γ_1 *is equivalent to* Γ_2 *iff* $n_1 = n_2$ *and* $m_1 = m_2$*.*

Remark **46** In Theorem 25 the equivalence between optimal syntheses is the one given in Definition 36, p. 88).

2.10.2 Example: the Duffin Equation

In the following we show the qualitative shape of the synthesis of the controlled Duffin Equation. More precisely we want to determine the value of the couple (m,n) of Theorem 25.

The Duffin equation is given by the formula $\ddot{y} = -y - \varepsilon(y^3 + 2\mu\dot{y}),\ \varepsilon, \mu > 0$, ε small. By introducing a control term and transforming the second order equation in a first order system, we have:

$$\dot{x}_1 = x_2 \tag{2.99}$$
$$\dot{x}_2 = -x_1 - \varepsilon(x_1^3 + 2\mu x_2) + u. \tag{2.100}$$

To know the shape of the synthesis we need to know where $(F+G)_2(x) = 0$. If we set $a = \frac{1}{2\varepsilon\mu}$, this happens where

$$x_2 = a(1 - x_1 - \varepsilon x^3). \tag{2.101}$$

From (2.99) and (2.100) we see that, after meeting this curve, the trajectory γ^+ moves with $\dot{\gamma}_1^+ > 0$ and $\dot{\gamma}_2^+ < 0$. Then it meets the x_1-axis because otherwise, if $\gamma^+(t) \in \Omega := \{(x_1, x_2),\ x_1, x_2 > 0\}$, then we necessarily have (for $t \to \infty$) $\gamma_1^+ \to \infty$, $\dot{\gamma}_2^+ \to 0$, that is not permitted by (2.100). The behavior of the trajectory γ^- is similar.

In this case, the numbers (n, m) are clearly (∞, ∞) because the $+1$ trajectory that starts at z_1 meets the curve (2.101) exactly one time and behaves like γ^+. So the C-curve that starts at y_1 meets again the x_1 axis. The same happen for the -1 curve that starts at y_1. In this way an infinite sequence of y_i and z_i is generated.

2.10.3 Generalization to Bolza Problems

Quite easily we can adapt the previous program to obtain information about the optimal syntheses associated (in the previous sense) to second order differential equations, but for more general minimizing problems.

We have the well known:

Lemma 16 *Consider the control system:*

$$\dot{x} = F(x) + uG(x), \quad x \in \mathbb{R}^2,\ F, G \in C^\infty(\mathbb{R}^2, \mathbb{R}^2),\ F(0) = 0,\ |u| \leq 1. \tag{2.102}$$

Let $L : \mathbb{R}^2 \to \mathbb{R}$ be a C^3 bounded function such that there exists $\delta > 0$ satisfying $L(x) > \delta$ for any $x \in \mathbb{R}^2$.
Then, for every $x_0 \in R^2$, the problem: $\min \int_0^\tau L(x(t))dt$ *s.t.* $x(0) = 0,\ x(\tau) = x_0$, *is equivalent to the* minimum time problem *(with the same boundary conditions) for the control system $\dot{x} = F(x)/L(x) + uG(x)/L(x)$.*

By this Lemma it is clear that if we have a second order differential equation with a bounded–external force $\ddot{y} = f(y, \dot{y}) + u$, $f \in C^3(\mathbb{R}^2)$, $f(0,0) = 0$, $|u| \leq$

1, then the problem of reaching a point in the configuration space (y_0, v_0) from the origin, minimizing $\int_0^T L(y(t), \dot{y}(t))dt$, (under the hypotheses of Lemma 16) is equivalent to the minimum time problem for the system: $\dot{x}_1 = x_2/L(x)$, $\dot{x}_2 = f(x)/L(x) + 1/L(x)u$. By setting: $\alpha : \mathbb{R}^2 \to]0, 1/\delta[$, $\alpha(x) := 1/L(x)$, $\beta : \mathbb{R}^2 \to \mathbb{R}$, $\beta(x) := f(x)/L(x)$, we have: $F(x) = (x_2\alpha(x), \beta(x))$, $G(x) = (0, \alpha(x))$. From these it follows: $\Delta_A(x) = x_2\alpha^2$, $\Delta_B(x) = \alpha^2(\alpha + x_2\partial_2\alpha)$. The equations defining turnpikes are: $\Delta_A \neq 0$, $\Delta_B = 0$, that with our expressions becomes the differential condition $\alpha + x_2\partial_2\alpha = 0$ that in terms of L is:

$$L(x) - x_2\partial_2 L(x) = 0. \tag{2.103}$$

Remark **47** Since $L > 0$ it follow that the turnpikes never intersect the x_1-axis. Since (2.103) depends on $L(x)$ and not on the control system, all the properties of turnpikes depend only on the Lagrangian.

Now we consider some particular cases of Lagrangians.

L=$L(y)$ In this case the Lagrangian depends only on the position y and not on the velocity $\dot{y}$ (i.e. $L = L(x_1)$). (2.103) is never satisfied so there are no turnpikes.

L=$L(\dot{y})$ In this case the Lagrangian depends only on velocity and the turnpikes are horizontal lines.

L=$V(y) + \frac{1}{2}\dot{y}^2$ In this case we want to minimize an energy with a kinetic part $\frac{1}{2}\dot{y}^2$ and a positive potential depending only on the position and satisfying $V(y) > 0$. The equation for turnpikes is $(x_2)^2 = 2V(x_1)$.

2.10.4 The Non-locally Controllable Case

If we relax the condition $F(0) = 0$, that assures local controllability, there are some differences in the shape of the optimal syntheses in a neighborhood of $\gamma^+ \cup \gamma^-$. We briefly discuss what happen in a neighborhood of the origin and we leave the study of the other singularities as an exercise.

First observe that if $F(0) \neq 0$, but there exists $u \in [-1, 1]$ such that $F(0) + uG(0) = 0$ (this is a nongeneric case), then the shape of the optimal syntheses is exactly as in the case $F(0) \neq 0$. In the generic case, $\Delta_A(0) \neq 0$ ($F(0)$ and $G(0)$ are not parallel) and we can assume $\Delta_B(0) \neq 0$. According to Theorem 11, p. 44, only one switching is permitted depending on the sign of f. More precisely:

- if $f(0) = -\Delta_B(0)/\Delta_A(0) > 0$ then every optimal trajectory is of the type X, Y or $Y * X$–trajectory.
- if $f(0) = -\Delta_B(0)/\Delta_A(0) < 0$ then every optimal trajectory is of the type X, Y or $X * Y$–trajectory.

The shape of the local optimal syntheses is drawn in Figure 2.37.

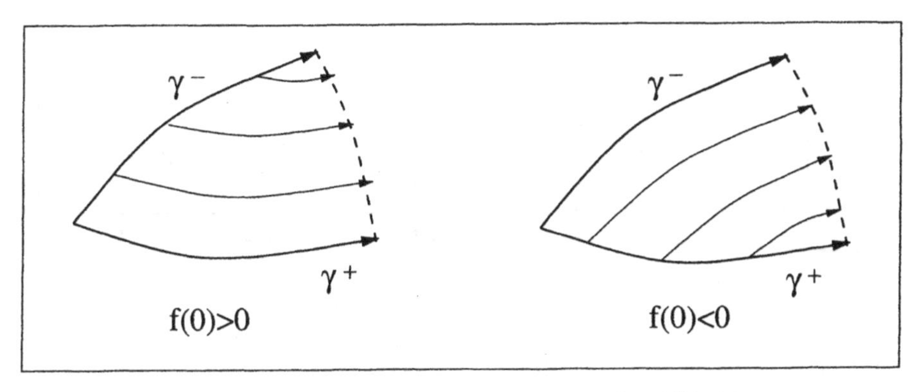

Fig. 2.37. Generic local optimal synthesis for $F(0) \neq 0$

Bibliographical Note

The concept of synthesis was introduced in the pioneering work of Boltianski [27]. Properties of extremal trajectories for single input affine control systems on the plane were first studied by Baitman in [19, 20] and later by Sussmann in [125, 126].

The existence of an optimal synthesis for analytic systems and under generic conditions for $\mathcal{C}^\infty$ systems was proved by Sussmann in [127] and by Piccoli in [109] respectively. Structural stability of generic syntheses was proved in [43]. Finally, a complete classification of generic singularities and generic optimal synthesis is given by Bressan and Piccoli in [44, 110]. Extensions to two dimensional compact manifolds (see Section 2.9, p. 113) are presented in [35, 104]. For graph theory, used in Section 2.9, p. 113 see [68].

Some example of optimal syntheses for two dimensional systems can be found in the book [69]. Jakubczyk and Respondek studied feedback equivalence of systems as our model problem in [73, 74], and more recently also bifurcations in [75].

The problem of the construction of the Time Optimal Synthesis for the system $\dot{x} = F(x) + uG(x), \quad |u| \leq 1$ where $x \in \mathbb{R}^3$, under generic conditions, is very difficult and still open. Results on the local structure of the reachable set can be found in [42, 90, 115, 116]. Results about the bound on the number of switchings, using higher order techniques, can be found in the recent interesting paper [11].

For the difficulties arising in extending our results to infinite time, we refer the reader to the works of Davydov [54, 55, 56, 57]. Related to our problem is also the interesting paper by Butenina [47].

The classification of the class of systems coming from second order differential equations, developed in Section 2.10.1, was published by the authors in [36].

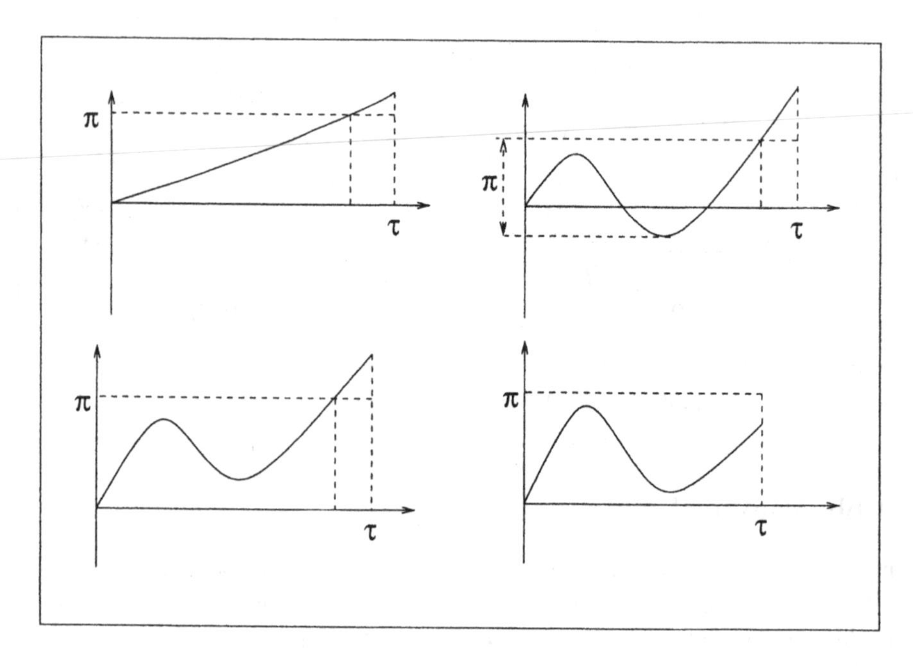

Fig. 2.38. Exercise 10

Exercises

Exercise 10 Draw the possible singularities along γ^+ for the θ^+ functions given in Figure 2.38.

Exercise 11 Recall Section 2.10.1, p. 118. Find the couple (m,n) for the Van der Pol equation: $\ddot{y} = -y + \epsilon(1-y^2)\dot{y} + u$, $\varepsilon > 0$ and small. Study the reachable set $\mathcal{R}(T)$ for $T \to \infty$.

Exercise 12 Recall Section 2.10.1, p. 118. Find the couple (m,n) for the equation $\ddot{y} = -e^y + \dot{y} + 1$.

Exercise 13 Prove that if X_1, X_2 are smooth vector fields on R^n and ψ a smooth function on R^n, then $[X_1, \psi X_2] = L_{X_1}\psi\, X_2 + \psi[X_1, X_2]$ where $L_{X_1}\psi = \nabla\psi \cdot X_1$ is the Lie derivative of ψ along X_1.

Exercise 14 Prove Lemma 12, p. 47 using the generalized Legendre–Clebsch condition of Theorem 7, p. 25.
[Hint: using Definition 20, p. 44, Exercise 13, PMP and Lemma 6, p. 40 (in this order), write the generalized Legendre–Clebsch condition as $L_G f \geq 0$.]

Exercise 15 Prove formula (2.18) of the proof of Lemma 14, p. 52.
[Hint: use mean–value theorem and the differentiability of ψ.]

Exercise 16 Draw the optimal syntheses in a neighborhood of the origin in the nongeneric case in which γ^+, γ^- and $\Delta_B^{-1}(0)$ are as in Figure 2.39.

Exercise 17 Find all the singularities along $\gamma^+ \cup \gamma^-$ in the case $F(0) \neq 0$, for the generic situation.

Exercise 18 After reading Appendix B, compute the time optimal synthesis for Example A of the Introduction with target the unit closed ball.

Exercise 19 After reading Appendix B, compute the time optimal synthesis for Example 1 of Section 2.6.4, p. 62 with target the ball of radius $\sqrt{10}/3$. Do the same for Example 3 of Section 2.6.4, p. 62 with target the ball of radius 1/3.

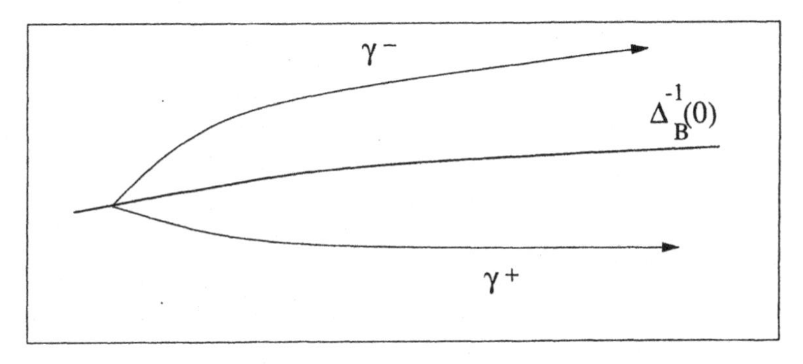

Fig. 2.39. Exercise 16

3

Generic Properties of the Minimum Time Function

Let us first introduce the concept of value function:

Definition 46 *Given a family of control problems $\{\mathcal{P}_{x_1}\}_{x_1 \in M}$ of type (1.1)-(1.7), one defines the value function $V(x_1)$ to be the value of the minimum of the problem $\mathcal{P}_{x_1}$.*

It is well known that, for a general optimal control problem of the form (1.7), p. 19, under suitable assumptions, the function V, that in general is not differentiable, satisfies the Hamilton-Jacobi-Bellman equation:

$$-\min_{u \in U}\{\nabla V(x) \cdot f(x,u) + L(x,u)\} = 0.$$

in viscosity sense [23].
Recall now our model problem:

$$\dot{x} = F(x) + uG(x), \quad x \in M, \quad |u| \leq 1 \tag{3.1}$$

where M is a smooth two dimensional manifold, we assume $F(x_0) = 0$ and consider the problem of reaching every point of M in minimum time from x_0. For this problem the value function is the minimum time function.

In this Chapter, we treat the problem of topological regularity of minimum time function. We recall that a smooth function is a Morse function if it has isolated critical points with nondegenerate second derivatives at these points. We say that a continuous function, not necessarily smooth, is topologically a Morse function if its level sets are homeomorphic to the level sets of a Morse function. This is sufficient, for example, to derive Morse inequalities, see [101]. In Figure 3.1 we represent the level sets of a Morse function on the plane.

It is known that a function that is the minimum of a finite number of smooth functions in generic position is a Morse function in topological sense [8, 98]. We give a positive answer to the question of V.I. Arnold: is the minimum time function generically a Morse function in topological sense? This result is

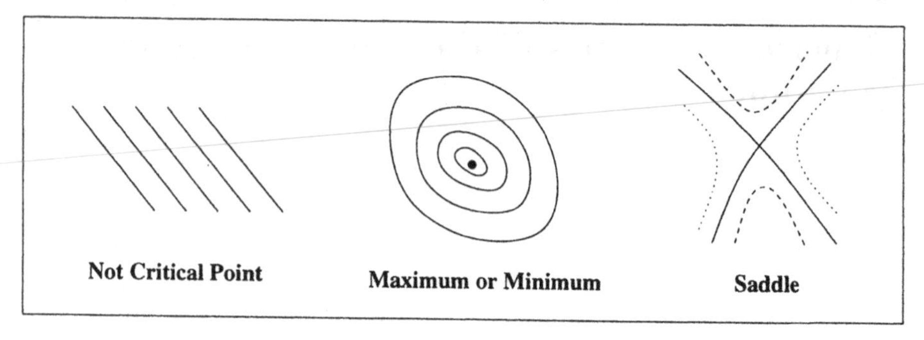

Fig. 3.1. Level curves of a Morse function.

in the same spirit of [8, 98], being now the minimization taken over a finite dimensional family of functions. In order to prove our result, we analyze the Minimum Time Fronts (briefly MTF) that are the level sets of the minimum time function. The properties of the MTFs are given in Theorem 27.

This analysis permits to understand the evolution of the topology of the reachable set. More precisely, the topology of the reachable set $\mathcal{R}(\tau)$ generically changes only at a discrete set of times when an overlap curve either is formed or ends. More precisely, at these times the reachable set may change topology because either a handle is added or a hole is closed.

In the following we give the definition of topological Morse function and state our main result (Theorem 26). We recall that the reachable set within time $T > 0$ is given by:

$$\mathcal{R}(T) := \{x \in M : \exists t \in [0,T], \exists \text{ a trajectory } \gamma : [0,t] \to M \text{ of (3.1)} \\ \text{such that } \gamma(0) = x_0,\ \gamma(t) = x\}, \tag{3.2}$$

and we call simply reachable set the set:

$$\mathcal{R}(\infty) := \{x \in M : \exists t \geq 0, \exists \text{ a trajectory } \gamma : [0,t] \to M \text{ of (3.1)} \\ \text{such that } \gamma(0) = x_0,\ \gamma(t) = x\}. \tag{3.3}$$

Results about the structure of $\mathcal{R}(\infty)$ for (3.1) can be found in [54], while here we are interested in the evolution of $\mathcal{R}(T)$. Recall that the minimum time function $\mathbf{T}(\cdot) : \mathcal{R}(\infty) \to \mathbb{R}^+$ (now we drop the label x_0) is by definition:

$$\mathbf{T}(x) := \inf\{t \geq 0 : \text{ there exists a trajectory } \gamma \text{ of (3.1) s.t.} \\ \gamma(0) = x_0,\ \gamma(t) = x\}. \tag{3.4}$$

In general $\mathbf{T}(\cdot)$ is not C^1, but in our case, under generic hypotheses, it is C^∞ outside a rectifiable set of codimension 1.

Let us introduce the following:

Definition 47 *We say that a set $Q \subset M$ is* <u>*thin*</u> *if $Q = \cup_{i=1}^{n} M_i$, with $n \in \mathbb{N}$, M_i connected embedded smooth one dimensional manifold with boundary. We say that a function $f : A \subset M \to \mathbb{R}$, A open, is* <u>*piecewise C^∞*</u> *if it is C^∞ except a thin set Q.*

A smooth function is said Morse function if at critical points its Hessian is non singular. To treat the case of a piecewise C^∞ function, we introduce the following:

Definition 48 *Let $f : A \subset M \to \mathbb{R}$ be a piecewise C^∞ function and $Q \subset A$ the corresponding thin set. We say that f is* <u>*topologically a Morse function*</u> *if, taking Q minimal, the following holds. On the set $A \setminus Q$, f is a Morse function and for each $x \in Q$ there exist a neighborhood $I(x)$ of x and an homeomorphism $\psi : I(x) \to I(x)$ such that $f \cdot \psi : I(x) \to \mathbb{R}$ is a Morse function.*

Thus a topologically Morse function is a function whose level sets are homeomorphic to one of the cases of Figure 3.1.

Theorem 26 **(Morse Property)** *Fix $\tau > 0$. Under generic conditions on F and G, on the interior of the set $\mathcal{R}(\tau)$, the minimum time function $\mathbf{T}(\cdot)$ is a piecewise C^∞ function and topologically a Morse function.*

In Section 3.1 we refine the definition of Frame Curves, recall the shape of the singularities of the optimal synthesis (classified in Chapter 2) and define a suitable set of bad points. In Section 3.2 the definition of <u>optimal strip</u> is introduced and the local properties of the <u>Minimum Time Front</u>, briefly MTF, are studied. In Section 3.3 we prove that the minimum time function is topologically a Morse function. The case of a two-dimensional initial manifold is discussed in Appendix B. The results developed in this Chapter were obtained in [40].

3.1 Basic Definitions and Statements of Results

In Chapter 2 it was proved that, under generic assumptions, the problem of reaching in minimum time every point of the reachable set for the system (3.1) admits a regular synthesis and it is provided a complete classification of all types of Frame Points (briefly FPs) and Frame Curves (FCs).

We call $\gamma^\pm : [0, t_f^\pm] \to M$ the extremal trajectories exiting x_0 with constant control ± 1, where $t_f^\pm$ are the last times in which $\gamma^\pm$ are extremal (if they are less than τ) or τ (otherwise).

Moreover let $t_{op}^\pm$ the last times in which $\gamma^\pm$ are optimal, we have $t_{op}^\pm \le t_f^\pm$. We define $\gamma_{op}^\pm = \gamma^\pm|_{[0, t_{op}^\pm]}$.

Remark **48** Notice that in Chapter 2 $\gamma^{\pm}$ were the trajectories exiting x_0 with constant control ± 1 defined on the whole $[0, \tau]$.

We recall that all the possible FCs are the following:

- FCs of kind Y (resp. X), that correspond to subsets of γ^{+}_{op} (resp. subsets of γ^{-}_{op}),
- FCs of kind C, called switching curves, that are curves made of switching points,
- FCs of kind S, i.e. turnpikes,
- FCs of kind K, called overlaps, reached optimally by two trajectories coming from different directions.
- FCs that are arcs of optimal trajectories that start at FPs. These trajectories "transport" special information. Here we need to distinguish three kinds of these curves:
  - curves of kind γ_A that are abnormal extremals (recall that they are bang-bang, see Proposition 2, p. 49),
  - curves of kind γ_k that are curves passing through an endpoint of an overlap (i.e. that start at the FPs of kind $(Y, K)_3$ $(S, K)_2, (C, K)_1$, see below)
  - curves of kind γ_0 that are the other arcs of optimal trajectories that start at FPs.

There are eighteen topological equivalence classes of FPs: (X, Y), $(Y, C)_{1,2,3}$, (Y, S), $(Y, K)_{1,2,3}$, $(C, C)_{1,2}$, $(C, S)_{1,2}$, $(C, K)_{1,2}$, $(S, K)_{1,2,3}$, (K, K). The optimal synthesis near each Frame Point is showed in Figure 2.9, p. 61.

Definition 49 *We define the following sets of bad points:*

- $bad_k := \{x \in \mathcal{R}(\tau) :$ *there exist two optimal trajectories* $\gamma_1, \gamma_2 : [0, T] \to M$, $T > 0$ *and* $\varepsilon > 0$ *such that* $\gamma_1(0) = \gamma_2(0) = x_0$, $\gamma_1(T) = \gamma_2(T) = x$, $\gamma_1(t) \neq \gamma_2(t)$ *for every* $t \in [T - \varepsilon, T[\}$.
- $bad_\gamma := \{x \in \mathcal{R}(\tau) :$ *there exist two optimal trajectories* $\gamma_1, \gamma_2 : [0, T] \to M$, $T > 0$, $a \in]0, T[$ *and* $0 < \varepsilon < a$ *such that* $\gamma_1(0) = \gamma_2(0) = x_0$, $\gamma_1(T) = \gamma_2(T) = x$, $\gamma_1(t) = \gamma_2(t)$ *for* $t \in [a, T]$ *and* $\gamma_1(t) \neq \gamma_2(t)$ *for* $t \in [a - \varepsilon, a[\}$.
- $bad_{CV} := bad_k \cup bad_\gamma$.
- $bad_{CX} := Supp(\gamma^{+}_{op}) \cup Supp(\gamma^{-}_{op}) \setminus (\gamma^{+}(t^{+}_{f}) \cup \gamma^{-}(t^{-}_{f}))$.

Remark **49** Of course bad_k is the set of all the overlap FCs and bad_γ is the set of γ_k's. Moreover the set bad_{CX} is the optimal part of $Supp(\gamma^{\pm})$ if $t^{\pm}_{op} < t^{\pm}_{f}$, otherwise it is $Supp(\gamma^{\pm})$ without the terminal point. See Figure 3.2.

Definition 50 *The minimum time front in time* $0 < T < \tau$ *(in the following MTF) is the* T*-level surface of* $\mathbf{T}(x)$*:*

$$F_T = \{x \in \mathcal{R}(\tau) : \quad \mathbf{T}(x) = T\}. \tag{3.5}$$

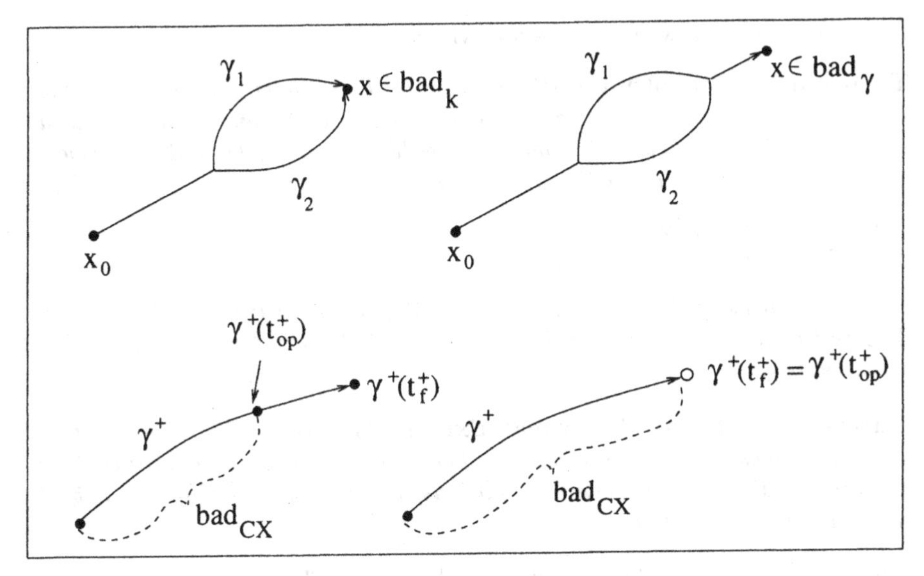

Fig. 3.2. Definition 49

We want to prove that, under generic conditions on (F, G) and for generic T, the corresponding MTF is a one dimensional piecewise C^1 compact manifold. More precisely the result is valid for every T such that the following conditions hold.

(C1) No FP of kind $(Y, K)_2$, $(Y, K)_3$ or (K, K) is on the front F_T.
(C2) The front F_T is not tangent to bad_k.

Definition 51 *Let $x \in F_T$. We define the contingent cone to the set $\mathcal{R}(\mathbf{T}(x)) = \mathcal{R}(T)$ at x as:*

$$C^C(\mathcal{R}(\mathbf{T}(x)), x) = \{v \in T_xM : \exists\ \xi :]-\varepsilon, \varepsilon[\to \mathcal{R}(\mathbf{T}(x)) \text{ smooth } s.t.\ \xi(0) = x, \dot{\xi}(0) = v\}.$$

and the normal cone as:

$$C^N(\mathcal{R}(\mathbf{T}(x)), x) = \{w \in T_x^*M : \ < w, v > \leq 0, \ \textit{for all } v \in C^C(\mathcal{R}(\mathbf{T}(x)), x)\}$$

Let $x \in \partial\mathcal{R}(\mathbf{T}(x))$. Notice that if $\partial\mathcal{R}(\mathbf{T}(x))$ is smooth at x then we have $C^N(\mathcal{R}(\mathbf{T}(x)), x) = \{\alpha \nabla \mathbf{T}(x) : \ \alpha \geq 0\}$. When $M = \mathbb{R}^2$ and $\partial\mathcal{R}(\mathbf{T}(x))$ has a corner at x, we have the following: $\mathcal{R}(\mathbf{T}(x))$ is convex at x iff $C^N(\mathcal{R}(\mathbf{T}(x)), x)$ is a cone with nonempty interior while $\mathcal{R}(\mathbf{T}(x))$ is concave iff $C^N(\mathcal{R}(\mathbf{T}(x)), x) = \{\emptyset\}$.

In the next section, we prove the following:

Theorem 27 (Smoothness Properties of F_T) *Under generic conditions on (F,G), for $0 < T < \tau$ satisfying conditions* **(C1)** *and* **(C2)**, *F_T is a one–dimensional piecewise C^1 compact embedded submanifold of M. Moreover, $x \in F_T$ is a point at which F_T is not C^1 iff:*

(case 1) $x \in bad_{CV}$ *or*
(case 2) $x \in bad_{CX}$.

For every such x we have $C^N(\mathcal{R}(T),x) = \{0\}$ (case 1) or $Int(C^N(\mathcal{R}(T),x)) \neq \emptyset$ (case 2), i.e. if $M = \mathbb{R}^2$ then $\mathcal{R}(T)$ is concave at x if $x \in bad_{CV}$ and convex if $x \in bad_{CX}$.

Theorem 28 (F_T on Abnormal Extremals) *Fix $0 < T < \tau$, let $\bar{x} \in F_T$ and suppose $\bar{x} \notin bad_{CV} \cup bad_{CX}$. Let $\bar{\gamma}$ be the unique optimal trajectory reaching $\bar{x}$. Then, under generic conditions, F_T is tangent to $Supp(\bar{\gamma})$ at $\bar{x}$ iff $\bar{\gamma}$ is an abnormal extremal.*

Similar properties to the ones described by this Theorem were studied in [32, 129].

3.2 Properties of the Minimum Time Front F_T

To prove Theorems 27, p. 132 and 28, p. 132, we first study the smoothness properties of the MTF in a neighborhood of $Supp(\gamma^+_{op}) \cup Supp(\gamma^-_{op})$. From now on, for simplicity, we consider the case $M = \mathbb{R}^2$ and $x_0 = 0$. The conclusions are valid in the general case mutatis mutandis.

3.2.1 Definition of Strips

Now we give the key definition of optimal strip. An optimal strip is essentially a one parameter continuous family of optimal trajectories formed by the same sequence of arcs.

Definition 52 *Let a, b be two real numbers such that $0 \leq a < b \leq \tau$, $x \in \mathcal{R}(\tau)$ and $f : [a,b] \to \mathbb{R}$ a function such that $f(\alpha) \geq \alpha$ for every $\alpha \in [a,b]$. A set of trajectories $\mathcal{S}^{a,b,x,f} = \{\gamma_\alpha : [0, f(\alpha)] \to M,\ \alpha \in [a,b],\ \gamma_a(a) = x\}$ is called an optimal strip if:*

i) $\forall\ \alpha \in [a,b]$, $\gamma_\alpha : [0, f(\alpha)] \to M$, *is an optimal trajectory for the control problem (3.1). Moreover there exists $\varepsilon > 0$ such that $\gamma|_{[\alpha,\alpha+\varepsilon]}$ corresponds to a constant control ± 1.*
ii) $\forall\ \alpha \in]a,b[$, γ_α *does not switch on $\Delta_A^{-1}(0) \cup \Delta_B^{-1}(0)$ after time α.*
iii) The set $\mathcal{B}^{a,b,x,f} = \{y \in \mathcal{R}(\infty) :\ \exists\ \alpha \in\ ]a,b[$ and $t \in]\alpha, f(\alpha)[$ such that $y = \gamma_\alpha(t)$, t is a switching time for $\gamma_\alpha\}$ is never tangent to X or Y.

iv) The map $\eta : \ \alpha \in \ [a,b] \mapsto \gamma_\alpha(\alpha) \in M$ *is a bang or singular arc and for* $a \leq \alpha' \leq \alpha \leq b$, *it holds* $\gamma_\alpha(t) = \gamma_{\alpha'}(t)$ *for* $t \in [0, \alpha']$.

The function η is called the base of the optimal strip, moreover $Int(S^{a,b,x,f}) := \{\gamma_\alpha : \ \alpha \in]a,b[\}$ is called an open optimal strip and $\partial S^{a,b,x,f} := \{\gamma_a, \ \gamma_b\}$ is called optimal strip border. The concept of strip is clarified in Figure 3.3.

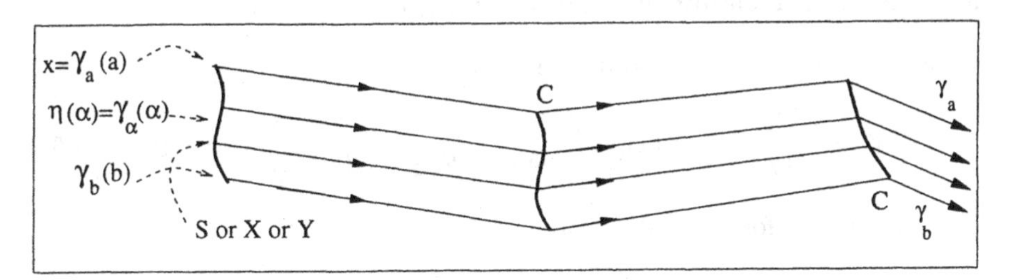

Fig. 3.3. optimal strip

The construction of the optimal synthesis, as in Chapter 2, can be done in the following way, with a new algorithm. One first constructs all optimal strips bifurcating from $\gamma_{op}^{\pm}$. Then one studies the evolution of each strip. It can happen that an optimal strip is divided into two strips when some trajectory of the strip enters a turnpike. This happens at Frame Points of kind $(C,S)_1$ and $(S,K)_1$ (moreover two strips with the turnpike as base start at these points). A strip can terminate on some K curve. Finally, a strip can glue together with another strip at Frame Points of kind $(Y,K)_3$, $(C,K)_1$ and $(S,K)_2$ (indeed the strip is divided into two parts, one of which ends on the K curve and the other glues together with another strip).
Let $\{S_M^i\}_{i \in I}$ be the set of all the maximal strips, i.e. the strips with maximal base and maximal time $f(\alpha)$. Clearly we may split the synthesis Γ as:

$$\Gamma = \bigcup_{i \in I} S_M^i.$$

This partition of the optimal synthesis permits to study the evolution of the MTF and the properties of the minimum time function separately for the base, the borders, the internal of the strips and for the overlap curves. Notice that if $S^{a,b,x,f}$ is a maximal strip then $f(\alpha) < \tau$ for some $\alpha \in]a,b[$ iff there exists an overlap curve K such that $\gamma_\alpha(f(\alpha)) \in K$. We can split the maximal strips in such a way that they satisfy the following:

v) either $\{\gamma_\alpha(f(\alpha)), \alpha \in [a,b]\}$ is a K curve or $f(\alpha) = \tau$ for every $\alpha \in [a,b]$.

Moreover notice that every FC of the optimal synthesis, that is not an overlap curve, is the support of a base or the support of a border of an optimal strip.

In Section 3.2.2 we study the evolution of F_T in a neighborhood of $x_0 = 0$, while in Section 3.2.3 in a neighborhood of $\gamma^+_{op} \cup \gamma^-_{op}$. Then in Section 3.2.4 we study the evolution of the MTF on the other bases of the strips (i.e the turnpikes), in Section 3.2.5 on the internal of the strips, in Section 3.2.6 on overlap curves and finally in Section 3.2.7 on the borders.

3.2.2 F_T in a Neighborhood of the Origin

Choose a local system of coordinates such that:

$$Y = \begin{pmatrix} 1 \\ 0 \end{pmatrix}, \qquad X(x) = \begin{pmatrix} -1 + O(x_1, x_2) \\ -b\,x_1 + b'x_2 + O(|x|^2) \end{pmatrix}, \qquad b < 0. \tag{3.6}$$

The expressions for γ^+_{op} and γ^-_{op} are:

$$\gamma^+_{op}(t) = \begin{pmatrix} t \\ 0 \end{pmatrix}, \quad \gamma^-_{op}(t) = \begin{pmatrix} -t + O(t^2) \\ \frac{1}{2}bt^2 + O(t^3) \end{pmatrix}. \tag{3.7}$$

Let us compute the MTF along γ^+_{op}, in the region where $x_2 < 0$. Consider a trajectory γ^α corresponding to constant control -1 on the interval $[0,\alpha[$ and to constant control +1 on the interval $[\alpha, T]$ where α belongs to the interval $[0, T]$. Using the expressions (3.7) it follows that $\gamma^\alpha(T)$ has coordinates $x_1(\alpha) = T - 2\alpha + O(\alpha^2)$, $x_2(\alpha) = \frac{1}{2}b\alpha^2 + O(\alpha^3)$, from which it follows the expression for the MTF:

$$x_2(x_1) = \frac{1}{8}b(x_1 - T)^2 + O(x_1^3). \tag{3.8}$$

Let us compute the MTF along γ^+_{op}, in the region where $x_2 > 0$. A point with $x_2 > 0$ is reached at time T by a trajectory corresponding to constant control +1 in the interval $[0, T-\alpha[$ and to constant control -1 in the interval $[T-\alpha, T]$ where α belong to the interval $[0, T]$. Consider now an optimal trajectory γ^-_η of (3.1), corresponding to constant control -1, and having initial condition $\gamma^-_\eta(0) = (\eta, 0)$. We have:

$$\gamma^-_\eta(t) = \begin{pmatrix} \eta - t + O(t^2, t\eta) \\ \frac{1}{2}bt^2 - b\eta t + O(t^3, \eta^2 t, \eta t^2) \end{pmatrix}. \tag{3.9}$$

Similarly to the previous case we obtain the expression for the MTF:

$$x_2(x_1) = \frac{1}{2}bT(x_1 - T) + \frac{3}{8}b(x_1 - T)^2 + O((x_1 - T)^2).$$

In the same way we may compute the MTF along γ^-_{op}, moreover the MTF is clearly smooth on $I(0) \setminus (Supp(\gamma^+_{op}) \cup Supp(\gamma^-_{op}))$, where $I(0)$ is a neighborhood of the origin.

This result can be summarized in the following (see Figure 3.4):

Proposition 7 *For $T > 0$, sufficiently small, F_T is a 1–dimensional piecewise C^1 embedded compact submanifold of M. The only two points in which F_T is not C^1 are the points $F_T \cap Supp(\gamma_{op}^+)$ and $F_T \cap Supp(\gamma_{op}^-)$. Moreover if x is a such point we have $Int(C^N(\mathcal{R}(T), x)) \neq \emptyset$, that is $\mathcal{R}(T)$ is convex at x (see Figure 3.4).*

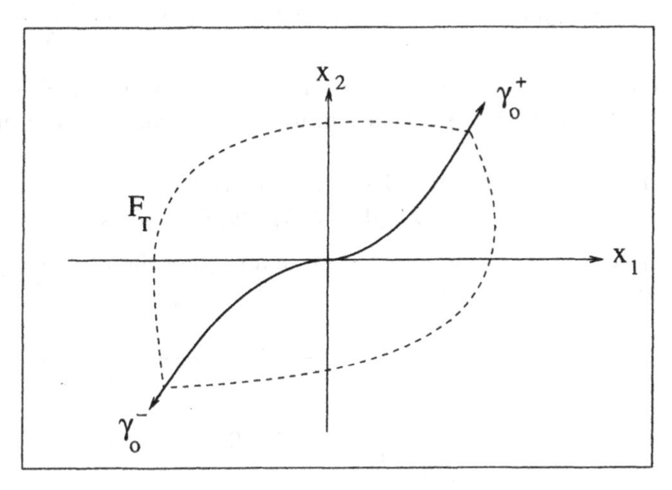

Fig. 3.4. Proposition 7

3.2.3 F_T in a Neighborhood of $\gamma_{op}^+ \cup \gamma_{op}^-$

Let $I(\gamma_{op}^\pm)$ be a neighborhood of $Supp(\gamma_{op}^+) \cup Supp(\gamma_{op}^-)$. In this subsection we study the smoothness properties of $F_T \cap I(\gamma_{op}^\pm)$. For this purpose we analyze F_T in a neighborhood of all the $(Y, \,.\,)$ FPs ordered by increasing time.

Analysis up to the first (Y, S) or $(Y, C)_1$ FP.

Assume $t_1'^+ > 0$ (being the case $t_1^+ > 0$ entirely similar) and suppose that at the point $\gamma_{op}^+(t_1'^+)$ a turnpike starts. In this case it bifurcates on the left of γ_{op}^+. Choose a system of coordinates such that:

$$\begin{cases} Y = \begin{pmatrix} 1 \\ 0 \end{pmatrix}, \\ X(x) = \begin{pmatrix} a_0 + a_1(x_1 - t_1'^+) + a_2 x_2 + O((x_1 - t_1'^+)^2, x_2^2) \\ b_0 + b_1(x_1 - t_1'^+) + b_2 x_2 + O((x_1 - t_1'^+)^2, x_2^2) \end{pmatrix}, \\ a_0 < 1, \quad b_0 > 0. \end{cases} \tag{3.10}$$

To have a (Y, S) point at $(t'_0, 0)$ the following conditions must hold:

1) $(t_1'^+, 0) \in \Delta_B^{-1}(0)$ that implies:

$$a_2 = \frac{a_1 b_0}{a_0 - 1}.$$

Under this condition, the set $\Delta_B^{-1}(0)$ is locally described by the equation:

$$x_2(x_1) = m_s(x_1 - t_1'^+) + O((x_1 - t_1'^+)^2) \text{ where } m_s = \frac{a_1 a_2 - a_1 b_1}{a_2^2 - a_1 b_2},$$

and we assume the generic conditions $a_2^2 - a_1 b_2 \neq 0$, $m_s \neq 0$. The case in which $a_2^2 - a_1 b_2 = 0$ is the case in which the tangent to the turnpike is vertical.

2) X and Y point to opposite sides of $\Delta_B^{-1}(0)$ that implies $arctan(b_0/a_0) > arctan(m_s)$ where we assume $[0, \pi[$ as range of the function $arctan$.

Let us compute the evolution of the MTF in a neighborhood I of $(t_1'^+, 0)$. We refer to Figure 3.5.

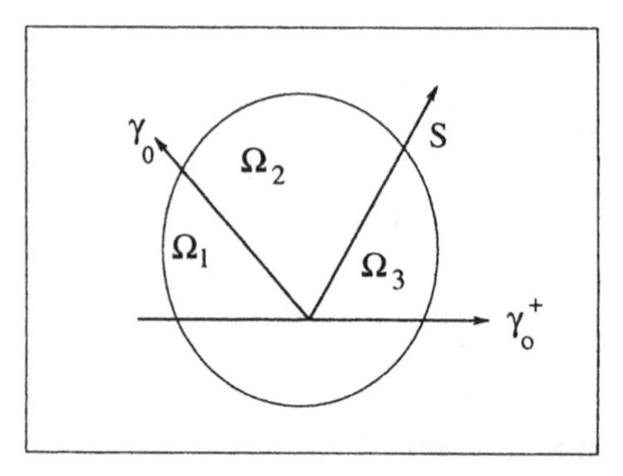

Fig. 3.5. Evolution of the MTF in a neighborhood I of $(t_1'^+, 0)$

We have to compute the front F_T in the region $\Omega_1 \subset I$ enclosed by γ_{op}^+ (before time $t_1'^+$) and the trajectory γ_0 corresponding to constant control -1 starting from $(t_1'^+, 0)$, on the region Ω_2 enclosed by γ_0 and the turnpike, finally on the region Ω_3 between the turnpike and γ_{op}^+ (after time $t_1'^+$).
In Ω_1 using the expression of $X(x)$ given by formula (3.10), we obtain for F_T the expression:

$$x_2(x_1) = \frac{b_0}{a_0 - 1}(x_1 - T) + O((x_1 - t_1'^+)^2). \tag{3.11}$$

To compute the locus F_T in Ω_2 and Ω_3 we need the explicit expression for the singular trajectory. The velocity of the singular trajectory at a point

$x \in S$ is the convex combination of $X(x)$ and $Y(x)$ corresponding to the direction of the turnpike. This guarantees the smoothness of the MTF on S (see Proposition 9, p. 141) and gives us the expression for the singular trajectory γ^s ($S = Supp(\gamma^s)$, $\gamma^s(0) = \gamma_{op}^+(t_1'^+)$):

$$\gamma^s(t) = \begin{pmatrix} t_1'^+ + v_1 t + O(t^2) \\ v_2 t + O(t^2) \end{pmatrix}$$

where:

$$v_1 = \frac{b_0 \left(a_1 b_0{}^2 - (-1 + a_0)^2 b_2\right)}{a_1 b_0 \left(-1 + 2 a_0 - a_0{}^2 + b_0{}^2\right) + (-1 + a_0)^2 \left((-1 + a_0) b_1 - b_0 b_2\right)} \tag{3.12}$$

$$v_2 = \frac{(-1 + a_0) b_0 (a_1 b_0 + b_1 - a_0 b_1)}{a_1 b_0 \left(-1 + 2 a_0 - a_0{}^2 + b_0{}^2\right) + (-1 + a_0)^2 \left((-1 + a_0) b_1 - b_0 b_2\right)} . \tag{3.13}$$

With similar arguments to the ones of the previous Section we obtain the expression for F_T in Ω_2:

$$x_2(x_1) = v_2(T - t_1'^+) + \frac{(x_1 - t_1'^+) - v_1(T - t_1'^+)}{(a_0 - v_1)}(b_0 - v_2) + O((x_1 - t_1'^+)^2).$$

Using (3.12) and (3.13), we obtain exactly the expression (3.11). This means that F_T is C^1 at the intersection point with $Supp(\gamma_0)$. Exactly the same result holds in the case when condition **2)** fails that is when at $(t_1'^+, 0)$ a switching curve bifurcates from γ_{op}^+. In this case, we have to use directly the PMP, but we omit this proof because it is entirely similar to the proof of Proposition 10 below. The smoothness properties of F_T in a neighborhood of $Supp(\gamma_{op}^+|_{]0,t_1'^+]})$ comes immediately from the shape of the synthesis in this region. All these results are summarized in the following:

Proposition 8 *Suppose $t_1'^+ > 0$, fix $y_0 \in Supp(\gamma_{op}^+|_{]0,t_1'^+]})$ and let t_0 be such that $y_0 = \gamma_{op}^+(t_0)$. There exists a ball B with center in y_0 such that if $\varepsilon > 0$ satisfies $\gamma_{op}^+(T) \in B$ for every $T \in]t_0 - \varepsilon, t_0 + \varepsilon[$, then $F_T \cap B$ is a one dimensional piecewise C^1 embedded submanifold of M. Moreover F_T is not C^1 only at the point $x = F_T \cap Supp(\gamma_{op}^+)$ and we have $Int(C^N(\mathcal{R}(T), x)) \neq \emptyset$, that is $\mathcal{R}(T)$ is convex at x.*

Analysis up to the First $(Y, K)_1$ or $(Y, C)_2$ FP.

Assume $s_1^+ > 0$ (being the case $s_1'^+ > 0$ entirely similar). In this case an abnormal extremal γ_A bifurcates on the right of $\gamma_{op}^+(s_1^+)$ and we can have two kind of FPs:

- $(Y, K)_1$ if an overlap curve starts at $\gamma_{op}^+(s_1^+)$,
- $(Y, C)_2$ if a switching curve starts at $\gamma_{op}^+(s_1^+)$.

The analysis of the first case is postponed to Section 3.2.6, hence let us consider the second case. Set $x = (x_1, x_2)$, and choose a local system of coordinates as done in formulas (2.66) and (2.67) p. 94:

$$Y \equiv \begin{pmatrix} 1 \\ 0 \end{pmatrix} \quad X(x) = \begin{pmatrix} -1 + a_1 x_1 + a_2 x_2 + O(|x|^2) \\ b_1 x_1 + b_2 x_2 + O(|x|^2) \end{pmatrix}, \quad b_1 > 0, \tag{3.14}$$

$$X(x) = \begin{pmatrix} c_0 + c_1(x_1 - s_1^+) + c_2 x_2 + O(|x - (s_1^+, 0)|^2) \\ d_1(x_1 - s_1^+) + d_2 x_2 + O(|x - (s_1^+, 0)|^2) \end{pmatrix}, \quad \begin{cases} c_0 \in]0,1[, \\ d_1 < 0. \end{cases} \tag{3.15}$$

For the signs of b_1 and d_1 and the condition $c_0 < 1$ see the discussion after formula (2.67) p. 94, while the condition $c_0 > 0$ follows from the fact that a C curve originates. The expression for X in formula (3.14) is useful in a neighborhood of $(0,0)$, while the expression (3.15) gives information in a neighborhood I of $(s_1^+, 0)$. In Section 2.8, p. 86, we proved that at $(s_1^+, 0)$ a C curve starts if:

$$\delta := c_0 d_1 - b_1 \left(\frac{1 - c_0}{2} \right)^2 < 0. \tag{3.16}$$

Indeed, following Chapter 2, one can compute the expression of the curve D reached at the same time by trajectories leaving the origin with different controls. The above condition corresponds precisely to having that extremal trajectories reach the C curve before intersecting D. If the opposite happens D is an overlap curve of the optimal synthesis.

In Section 2.8, p. 86, we computed the expression of the C curve starting at $(s_1^+, 0)$ and of the abnormal extremal bifurcating from the same point:

$$x_2(x_1) = -\frac{2d_1^2}{b_1(1 - c_0)^2}(x_1 - s_1^+)^2 + O((x_1 - s_1^+)^3), \tag{3.17}$$

$$\gamma_A(t) = \begin{pmatrix} x_1(t) \\ x_2(t) \end{pmatrix} = \begin{pmatrix} s_1^+ + c_0 t + O(t^2) \\ d_1(\frac{1}{2} c_0 t^2) + O(t^3) \end{pmatrix}. \tag{3.18}$$

$Supp(\gamma_A)$ is described as:

$$x_2^{\gamma_A}(x_1) = \frac{1}{2} \frac{d_1}{c_0}(x_1 - s_1^+)^2 + O((x_1 - s_1^+)^3). \tag{3.19}$$

Divide I in the four regions Ω_A, Ω_B, Ω_C, Ω_D delimited by γ_{op}^+, C and γ_A as in Figure 3.6. In the following we compute F_T ($T = \tau_1 + \tau_2$) in these regions and we check that on C and γ_A the tangents to F_T from both sides coincide.

In Ω_A the expression for F_T is obtained similarly to (3.8):

$$x_2^A(x_1) = \frac{1}{8} b_1 (x_1 - T)^2 + O(x_1^3). \tag{3.20}$$

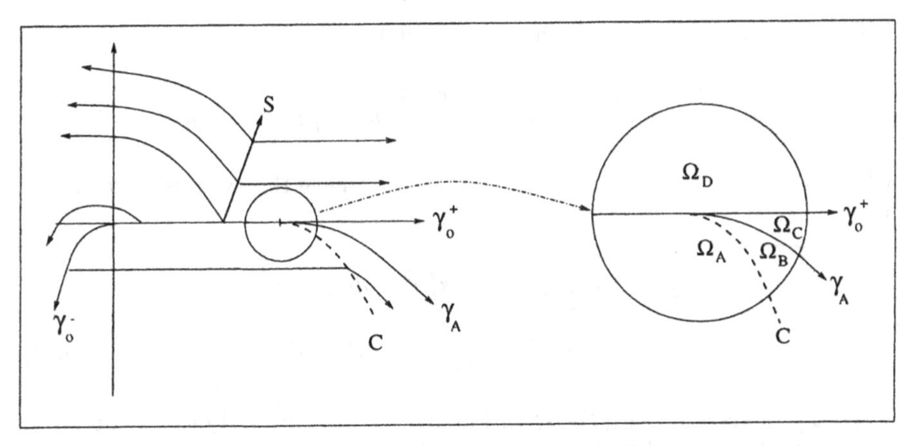

Fig. 3.6. Analysis of F_T at the $(Y, C)_2$ FP

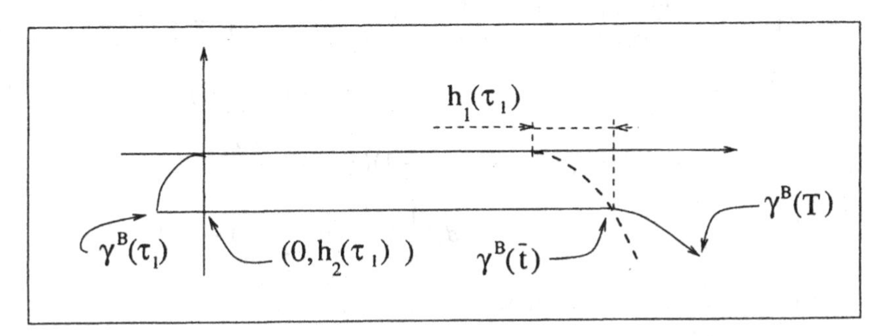

Fig. 3.7. Analysis of F_T at the $(Y, C)_2$ FP

A point $x^B \in \Omega_B$ is reached by an optimal trajectory γ^B corresponding to constant control -1 in the interval $[0, \tau_1]$, to constant control +1 in the interval $[\tau_1, \bar{t}]$ (where $\bar{t}$ is the time in which this trajectory intersects the C curve) and to constant control -1 in the interval $[\bar{t}, T]$ (see Figure 3.7).
Let us parameterize the curve C with τ_1. Using (3.17) we obtain the expression:

$$\begin{cases} x_1(\tau_1) = s_1^+ + h_1(\tau_1) + O(\tau_1^2) \\ x_2(\tau_1) = h_2(\tau_1) + O(\tau_1^3), \end{cases}$$

$$\text{where } h_1(\tau_1) := \frac{b_1(1 - c_0)}{2d_1}\tau_1, \qquad h_2(\tau_1) := \frac{1}{2}b_1\tau_1^2.$$

The expression for an optimal trajectory γ^h of (3.1) corresponding to constant control -1, and having initial condition $\gamma^h(0) = (s_1^+ + h_1, h_2) \in I$ is:

$$\gamma^h(t) = \begin{pmatrix} x_1(t) = s_1^+ + h_1 + c_0 t + O(t^2, h_1 t, h_2 t) \\ x_2(t) = h_2 + d_1(\frac{1}{2}c_0 t^2 + h_1 t) + O(t^3, h_1 t^2, h_1^2 t, h_2 t^2, h_2^2 t) \end{pmatrix}. \quad (3.21)$$

It follows that $x^B \in F_T \cap \Omega_B$ has coordinates:

$$\begin{cases} x_1^B(\tau_1) = s_1^+ + h_1(\tau_1) + c_0\beta(\tau_1) + O(\tau_1^2) \\ x_2^B(\tau_1) = h_2(\tau_1) + d_1(\frac{1}{2}c_0\beta(\tau_1)^2 + h_1(\tau_1)\beta(\tau_1)) + O(\tau_1^3), \end{cases}$$

where $\beta(\tau_1) := T - \bar{t} = T - h_1(\tau_1) - s_1^+ - 2\tau_1$. Hence the expression for F_T in Ω_B:

$$\begin{aligned} x_2^B(x_1) = &\frac{1}{2}c_0 d_1(T - s_1^+)^2 + d_1(T - s_1^+)x_1' \\ &+ d_1\frac{(2-c_0)b_1 + 4d_1}{8\delta}x_1'^2 + O(x_1'^3), \end{aligned} \tag{3.22}$$

where $x_1' := (x_1 - s_1^+ - c_0(T - s_1^+))$. Finally, the front $F_T \cap \Omega_C$ can be obtained by lengthy but straightforward computations:

$$\begin{aligned} x_2^C(x_1) &= d_1\frac{x_1'}{1-c_0}(T - s_1^+ - \frac{x_1'}{1-c_0}) + \frac{1}{2}c_0 d_1(T - s_1^+ - \frac{x_1'}{1-c_0})^2 \\ &\qquad + O((T - s_1^+ - \frac{x_1'}{1-c_0})^3) \qquad (3.23) \\ &= \frac{1}{2}c_0 d_1(T - s_1^+)^2 - d_1(\frac{1}{2}c_0 - 1)(T - s_1^+)\frac{x_1'}{1-c_0} \\ &\qquad + d_1(\frac{1}{2}c_0 - 1)(\frac{x_1'}{1-c_0}) + O((\frac{x_1'}{1-c_0})^3). \end{aligned}$$

Now it is easy to check the following:

- $\dot{x}_2^A(\tilde{x}_1) = \dot{x}_2^B(\tilde{x}_1)$ where $\tilde{x}_1$ is the intersection of the two fronts with the C curve. This means that F_T is C^1 in $\tilde{x}_1$.
- $\dot{x}_2^B(\bar{x}_1) = \dot{x}_2^C(\bar{x}_1) = \dot{x}_2^{\gamma_A}(\bar{x}_1)$ where $\bar{x}_1$ is the intersection of the two fronts with $Supp(\gamma_0)$. This means that F_T is C^1 in $\bar{x}_1$ and it is tangent to γ_A in $\bar{x}_1$.

Analysis of the other Frame Points.
The front F_T at the other singular points along $\gamma_{op}^{\pm}$ can be analyzed by arguments completely similar to those used above. The only delicate points are $\gamma_{op}^{\pm}(t_{op}^{+})$. At these points we can have a $(Y,K)_{2,3}$ or a $(Y,C)_3$ FP. The first two cases are analyzed in details in the proof of Theorem 26. In the case $(Y,K)_3$ that is the only case in which a γ_k trajectory generates from $\gamma_{op}^{\pm}$, it is easy to check that the MTF forms a concave corner on γ_k. The last case, can be studied as the interior of strips (see below). We obtain that F_T is always a piecewise C^1 manifold with corners only on $\gamma_{op}^{\pm}$. Notice that an abnormal extremal may bifurcates from a $(Y,C)_3$ points and in this case we can prove that the front is tangent to the abnormal extremal by arguments similar to those used above.

3.2.4 The MTF on Other Bases

The front F_T on the singular trajectories is described by the following:

Proposition 9 *Suppose that there exists a point $y^0 \in F_T \cap S$ where S is a turnpike of the optimal synthesis. Then F_T is C^1 at y^0.*

Proof. It is a consequence of the fact that the velocity of the singular trajectory at a point $x \in S$ is the convex combination of $X(x)$ and $Y(x)$ corresponding to the direction of the turnpike. Indeed, fix $y \in F_T$, $y = \gamma_s(T)$ and for every ε small let $x = \gamma_s(T-\varepsilon)$ where γ_s runs the turnpike. Notice that, generically, $\dot\gamma_s(T) = \lambda X(y) + (1-\lambda)Y(y)$, for some $0 < \lambda < 1$. Let γ_X^ε be the X–trajectory such that $\gamma_X^\varepsilon(0) = x$ and define γ_Y^ε in the same way with X replaced by Y. We have:

$$\gamma_X^\varepsilon(\varepsilon) = x + X(y)\varepsilon + O(\varepsilon^2) \in F_T$$
$$\gamma_Y^\varepsilon(\varepsilon) = x + Y(y)\varepsilon + O(\varepsilon^2) \in F_T.$$

Now the tangent from the left and from the right to F_T at y are computed using $y - x = \dot\gamma_s(T)\varepsilon + O(\varepsilon^2)$, by:

$$\begin{aligned}\frac{d}{d\varepsilon}(\gamma_X^\varepsilon(\varepsilon) - y) &= X(y) - \dot\gamma_s(T) = (1-\lambda)(X-Y)(y)\\ \frac{d}{d\varepsilon}(\gamma_Y^\varepsilon(\varepsilon) - y) &= Y(y) - \dot\gamma_s(T) = -\lambda(X-Y)(y),\end{aligned} \tag{3.24}$$

proving that F_T is C^1 at y. ■

3.2.5 The MTF on Open Strips

In this Section we prove that the MTF restricted to an open strip is always C^1.

Remark **50** Notice that the C^1 property of the MTF can also be obtained reasoning on the position of the covectors with respect to the MTF. More precisely, each covector of the PMP has negative scalar product with all variational vectors corresponding to needle variations. From the structure of the optimal synthesis, it is clear that, fixed an optimal trajectory, the neighboring trajectories of the optimal synthesis can be obtained by needle variations. This implies that the MTF of a strip can not form angles in such a way that the reachable set is locally concave, otherwise this would contradict optimality. On the other side, after the first switching the covector is determined up to the multiplication by a positive constant, hence also angles of the MTF that render the reachable set locally convex are prohibited.

Fix an optimal strip $\mathcal{S}^{a,b,x,f}$ and let $T \in]a, \sup\{f(\alpha), \ \alpha \in]a,b[\}[$. We define the strip-MTF (in the following s-MTF) to be the set:

$$F_T^{a,b,x,f} := \{y \in F_T : \ y = \gamma_\alpha(T) \text{ for some } \gamma_\alpha \in \mathcal{S}^{a,b,x,f}, \ \alpha \in]a,b[\}.$$

Proposition 10 *Let $\mathcal{S}^{a,b,x,f}$ be an optimal strip, $T \in]a, \sup\{f(\alpha), \ \alpha \in]a,b[\}[$ and F_T the corresponding s-MTF. Then $F_T^{a,b,x,f}$ is a one-dimensional C^1 embedded submanifold of M. Moreover if $T \in]a, \inf\{f(\alpha), \ \alpha \in]a,b[\}[$ then $F_T^{a,b,x,f}$ is also connected.*

The proof follows from the next two Lemmas. With the first we prove that the s-MTF is smooth when it intersects the "first" switching locus. With the second we prove that if the s-MTF is C^1 when it intersect a switching locus, then it is C^1 when it intersects the subsequent. By induction we reach the conclusion.

Lemma 17 *Let $Int(\mathcal{S}^{a,b,x,f})$ be an open optimal strip. Fix $\alpha_0 \in]a,b[$ and suppose that $T_0 \in]\alpha_0, f(\alpha_0)[$ is the first switching time of γ_{α_0} after α_0. Then $F_{T_0}^{a,b,x,f}$ is C^1 in $y^0 := \gamma_{\alpha_0}(T_0)$.*

Proof. Suppose that γ_{α_0} corresponds to constant control $+1$ in the interval $]\alpha_0, T_0[$. Choose a local system of coordinates in such a way that $Y \equiv (1,0)$, $\gamma_{\alpha_0}(\alpha_0) = (0,0)$ and the expression for the base of the strip is:

$$\eta(t) = \begin{pmatrix} 0 \\ t - \alpha_0 \end{pmatrix}, \quad t \in]a,b[.$$

In this system of coordinates we have $y^0 = (T_0 - \alpha_0, 0)$ and the expression of G on the base is ($t \in]a,b[$):

$$G(\eta(t)) = \begin{cases} \frac{1}{2}\begin{pmatrix} 1 \\ -1 \end{pmatrix} & \text{if the base is a } X\text{–trajectory, so } X(0,t) = (0,1) \\ \frac{1}{1-\varphi(\eta(t))}\begin{pmatrix} 1 \\ -1 \end{pmatrix} & \text{if the the base is a } Z\text{–trajectory} \\ & \text{so } (F + \varphi G)(0,t) = (0,1). \end{cases} \tag{3.25}$$

Let λ_α be the covector associated to γ_α ($\alpha \in]a,b[$). From the condition $\lambda_\alpha(\alpha) \cdot G(\gamma_\alpha(\alpha)) = 0$ we obtain, up to multiplication by a constant $\lambda_\alpha(\alpha) = \pm(1,1)$. The switching curve C passing from y^0 is determined by the equation:

$$\lambda_\alpha(t) \cdot G(\gamma_\alpha(t)) = 0. \tag{3.26}$$

Let t_α be the first switching time of γ_α after α, then $\dot{\lambda}_\alpha(t) = 0$ for $t \in [\alpha, t_\alpha]$. We get that condition (3.26) is equivalent to:

$$G_1(x) + G_2(x) = 0, \tag{3.27}$$

where $G_1(x)$ and $G_2(x)$ are the two components of G.

In terms of components of the vector field X, (3.27) can be written as:

$$X_1(x) + X_2(x) = 1. \tag{3.28}$$

We suppose that at y^0, $F_T^{a,b,x,f}$ and C are not tangent otherwise the conclusion follows easily. Assume that we are in the situation of Figure 3.8 (the other case being similar). In a neighborhood of the point y^0 we may parameterize the C curve by a function:

$$y(\alpha) = \begin{pmatrix} T_0 - \alpha_0 - a(\alpha) \\ \alpha \end{pmatrix}, \tag{3.29}$$

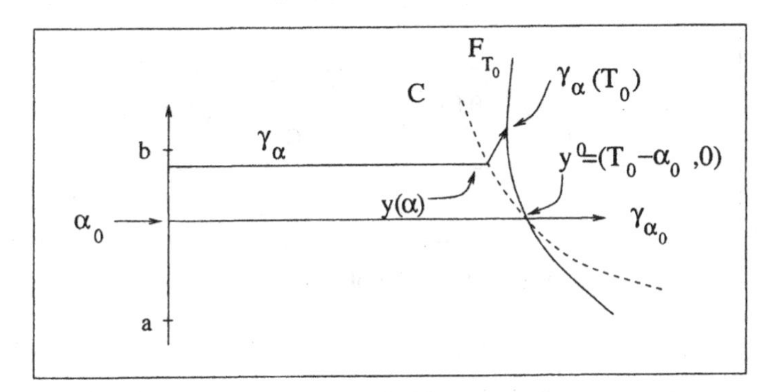

Fig. 3.8. Proof of Lemma 17

where $a(\cdot)$ is a C^1 function defined in a neighborhood of α_0 and satisfying $a(\alpha_0) = 0$. Define for $\alpha > \alpha_0$:

$$z(\alpha) := \gamma_\alpha(T_0) = y(\alpha) + (a(\alpha) + \alpha_0 - \alpha)X(y^0) + O(\alpha^2). \tag{3.30}$$

It follows

$$\dot{z}(\alpha_0) = \begin{pmatrix} -\dot{a}(\alpha_0) + (\dot{a}(\alpha_0) - 1)X_1(y^0) \\ 1 + (\dot{a}(\alpha_0) - 1)X_2(y^0) \end{pmatrix}.$$

While in the region such that $x_2 < 0$ we have that the s–MTF has direction $(-1, 1)$. The front $F_T^{a,b,x,f}$ is C^1 in y^0 iff $\dot{z}(\alpha_0)$ is parallel to $(-1, 1)$. This condition is satisfied since (3.28) holds. ■

Lemma 18 *Let $Int(\mathcal{S}^{a,b,x,f})$ be an open optimal strip. Fix $\alpha_0 \in]a, b[$ and suppose that $T_0, T_1 \in]\alpha_0, f(\alpha_0)[$ are two subsequent switching times of γ_{α_0} after α_0. Suppose that $F_{T_0}^{a,b,x,f}$ is C^1 in $y^0 := \gamma_{\alpha_0}(T_0)$. Then $F_{T_1}^{a,b,x,f}$ is C^1 in $y^1 := \gamma_{\alpha_0}(T_1)$.*

Proof. Suppose that γ_{α_0} corresponds to constant control $+1$ in the interval $]T_0, T_1[$. Set $x = (x_1, x_2)$, choose a local system of coordinates in such a way $Y \equiv (1,0)$, $y_0 = \gamma_{\alpha_0}(T_0) = (0,0)$, the expression for the switching curve C_0 passing from y^0 is:

$$C_0 = \{x \in M : x_1 = 0, \quad x_2 \in]a', b'[\quad a' < 0, \quad b' > 0\},$$

and $X(0, x_2) = (\beta, \beta)$ $(\beta \in \mathbb{R} \setminus \{0\})$ for $x_2 \in]a', b'[$ (see Figure 3.9). For simplicity set $\alpha_0 = 0$.

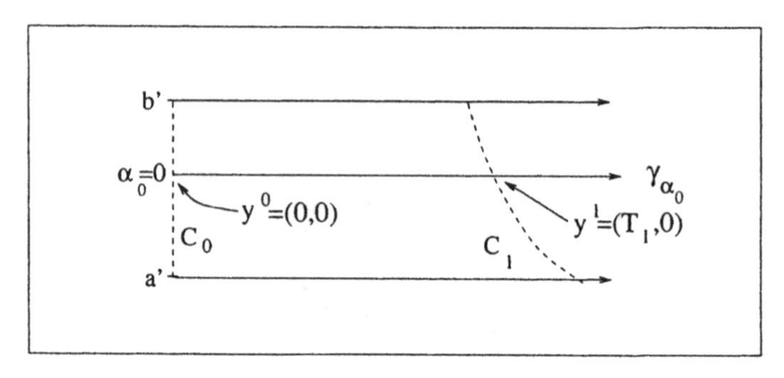

Fig. 3.9. Proof of Lemma 18

In this system of coordinates we have $y^1 = (T_1 - T_0, 0)$ and the expression of G on C_0 is:

$$G((0, x_2)) = \frac{1}{2}\begin{pmatrix} 1-\beta \\ -\beta \end{pmatrix} \text{ for every } x_2 \in]a', b'[. \tag{3.31}$$

Using the same reasoning as to the proof of the previous Lemma, we have that the switching curve C_1 passing through y^1 is determined by the condition:

$$X_1(x) + \frac{1-\beta}{\beta} X_2(x) = 1. \tag{3.32}$$

Suppose that we are in the same situation of the proof of Lemma 17 (see Figure 3.8). Let $T(\alpha)$ and $\mathcal{T}(\alpha)$ be the switching times of $\gamma(\alpha)$ on respectively switching curves C_1 and C_2. Define $a(\alpha) = (T_1 - T_0) - T(\alpha) - \mathcal{T}(\alpha)$ and notice that $a(\cdot)$ is C^1 with $a(0) = 0$. At time T_1 a trajectory with $\alpha > 0$ reaches the point:

$$z(\alpha) := \gamma_\alpha(T_1) = y(\alpha) + (a(\alpha) - \alpha \dot{T}(0))X(y^1) + O(\alpha^2), \tag{3.33}$$

where:

$$y(\alpha) = \begin{pmatrix} T_1 - T_0 - a(\alpha) \\ \alpha \end{pmatrix},$$

(being $a(\cdot)$ a C^1 function defined in a neighborhood of $\alpha = 0$ and satisfying $a(0) = 0$) and $T(\alpha)$ is the switching time of γ_α on the switching curve passing from y^0. It follows:

$$\dot{z}(0) = \begin{pmatrix} -\dot{a}(0) + (\dot{a}(0) - \dot{T}(0))X_1(y^1) \\ 1 + (\dot{a}(0) - \dot{T}(0))X_2(y^1) \end{pmatrix}.$$

This vector must be parallel to the vector $(-\dot{T}(0), 1)$, i.e. the following conditions must hold:

(c1) $\dot{T}(0) = \frac{1-\beta}{\beta}$ if $\dot{T}(0) \neq 0$,

(c2) $\beta = 1$ if $\dot{T}(0) = 0$.

Imposing that $F_{T_0}^{a,b,x,f}$ is C^1 in $y^0 := \gamma_{\alpha_0}(T_0)$, it is easy to check that **(c1)** and **(c2)** are verified. ■

3.2.6 F_T on Overlap Curves

The MTF along overlap curves is described by the following:

Proposition 11 *Let $y^0 = F_T \cap K$ where K is some overlap curve of the optimal synthesis and suppose that condition* **(C2)** *(at the end of Section 3.1, p. 129) holds. Then there exists a ball B centered at y^0 such that $F_T \cap B$ is not C^1 only at y^0. Moreover, generically, $C^N(\mathcal{R}(T), y^0) = \{0\}$, that is $\mathcal{R}(\tau)$ is concave at y^0.*

Proof. Choosing B sufficiently small, we have that y^0 is the only possible point in which F_T is not C^1, indeed $(F_T \cap B) \setminus \{y^0\}$ is contained in interiors of strips. By condition **(C2)**, the two connected components of $(F_T \cap B) \setminus \{y^0\}$ cannot be tangent to K. Moreover the K curve is reached from the two sides by trajectories with constant control $+1$ and -1 respectively. By contradiction suppose that we have $Int(C^N(\mathcal{R}(T), y^0)) \neq \emptyset$. This means that we have the situation of Figure 3.10. The point $\bar{x}$ is reached by a trajectory γ^{-1} (corresponding to constant control -1) at time T, while it is reached at time $T' < T$ by a trajectory γ^{+1} corresponding to constant control $+1$. This is due to the fact that γ^{+1} reaches x' at time T. This contradicts the optimality of γ^{-1}. The situation in which $(F_T \cap B)$ is C^1 in y^0 is not generic. Thus generically $\mathcal{R}(T)$ is concave at y_0, hence $C^N(\mathcal{R}(T), y^0) = \{0\}$. ■

3.2.7 F_T on Borders

In this section we study the smoothness property of the MTF on strip borders. In the first and second Sections we analyze respectively the strip borders of kind γ_0 and γ_A. In the third Section we analyze the strip borders of kind

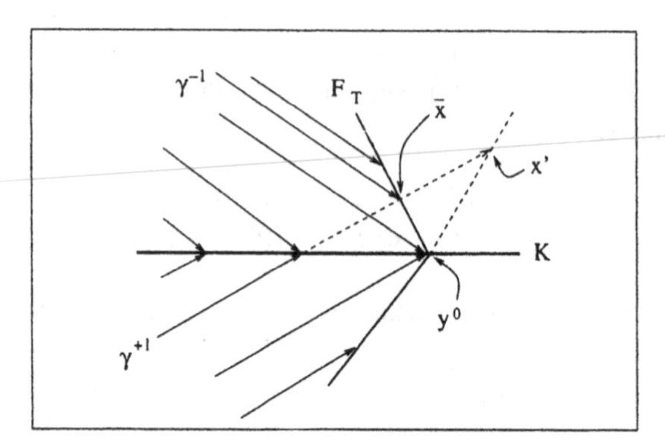

Fig. 3.10. Proof of Proposition 11

γ_k. Since at the two sides of a border of a strip the trajectories have the same control, we have to analyze only the FPs at which this border generates. The FPs of kind $(Y,.)$ are already studied in Section 3.2.3, p. 135. Here we complete the analysis.

F_T on Strip Border of Kind γ_0

The strip borders of kind γ_0, that do not bifurcate from $\gamma^{\pm}_{op}$, are generated at the following FPs: $(C,C)_1$, $(C,S)_{1,2}$, $(S,K)_1$, and at $(C,C)_2$ if $\Delta_A \neq 0$ (see Figure 2.9, p. 61). We analyze in details the case $(C,S)_1$ and $(S,K)_1$. The other cases can be treated by the same argument used in Section 3.2.4.

Assume to have a $(C,S)_1$ or a $(S,K)_1$ FP at $y^1 := \gamma_0(t_1)$ and suppose the following:

- the previous switching time of γ_0 is $t_0 < t_1$;
- γ_0 corresponds to constant control $+1$ in the interval $[t_0,t_1]$;
- $y^0 := \gamma_0(t_0)$ belongs to a X FC (being similar the cases in which it belong to a switching curve or to a turnpike).

Set $T_0 = t_1 - t_0$. Choose a system of coordinates and shift the time in such a way that $y^0 = \gamma_0(0) = 0$, and:

$$Y = \begin{pmatrix} 1 \\ 0 \end{pmatrix}, \qquad X(0,x_2) = \begin{pmatrix} 0 \\ 1 \end{pmatrix}$$

$$X(x) = \begin{pmatrix} a_0 + a_1(x_1 - T_0) + a_2 x_2 + O((x_1-T_0)^2, x_2^2) \\ b_0 + b_1(x_1 - T_0) + b_2 x_2 + O((x_1-T_0)^2, x_2^2) \end{pmatrix}, \quad a_0 < 1, \ \ b_0 > 0.$$

To have a $(C,S)_1$ or a $(S,K)_1$ FP at y^1 the following condition must hold (cfr. Section 3.2.3, p. 135):

$$a_2 = \frac{a_1 b_0}{a_0 - 1}. \tag{3.34}$$

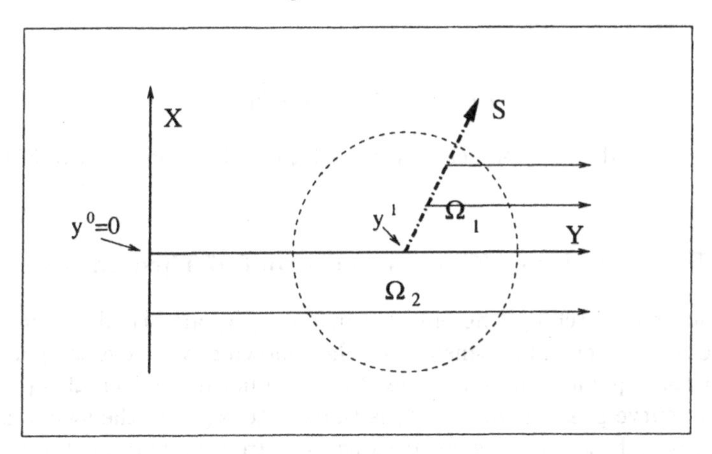

Fig. 3.11. Analysis of the minimum time front at frame points of type $(C,S)_1$ or $(S,K)_1$

Exactly as in Section (3.2.3), we obtain the equation of the singular trajectory starting from y_1 at time 0:

$$\gamma^s(t) = \begin{pmatrix} T_0 + v_1 t + O(t^2) \\ v_2 t + O(t^2) \end{pmatrix},$$

where v_1 and v_2 are given by formulas (3.12) and (3.13). Let I be a neighborhood of y^1 and let us compute the MTF in the region Ω_1 enclosed by γ_0 (after T_0) and S (see Figure 3.11). Consider a trajectory corresponding to constant control $+1$ in $[0, T_0[$, to the singular control $\varphi(\gamma^s(t-T_0))$ on $[T_0, T_0+\alpha[$ and to constant control $+1$ on $[T_0+\alpha, T]$. At time T this trajectory reaches the point:

$$z(\alpha) = (T_0, 0) + \alpha\dot{\gamma}^s(0) + (T - T_0 - \alpha)\begin{pmatrix} 1 \\ 0 \end{pmatrix} + O(\alpha^2) \in \Omega_1.$$

It follows:

$$\dot{z}(0) = \dot{\gamma}^s(0) - \begin{pmatrix} 1 \\ 0 \end{pmatrix}.$$

The MTF is C^1 at $\gamma_0(T)$ iff $\dot{z}(0)$ is parallel to $(1,-1)$, so we have to check the condition:

$$v_1 + v_2 = 1. \tag{3.35}$$

Let $\lambda(\cdot)$ be the covector associated to $\gamma_0(\cdot)$. From the following equations:

$$\begin{aligned}
&\lambda(0)\cdot G(\gamma_0(0)) = 0, \\
&\dot{\lambda}(t) = 0 \ \ \forall\, t \in [0, T_0], \\
&\lambda(T_0)\cdot G(\gamma_0(T_0)) = 0, \\
&\lambda(T_0)\cdot [F,G](\gamma_0(T_0)) = 0,
\end{aligned}$$

we obtain:

$$a_0 + b_0 = 1, \qquad a_1 + b_1 = 0. \tag{3.36}$$

Using (3.36) and (3.34) one checks immediately that condition (3.35) is verified.

F_T on Strip Border of Kind γ_A and Proof of Theorem 28

After the first switching, the only FP involving an abnormal extremal is the point $(C,C)_2$ where Δ_A vanishes. In the following we prove that if t_n, t_{n+1} are two consequent switching times for an abnormal extremal γ_A, and the switching curve passing through t_n is tangent to γ_A then the switching curve passing through t_{n+1} is tangent to γ_A as well. From this fact, with an analysis completely similar to that of Section 3.2.5, it follows that the MTF is C^1 on γ_A and it is always tangent to $Supp(\gamma_A)$. This proves Theorem 28.

Let us consider only the case in which γ_A corresponds to constant control $+1$ on $[t_n, t_{n+1}]$, the opposite case being similar. Choose a local system of coordinates and rescale the time in such a way that:

$$\begin{aligned}
&t_n = 0, \quad \gamma_A(t_n) = 0, \\
&Y \equiv \begin{pmatrix} 1 \\ 0 \end{pmatrix} \quad X(x) = \begin{pmatrix} e_0 + e_1 x_1 + e_2 x_2 + O(|x|^2) \\ f_1 x_1 + f_2 x_2 + O(|x|^2) \end{pmatrix}, \\
&X(x) = \begin{pmatrix} g_0 + g_1(x_1 - t_{n+1}) + g_2 x_2 + O(|x - (t_{n+1}, 0)|^2) \\ h_1(x_1 - t_{n+1}) + h_2 x_2 + O(|x - (t_{n+1}, 0)|^2) \end{pmatrix}.
\end{aligned}$$

We can parameterize the switching curve C_n through $\gamma_A(t_n)$ in the following way. Given $x \in C_n$ there exists (γ_x, λ_x) extremal that switches at x at time $\tau_n(x)$. By induction, we can assume that τ_n is invertible for x near $\gamma(t_n)$ and parameterize C_n by τ_n:

$$\begin{aligned}
x_1(\tau_n) &= \alpha_n \tau_n + O(\tau_n^2), \\
x_2(\tau_n) &= \beta_n \tau_n^2 + O(\tau_n^3),
\end{aligned}$$

for some α_n, β_n that by genericity we may assume different from zero.
Let $(\tilde{\gamma}, \tilde{\lambda}) := (\tilde{\gamma}, \tilde{\lambda})_{\tau_n}$ be the extremal trajectory of (3.1) switching at τ_n on C_n and let τ_{n+1} be the next switching time. We have:

$$\tilde{\gamma}(\tau_n) = \begin{pmatrix} \alpha_n \tau_n + O(\tau_n^2) \\ \beta_n \tau_n^2 + O(\tau_n^3) \end{pmatrix}, \quad \tilde{\gamma}(\tau_{n+1}) = \begin{pmatrix} \alpha_n \tau_n + \tau_{n+1} + O(\tau_n^2) \\ \beta_n \tau_n^2 + O(\tau_n^3) \end{pmatrix}.$$

and:

$$\tilde{\lambda}(\tau_n) \cdot G(\tilde{\gamma}(\tau_n)) = 0, \tag{3.37}$$

$$\tilde{\lambda}(\tau_{n+1}) \cdot G(\tilde{\gamma}(\tau_{n+1})) = 0. \tag{3.38}$$

With computations entirely similar to the those of Section 3.2.7 and using similar generic conditions we conclude that the switching curve passing through $\gamma(t_{n+1})$ has the expression:

$$x_1(\tau_{n+1}) = t_{n+1} + \alpha_{n+1}(\tau_{n+1} - t_{n+1}) + O((\tau_{n+1} - t_{n+1})^2),$$
$$x_2(\tau_{n+1}) = \beta_{n+1}(\tau_{n+1} - t_{n+1})^2 + O((\tau_{n+1} - t_{n+1})^3),$$

for some $\alpha_{n+1}, \beta_{n+1} \neq 0$. This concludes the proof.

F_T on Strip Border of Kind γ_k

The strip borders of kind γ_k, that do not bifurcate from $\gamma_{op}^{\pm}$, are generated at the following FPs: $(C,K)_1, (S,K)_2$. Again it is easy to check that the MTF forms a corner on γ_k. Moreover, for every $x \in \gamma_k$, one has $C^N(\mathcal{R}(\mathbf{T}(x)), x) = \{0\}$ i.e. the reachable set is locally concave at x. For example in the $(C,K)_1$ case, using again the PMP, one checks that the front is not deflected when passing through the C curve. Since the front forms a concave angle on K, the same happens on γ_k.

3.2.8 Proof of Theorem 27

The front at points in the interior of a strip or a strip border has been analyzed in the previous sections.

Assume that x is a Frame Point and **(C1)** holds, so x is not of kind $(Y,K)_{2,3}$ or (K,K). The cases (Y,X), $(Y,C)_{1,2,3}$, (Y,S) and $(Y,K)_1$ have been treated in Section 3.2.2, p. 134. The cases $(C,C)_1$, $(C,S)_{1,2}$, $(S,K)_1$ and $(C,C)_2$ with $\Delta_A \neq 0$, were analyzed in Section 3.2.7, p. 146. The case $(C,C)_2$ with $\Delta_A = 0$ is in Section 3.2.7, p. 148. Finally, the cases $(C,K)_1$ and $(S,K)_2$ are in Section 3.2.7, p. 149. The remaining case $(C,K)_2$ is easily treated by using Proposition 10, p. 142 and 11, p. 145.

Finally, if x is on a frame curve of kind K, that is where a strip may end, and the condition **(C2)** holds then we can apply Proposition 11.

This concludes the proof of Theorem 27. ■

3.3 Proof of Theorem 26

We are going to prove Theorem 26, p. 129 analyzing the MTF first at points inside a strip, then at points on Frame Curves and finally at Frame Points.

If x is a point inside a strip, not on a C curve, the wave fronts are locally smooth one dimensional manifolds. Moreover, the optimal synthesis is locally a flow box of one of the vector fields X and Y. It easily follows that the minimum time function $\mathbf{T}$ is smooth and x is not a critical point (see Figure 3.1, p. 128), since $\mathbf{T}$ increases in the direction of X or Y.

Assume now that x belongs to a Frame Curve (but is not a Frame Point). If the Frame Curve is of kind X, Y, γ_0 or S then the front is a piecewise C^1 manifold and the minimum time is increasing along the Frame Curve, hence the function $\mathbf{T}$ is locally topologically equivalent to a nonzero linear function and x is not a critical point. Notice that indeed the front is C^1 in case of γ_0 and S curves. If the Frame Curve is of kind C then the front is C^1 and there is an increasing direction (indeed two directions X and Y) so $\mathbf{T}$ is C^1 and topologically equivalent to a nonzero linear function and x is not a critical point. We are left with the cases of K and γ_k Frame Curves. Given a Frame Curve D of K type, there exist two Frame Curves D_1 and D_2 such that the curve D is determined by the solution of the following system, see Chapter 2:

$$\begin{cases} (exp\ tX)\,\alpha_1(s) = (exp\ t'Y)\,\alpha_2(s') \\ s + t = s' + t' \end{cases},$$

where $(exp\ tV)\bar{x}$ indicates the integral trajectory at time t of the vector field V starting from $\bar{x}$ and α_1 and α_2 are two smooth parameterizations of the curves D_1 and D_2 respectively. It is clear that under generic assumptions on X and Y, hence on F and G, the above system determines a smooth curve D such that the minimum time function $\mathbf{T}$ restricted to D is a Morse function. We have now three cases:

a) x is not critical for $\mathbf{T}$ on D,
b) x is a maximum for $\mathbf{T}$ on D,
c) x is a minimum for $\mathbf{T}$ on D.

In case a), we are in the situation of Proposition 11, the MTFs near x are piecewise C^1 manifolds. The function $\mathbf{T}$ is increasing (or decreasing) in the direction tangent to K and is topologically equivalent to a nonzero linear function. Therefore x is not a critical point.

Assume case b) holds, then x is not a Frame Point and condition **(C2)** is violated. Since $\mathbf{T}$ is increasing along the two strips ending at K, x is a maximum for $\mathbf{T}$. The MTF through x reduces to a point and is homeomorphic to a circle for points in a neighborhood of x. Therefore $\mathbf{T}$ is topologically equivalent to $f(x_1, x_2) = -x_1^2 - x_2^2$ at the origin, that is x is topologically a maximum.

If c) holds, then either x is a Frame Point or x is not a Frame Point and condition **(C2)** does not hold. Assume the latter occurs. Again, $\mathbf{T}$ is increasing along the two strips ending at K. The MTF at x is formed by two tangent parabolas through x, for points in a neighborhood is formed by two branches that are smooth if they do not intersect the K curve. Therefore $\mathbf{T}$ is topologically equivalent to $f(x_1, x_2) = x_1^2 - x_2^2$ at the origin, that is x is topologically a regular saddle.

If the Frame Curve is of kind γ_k then we are in case a) above, indeed $\mathbf{T}$ is increasing along γ_k, and we conclude in the same way.

We are left with the case of x Frame Point. The points along $\gamma^{\pm}$ have been analyzed above and the minimum time function $\mathbf{T}$ is topologically equivalent to a linear function at those points. The same happens for the Frame Points of kind $(C,C)_{1,2}$, $(C,S)_{1,2}$, $(C,K)_{1,2}$, $(S,K)_{1,2,3}$. It remains to analyze the cases $(Y,K)_{2,3}$ and (K,K).

Assume that $x = \gamma^+(\mathbf{T}(x))$ is a $(Y,K)_2$ Frame Point (being the case of γ^- entirely similar). For $\varepsilon > 0$ sufficiently small, there are two strips $\mathcal{S}^{1,2}$ containing $\gamma^+|_{[\mathbf{T}(x)-\varepsilon,\mathbf{T}(x)]}$ and one strip $\mathcal{S}^3$ reaching x from the other side. We indicate by $\mathcal{R}_+$ the reachable set corresponding to trajectories of $\mathcal{S}^{1,2}$ and by $\mathcal{R}_*$ the reachable set corresponding to trajectories of $\mathcal{S}^3$. Generically, we have that $Int(C^N(\mathcal{R}_+(\mathbf{T}(x)),x)) \neq \emptyset$ and $C^N(\mathcal{R}_*(\mathbf{T}(x)),x)$ is a half line. Hence x is the first point reached on the overlap curve K and is a minimum for $\mathbf{T}$ restricted to K. The front $F_{\mathbf{T}(x)}$ is formed by a smooth branch on the strip $\mathcal{S}^3$ and by a piecewise $\mathcal{C}^1$ branch with an angle at x on the strips $\mathcal{S}^{1,2}$. For times $t < \mathbf{T}(x)$, near $\mathbf{T}(x)$, the front F_t consists of one smooth branch on $\mathcal{S}^3$ and one piecewise smooth on $\mathcal{S}^{1,2}$, while for $t > \mathbf{T}(x)$, near $\mathbf{T}(x)$, consists of two piecewise smooth branches crossing K on opposite sides of x. We conclude that x is topologically a regular saddle for $\mathbf{T}$.

The analysis of the point $(Y,K)_3$ can be done in a similar way. Notice that x is an endpoint of K and one easily checks that x is a minimum for $\mathbf{T}$ restricted to K. However, now there are four strips intersecting at x. Two of them end at K coming from different directions, while the other two have the γ_k curve through x as border with γ^+ being the base of one of the two. Again, we conclude that x is topologically a regular saddle for $\mathbf{T}$.

Assume x is of kind (K,K). If x is reached by a γ_0 Frame Curve then the fronts on each side of K are $\mathcal{C}^1$ and generically we are in case a) above, where D now is the union of the two K curves, and $\mathbf{T}$ is equivalent to a nonzero linear function.

If x is reached by a γ_k Frame Curve, then the corresponding front has an angle on γ_k and the reachable set is locally concave, see Section 3.2.7. The function $\mathbf{T}$ has a maximum at x and x is topologically a maximum.

Thus, on $\mathcal{R}(\tau)$, $\mathbf{T}$ is topologically a Morse function with the set Q being the collection of all Frame Curves of $Int(\mathcal{R}(\tau))$.

3.4 Topology of the Reachable Set

From the the analysis of Chapter 2, it easily follows:

Theorem 29 (Topological Properties of $\mathcal{R}(T)$) *For every T, $\mathcal{R}(T)$ is connected. Moreover, if bad_k is empty, $\mathcal{R}(T)$ is simply connected. At the times T_0 when* **(C1)** *or* **(C2)** *fails, the topology of the reachable set changes. In the planar case, there exists $n \in \mathbb{N}$ such that for t in a left neighborhood of T_0 the reachable set $\mathcal{R}(t)$ is homeomorphic to a disk with n holes and for t in a right neighborhood of T_0, $\mathcal{R}(t)$ is homeomorphic to a disk with $n+1$ holes or $n-1$ holes. For a general manifold it may also happen that we add a handle.*

Notice that if the first part of Theorem 29 applies, that is $bad_k = \emptyset$, then the minimum time function has no critical point except x_0 and thus the level sets are homeomorphic to circles. More precisely, outside $Supp(\gamma_{op}^{\pm})$ the level sets are diffeomorphic to circles. In this case, trivially one concludes that the minimum time function topologically is a Morse function.

Exercises

Exercise 20 Define the generalized gradients of the value function:

$$\partial^- V(x) = \{l \in \mathbb{R}^2 : V(x) + l \cdot (y - x) \leq V(y)\}$$
$$\partial^+ V(x) = \{l \in \mathbb{R}^2 : V(x) + l \cdot (y - x) \geq V(y)\}.$$

Compute $\partial^{\pm} V(x)$ for $x \in Supp(\gamma^{\pm})$, for points on switching curves and on turnpikes.

Exercise 21 With the definitions of Exercise 20, compute $\partial^{\pm} V(x)$ for $x \in K$ in Example 4 of Chapter 2, see pp. 68. Is there any point at which both are empty?

Exercise 22 Determine the sets $bad_k, bad_\gamma, bad_{CV}, bad_{CX}$ of Definition 49, p. 130 for Example 13 of Chapter 2, see pp. 80.

Exercise 23 For Examples 1,2,3 and 4 of Chapter 2 determine the topology of the reachable set for any time. Compute the times for which the topology changes.

Exercise 24 Given a second order controlled equation, as in Section 2.10.1, p. 118, can the corresponding reachable set change topology?

4

Extremal Synthesis

In Chapter 2 an algorithm was defined to construct, under suitable conditions, a structurally stable time optimal synthesis from x_0 for the control system:

$$\dot{x} = F(x) + uG(x), \quad x, x_0 \in M, \quad |u| \leq 1 \tag{4.1}$$

where M is a smooth two dimensional manifold, and we assume $F(x_0) = 0$. Roughly speaking, at step n the algorithm constructs the extremal arcs made of n pieces such that each of them is either a bang arc or a singular arc. Moreover, the not optimal arcs are cut by the algorithm. However, the whole set of extremals happens to share a nice structure, so in **Step 3** of the geometric control approach, see Section 1.4, p. 27, an alternative method for the construction of the synthesis is indicated, namely, to construct all extremals in the cotangent bundle and then project on the base space. This second method is more involved, but induces a more clear idea of the properties of extremal trajectories and of the relationships between the optimal synthesis and the minimum time function. In particular it shows that the overlaps appearing in the optimal synthesis are related to projection singularities of the set of extremals. Thus we now describe a new algorithm to construct a regular extremal synthesis that is the collection of extremals in the cotangent bundle.

We start by constructing all the extremal pairs via an underlinealgorithm that stops (under generic conditions on the pair (F, G)) in a finite number of steps. Generically all extremal trajectories leave the origin with control $+1$ or -1. Thus, if $\gamma^{\pm}$ are the extremal arcs corresponding to constant controls ± 1 and exiting x_0, then each extremal trajectory bifurcates from $\gamma^{\pm}$.

The set of bifurcating extremal trajectories are divided into extremal strips, that are one dimensional families presenting the same sequence of arcs (bang and singular). Then the evolution of "extremal strips" is studied under generic conditions. The obtained set of extremal pairs has some regularity properties summarized in the definition of extremal synthesis. The set $\mathcal{N} \subset T^*M$ of pairs that are reached by an extremal trajectory fails to be a manifold but it is still a Whitney stratified subset of T^*M of dimension 3. Moreover, the extremal trajectories (in finite time) are again finite concatenations of bang and singular arcs and the set formed by points at which $\mathcal{N}$ fails to be a manifold is a finite union of two dimensional strata.

The covector associated to an extremal pair is defined up to the multiplication by a positive constant, hence we project $\mathcal{N}$ onto $M \times S^1$ obtaining a stratified set N of dimension two. The strata of dimension one and zero of N are called, also in this case, Frame Curves and Points. As a byproduct we get a complete classification of Frame Curves and Points.

Abnormal Extremals

A key role in the construction of the extremals synthesis, is played by abnormal extremals, that are trajectories with vanishing Hamiltonian. We already proved that the optimal ones are finite concatenations of bang arcs with switchings happening on the set of zeroes of the function Δ_A (that is where the two vector fields F and G are collinear), see Proposition 2, p. 49. In Section 4.3, p. 171, we study abnormal extremals in more details. In particular we show that the singularities of the synthesis involving abnormal extremals present some special features. Moreover, we classify all possible generic 28 singular points of the synthesis occurring along (projections of) abnormal extremals.

Also the minimum time front and the extremal front have some important properties at points reached by abnormal extremals. Namely they are always tangent to abnormal extremals (see Theorem 28, p. 132 and the analysis of the extremal front in Chapter 5).

From now on, for simplicity, we consider the case $M = \mathbb{R}^2$ and $x_0 = 0$. The conclusions are valid in the general case mutatis mutandis. Let us now introduce the precise concept of extremal synthesis and state the main result. First we need some definitions.

Definition 53 *Let $\Lambda \subset \mathbb{R}^n$ be a set such that $\Lambda = \cup_{j\in J} M_j$, where $J \subset \mathbb{N}$ and M_j are disjoint embedded connected submanifolds of $\mathbb{R}^n$. Then Λ is a Whitney stratified set if the collection $\mathcal{P} := \{M_j\}_{j\in J}$, called the stratification of Λ, is locally finite and the following holds:*

- *if $M_j \cap Clos(M_k) \neq \emptyset$, $j \neq k$, then $M_j \subset Clos(M_k)$ and $dim(M_j) < dim(M_k)$;*
- *let $x_n, y_n \in M_j$, $n \in \mathbb{N}$, $x_n, y_n \to \bar{x} \in M_k \subset Clos(M_j)$ and denote by ℓ_n the direction of $\mathbb{R}^n$ containing the segment joining x_n with y_n. If $T_{x_n} M_j \to T$ (affine subspace of $\mathbb{R}^n$) and $\ell_n \to \ell$, then $\ell \subset T$ and $T_{\bar{x}} M_k \subset T$.*

We define the dimension of Λ by $dim(\Lambda) = \max_j dim(M_j)$ and say that Λ is regular if every $z \in \Lambda$ belongs to the closure of a stratum of dimension $dim(\Lambda)$.

Given an extremal pair (γ, λ) one can easily check that for every $\alpha \in \mathbb{R}$, $\alpha > 0$, the pair $(\gamma, \alpha\lambda)$ is also extremal. It is then natural to introduce the following projection:

$$\pi_* : \mathbb{R}^2 \times (\mathbb{R}^2)_* \setminus \{0\} \to \mathbb{R}^2 \times S^1, \qquad \pi_*(x,p) = \left(x, \frac{p}{|p|}\right),$$

that is the projectivization of $T^*\mathbb{R}^2$. A key role is played by the trajectory-covector pairs $(\gamma^\pm, \lambda^\pm)$ starting from the origin and corresponding to constant control ± 1.

Definition 54 *A regular extremal synthesis Γ^* for (4.1) in time τ is a collection of trajectory-covector pair sets $\{\Gamma^*_{(x,p)} : (x,p) \in T^*(\mathbb{R}^2) = \mathbb{R}^4\}$ satisfying the following properties:*

1. *For every $(x,p) \in \mathbb{R}^4$, $\Gamma^*_{(x,p)}$ is the collection of all trajectory-covector extremal pairs $(\gamma, \lambda) : [0,a] \to \mathbb{R}^4$, $0 \le a \le \tau$, such that $\gamma(0) = 0$, $\gamma(a) = x$ and $\lambda(a) = p$.*
2. *For every $(x,p) \in \mathbb{R}^4$, $x \notin \gamma^\pm(Dom(\gamma^\pm))$, the set $\{\pi_*(\gamma,\lambda) : (\gamma,\lambda) \in \Gamma^*_{(x,p)}$, is finite.*
3. *For every $(x,p) \in \mathbb{R}^4$ and every $(\gamma,\lambda) \in \Gamma^*_{(x,p)}$, γ is a finite concatenation of bang and singular arcs.*
4. *The set $\mathcal{N} = \{(x,p) : \Gamma^*_{(x,p)} \neq \emptyset\}$ is a regular Whitney stratified subset of $\mathbb{R}^4$ and $dim(\mathcal{N}) = 3$.*
5. *The set of discontinuities of $\#(\Gamma^*_{(x,p)})$ on $\mathcal{N}$ (i.e. where $\mathcal{N}$ fails to be a manifold) is a finite collection of two dimensional embedded submanifolds of $\mathbb{R}^4$ (more precisely it is a finite collection of strata of the stratification $\mathcal{P}$ of $\mathcal{N}$.)*

For every $\tau > 0$ the algorithm, under generic assumptions on (F,G), constructs in a finite number of steps all the extremal pairs defined on the interval $[0,\tau]$. Under generic assumptions on τ the collection of these pairs has all the properties required by the definition of regular extremal synthesis. Using generic conditions similar to those of Chapter 2, we are able to prove structural stability of the extremal synthesis for fixed time. For the difficulties arising in extending these results for infinite time, we refer the reader to [54].

Theorem 30 *Fix $\tau > 0$. There exists an open dense subset Π_τ of Ξ such that for every $(F,G) \in \Pi_\tau$ there exists a regular structurally stable extremal synthesis Γ^*_τ in time τ.*

As a byproduct we obtain a detailed study of the singularities for the extremal regular synthesis. There exists a finite set of (equivalence classes of) stable singularities described by Theorems 36 and 37 (see Section 4.6, p. 195).

In Section 4.1.4 the local synthesis near $(\gamma^\pm, \lambda^\pm)$ is studied. In Section 4.2 the subdivision in extremal strips is operated and we make a detailed description of the algorithm. In Section 4.3 the case of abnormal extremals is treated. Finally in Section 4.4 the evolution of interior of strips complete the analysis.

4.1 Basic Facts

In this section we discuss some preliminary facts that are essential to state the new algorithm: generic conditions, anti-turnpikes, new frame curves, the set of extremals in a neighborhood of $\gamma^+ \cup \gamma^-$ and a special singularity.

4.1.1 Generic Conditions

In order to guarantee structural stability, in this Chapter, we assume the generic conditions **(GA1)÷(GA8)** stated in Section 2.8, p. 86. For the convenience of the reader, they are rewritten in the following.

(GA1) For every $t \in [0, t_f^+]$, $G(\gamma^+(t)) \neq 0$;
(GA2) $\dot\theta^+(0) \neq 0$, $\dot\theta^+(t_f^+) \neq 0$;
(GA3) If $\dot\theta^+(t) = 0$, then $\theta^+(t) \neq 0$, $\ddot\theta^+(t) \neq 0$;
(GA4) If $t \neq s$ and $\dot\theta^+(s) = \dot\theta^+(t) = 0$ then $\theta^+(s) \neq \theta^+(t)$;
(GA5) $(\nabla \Delta_B \cdot V)(\gamma^+(t)) \neq 0$, $V = X, Y$ at all points $t \in \{t_i^+,\ t_i'^+;\ \ i \geq 1\}$;
(GA6) If $t_f^+ = \tau$, then $\max\{|\theta^+(t) - \theta^+(\tau)|,\ \ t \in [0,\tau]\} < \pi$;
(GA7) Let v be the unit tangent vector at $\gamma^+(s_i^+)$ to the switching curve starting at $\gamma^+(s_i^+)$. We have $v \cdot X(\gamma^+(s_i^+)) \neq 0$, for each $i \geq 2$. And similarly for the points $\gamma^+(s_i'^+)$;
(GA8) Suppose $s_1^+ \neq 0$ and consider a local system of coordinates in a neighborhood of $\gamma^+(s_1^+)$ such that $Y = (1,0)$, $X(0) = (-1,0)$, $X(s_1^+,0) = (c_0, 0)$ and define $b_1 := \partial_{x_1} X_2(0)$, $d_1 := \partial_{x_1} X_2(s_1^+, 0)$. We have $c_0 \neq 0$ and $c_0 d_1 - b_1 \left(\frac{1-c_0}{2}\right)^2 \neq 0$. A similar condition is assumed if $s_1'^+ \neq 0$.

In addition, to avoid non generic self–intersections in the cotangent bundle, we assume:

(GA9) $\gamma^+(t) \neq (0,0)$ for every $t \in [0, t_f^+]$.

It is assumed that similar conditions hold for γ^-. Finally we consider a generic τ, i.e. we assume the generic condition:

(GAτ) Let $(\gamma, \lambda) : [0,\tau] \to \mathbb{R}^2 \times (\mathbb{R}^2)_*$ be en extremal pair. Recall Definition 16, p. 40, then the condition $\phi(\tau) = \lambda(\tau) \cdot G(\gamma(\tau)) = 0$ implies the following:

- if $\gamma = \gamma^+$, then $\tau \notin \{t_i^\pm, t_i'^\pm, s_i^\pm, s_i'^\pm\}$ (see Definition 38, p. 90);
- $\gamma(\tau) \notin \Delta_A^{-1}(0)$. The fact that this is generic follows from Theorem 18, p. 172 and Corollary 3, p. 176;

- if $\gamma(\tau) \in \Delta_B^{-1}(0)$ then there exists $\varepsilon > 0$ such that γ corresponds to the singular control φ on $[\tau - \varepsilon, \tau]$ (see (2.13)) and $|\varphi(\gamma(\tau))| < 1$;
- $X(\gamma(\tau))$ and $Y(\gamma(\tau))$ are not tangent to C or $\bar{C}$ curves.

Remark **51** Notice that **(GA6)** is a consequence of **(GA$_\mathcal{T}$)**.

4.1.2 Anti-Turnpikes

In Section 2.3, p. 40, we proved that an anti-turnipike is never optimal (see Lemma 12, p. 47). Anyway an anti-turnipike is extremal, thus, in principle, there may be an arc running an anti-turnpike and belonging to some extremal trajectory. The next Lemma shows that this is not the case for the extremal synthesis starting from the origin, since an extremal trajectory γ such that $\gamma(0) = 0$ never enters an anti-turnpike.

Lemma 19 *Let S be an anti-turnpike, (γ, λ) an extremal pair such that $\gamma(0) = 0$, and assume that there exist $\bar{t} \in Dom(\gamma)$ and $\varepsilon > 0$ such that $\gamma([\bar{t} - \varepsilon, \bar{t}[) \cap S = \emptyset$ and $\gamma(\bar{t}) \in S$. Then $\lambda(\bar{t}) \cdot G(\gamma(\bar{t})) \neq 0$.*

Proof. Let $\Omega, \Omega_X, \Omega_Y$ be as in Definition 22, p. 45. First suppose that γ reaches x from Ω with bang control, say $+1$, so that $\gamma([\bar{t}-\varepsilon, \bar{t}[) \subset \Omega_X$ (possibly choosing ε smaller). The function $\phi(t)$ satisfies the equation $\dot{\phi}(t) = \lambda(t) \cdot [F, G](\gamma(t))$. Since $\Omega \cap \Delta_A^{-1}(0) = \emptyset$, we can write for $x \in \Omega$, $[F, G](x) = f(x)F(x) + g(x)G(x)$. It follows $\dot{\phi}(t) = f(\gamma(t))\lambda(t) \cdot F(\gamma(t)) + g(\gamma(t))\lambda(t) \cdot G(\gamma(t))$ for a.e. $t \leq \bar{t}$. Now if γ was an abnormal extremal then the condition $\lambda(\bar{t}) \cdot G(\gamma(\bar{t})) = 0$ would imply $\gamma(\bar{t}) \in \Delta_A^{-1}(0)$ (see Proposition 1, p. 45), hence we consider the case:

$$\lambda(t) \cdot (F(\gamma(t)) + G(\gamma(t))) = cost > 0. \tag{4.2}$$

Assume by contradiction that $\lambda(\bar{t}) \cdot G(\gamma(\bar{t})) = 0$. From (4.2), taking $t' \leq \bar{t}$ sufficiently close to $\bar{t}$, one gets $\lambda(t) \cdot F(\gamma(t)) > 0$ for every $t \in [t', \bar{t}]$. Now for $t \in [t', \bar{t}[$ we have $f(\gamma(t)) > 0$, hence if k is such that $|g(x)| < k$ for x in a neighborhood of $\gamma(\bar{t})$, we get

$$\begin{aligned} &\dot{\phi}(t) = f(\gamma(t))\lambda(t) \cdot F(\gamma(t)) + g(\gamma(t))\lambda(t) \cdot G(\gamma(t)) > -k\phi(t) \quad \forall t \in [t', \bar{t}], \\ &\phi(t') = \phi_0 > 0. \end{aligned}$$

Hence ϕ is strictly positive on $[t', \bar{t}]$, which gives a contradiction. ■

Remark **52** Using the generic conditions **(P1)**–**(P7)** of Section 2.4, p. 48, one can check that an extremal trajectory γ can never enter an anti-turnpike even from a singular point. Indeed there are only two cases corresponding to conditions **(P6)** and **(P7)**. In the first case $\Delta_B^{-1}(0)$ is either formed by two turnpikes or by two anti-turnpikes, while in the second case the singular point is reached in infinite time along $\Delta_B^{-1}(0)$.

4.1.3 New Frame Curves

In Chapter 2 the one and zero dimensional strata (submanifold of $\mathbb{R}^2$) of the optimal synthesis are called respectively Frame Curves and Frame Points (in the following briefly FCs and FPs, respectively). As we see in next Sections, N (recall that we set $N := \pi_*(\mathcal{N})$) is a stratified subset of $\mathbb{R}^2 \times S^1$ of dimension two. With some abuse of notation we call also the one and zero dimensional strata of N Frame Curves and Frame Points respectively.

We use the same name for the curves and points in $\mathbb{R}^2 \times S^1$ and for their projections on $\mathbb{R}^2$, being clear from the context if we are treating the former or the latter. Finally the letters X, Y, C, S, K indicate the types of FCs as in Section 2.6.1, p. 58.

In the construction of the optimal synthesis at each step some extremal non optimal trajectories are cut by the algorithm generating the so called overlap curves. Now we keep all extremal trajectories thus the overlap curves are not constructed and some new FCs appears. The following facts illustrate the main differences with respect to the optimal case:

- we do not have any K (overlap) FC because these FCs are obtained as "cuts" of not optimal extremal trajectories (see Chapter 2 and 3);
- beside the C curves on which X and Y point to the same side, we also have a new type of FC (called $\bar{C}$) on which X and Y point to opposite sides (see Figure 4.1 Cases 1 and 2).

By definition switching curves are loci of conjugate points to other FCs and $\bar{C}$ curves were excluded from the optimal synthesis because trajectories switching on $\bar{C}$ are not locally optimal.

Notice that a C Frame Curve can be smoothly joined to a $\bar{C}$ Frame Curve if the switching locus becomes tangent to X or to Y. Other FCs that appear in the extremal synthesis are:

- FCs called W that are arcs of extremal trajectories characterized by the property that the projection of all the extremal trajectories close to them lie on the same side of W (see Figure 4.1 Case 3);
- FCs called γ_0 that are arcs of extremal trajectories that "transports" some special information, e.g. they switch every time they meet the locus $\Delta_A^{-1}(0)$ or they evolve into W FCs. The optimal synthesis close to a γ_0 FC is regular, but possibly the covectors do not depend in a smooth way from points on the plane.

More details are given later.

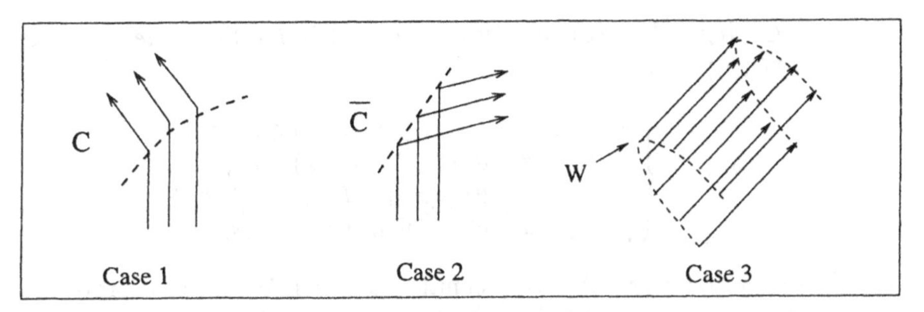

Fig. 4.1. FCs of kind C, $\bar{C}$ and W.

4.1.4 The Set of Extremals in a Neighborhood of $\gamma^+ \cup \gamma^-$

To explain better the situation, here we build the set of extremals in a neighborhood of $\gamma^+ \cup \gamma^-$. To this purpose, we first study the set of extremal in which the covectors are pulled back to their initial condition:

$$N_0 = \{(x,p) \in \mathbb{R}^2 \times S^1 : \text{there exists a time } t \in [0,\tau] \text{ and an extremal pair } (\gamma,\lambda) : [0,t] \to \mathbb{R}^2 \times S^1 \text{ such that } \gamma(0)=0,\ \gamma(t)=x,\ \lambda(0)=p\}.$$

One can easily see that in a neighborhood of the origin, the set of extremals is homeomorphic to N_0 (see Proposition 12).
Consider the projector $\pi : (x,p) \in \mathbb{R}^2 \times S^1 \mapsto x \in \mathbb{R}^2$. Clearly we have $\mathcal{R}(\tau) = \pi(N_0)$. If we set $\Lambda^0(x) = \{p \in S^1 : (x,p) \in N_0\}$, we have $(x, \Lambda^0(x)) = \pi^{-1}(x) \cap N_0$.

The structure of N_0 is described in the following Theorem (see Definitions 14, p. 37 and 38, p. 90 for the definitions of the function $\theta^+(\cdot)$ and of the times $t_f^+, t_i^+, t_i'^+, s_i^+, s_i'^+$):

Theorem 31 *Choose a local coordinate system such that* $G(0) = (0,1)$.

A) Let $x \in \mathcal{R}(\tau) \setminus (Supp(\gamma^+) \cup Supp(\gamma^-))$ *reached by a* $X * Y$ *trajectory* γ *at some time* $\bar{t}$:

$$\gamma : [0,\bar{t}] \to \mathcal{R}(\tau)$$
$$\gamma(0) = 0, \quad \gamma(\bar{t}) = x.$$

Then there exists a unique covector $\lambda : [0,\bar{t}] \to (\mathbb{R}^2)_*$ *such that* $|\lambda(0)| = 1$ *and* (γ,λ) *is an extremal pair on* $[0,\bar{t}]$. *Moreover, let* $t_0(\gamma) \in]0, t_f^+[$ *be the switching time of* γ, *then we have:*

$$\Lambda^0(x) = \begin{cases} \theta^+(t_0(\gamma)) & \text{if } t_0(\gamma) \in [s_i^+, t_i^+] \\ \pi + \theta^+(t_0(\gamma)) & \text{if } t_0(\gamma) \in [s_i'^+, t_i'^+] \end{cases} \tag{4.3}$$

B) $\Lambda^0(0) = [-\pi, \pi]$

C) Let $x \in$ Supp$(\gamma^+) \setminus \{0\}$ and let $t(x) \in]0, t_f^+]$ be the first time such that $x = \gamma^+(t(x))$. Then:

$$\Lambda^0(x) = \begin{cases} [\theta^+(t(x)), \pi + \theta^+(t'^+_{i-1})] & \text{if } t(x) \in]s_i^+, t_i^+], \\ [\theta^+(t_i^+), \pi + \theta^+(t'^+_{i-1})] & \text{if } t(x) \in]t_i^+, s'^+_i], \\ [\theta^+(t_i^+), \pi + \theta^+(t(x))] & \text{if } t(x) \in]s'^+_i, t'^+_i], \\ [\theta^+(t_i^+), \pi + \theta^+(t'^+_i)] & \text{if } t(x) \in]t'^+_i, s^+_{i+1}]. \end{cases}$$

Proof. Let us first prove **A)**. By definition p exists. We have to prove the uniqueness. Since γ is extremal, we know that there exists a non trivial cov-

160 4 Extremal Synthesis

C) Let $x \in$ Supp$(\gamma^+) \setminus \{0\}$ and let $t(x) \in]0, t_f^+]$ be the first time such that $x = \gamma^+(t(x))$. Then:

$$\Lambda^0(x) = \begin{cases} [\theta^+(t(x)), \pi + \theta^+(t'^+_{i-1})] & \text{if } t(x) \in]s_i^+, t_i^+], \\ [\theta^+(t_i^+), \pi + \theta^+(t'^+_{i-1})] & \text{if } t(x) \in]t_i^+, s'^+_i], \\ [\theta^+(t_i^+), \pi + \theta^+(t(x))] & \text{if } t(x) \in]s'^+_i, t'^+_i], \\ [\theta^+(t_i^+), \pi + \theta^+(t'^+_i)] & \text{if } t(x) \in]t'^+_i, s^+_{i+1}]. \end{cases}$$

Proof. Let us first prove **A)**. By definition p exists. We have to prove the uniqueness. Since γ is extremal, we know that there exists a non trivial cov-

160 4 Extremal Synthesis

C) Let $x \in$ Supp$(\gamma^+) \setminus \{0\}$ and let $t(x) \in]0, t_f^+]$ be the first time such that $x = \gamma^+(t(x))$. Then:

$[\theta^+(t(x)), \pi + \theta^+(t'^+_{i-1})]$ if $t(x) \in]s_i^+, t_i^+]$,

Now:

- if $t_0 \in [s_i^+, t_i^+[$ (resp. $t_0 \in [s_i'^+, t_i'^+[)$, then

  1. from Proposition 3, p. 92 at time t_0 we have a $Y * X$ switching;
  2. $\dot{\theta}^+(t_0) > 0$ (resp. < 0); (4.5)
  3. formula (4.4) is satisfied with the equality at time t_0, i.e. $\lambda(0) \cdot \bar{v}^+(t_0) = 0$; (4.6)

- if $t_0 = t_i^+$ (resp. $t_i'^+$), then

  1. from Proposition 3, p. 92, at time t_0 we have a $Y * X$ switching;
  2. θ has a maximum (resp. a minimum) at this time; (4.7)
  3. $\lambda(0) \cdot \bar{v}^+(t_0) = 0$. (4.8)

Notice that Proposition 3, p. 92, excludes that $t_0 \in]t_i^+, s_i'^+[$ or $t_0 \in]t_i'^+, s_{i+1}^+[$. Formulas (4.5), (4.6), in the first case, and (4.7), (4.8), in the second, determine exactly the $\lambda(0)$ given by the right hand side of formula (4.3) that includes both cases. Now, since the evolution equation of λ (see **i)** of Theorem 10, p. 35) satisfies the Caratheodory uniqueness conditions, $\lambda(t)$ is unique, and (4.3) follows.

Because of the request "a.e." in the statement of the PMP, at the origin every $\lambda(0)$ is allowed. By normalizing we have $\Lambda^0(0) = [-\pi, \pi]$. This proves **B)**.

By the same argument of the proof of formula (4.3) we have that for a.e. $t \in]0, t(x)]$:

$$0 \leq \lambda(0) \cdot \bar{v}^+(t). \tag{4.9}$$

From the definition of the times $t_i^+, s_i^+, t_i'^+, s_i'^+$, it follows the expression for $\Lambda^0(x)$ in the case $x \in \text{Supp}(\gamma^+) \setminus \{0\}$, that concludes the proof. ■

Remark **53** A similar Theorem can be written for the trajectory γ^- that starts from the origin with control -1.

Let $I(0)$ be a small neighborhood of the origin. The shape of $N_0|_{I(0)}$ can be obtained from Theorem 31 and it is drawn in Figure 4.3. Notice that N_0 in a neighborhood of the origin is homeomorphic to the set R described in Figure 4.4.
From the fact that in a neighborhood of the origin the extremal trajectories do not intersect, and from continuous dependence on initial data we get the following:

Proposition 12 *N_0 and N are homeomorphic in a neighborhood of the lift of origin.*

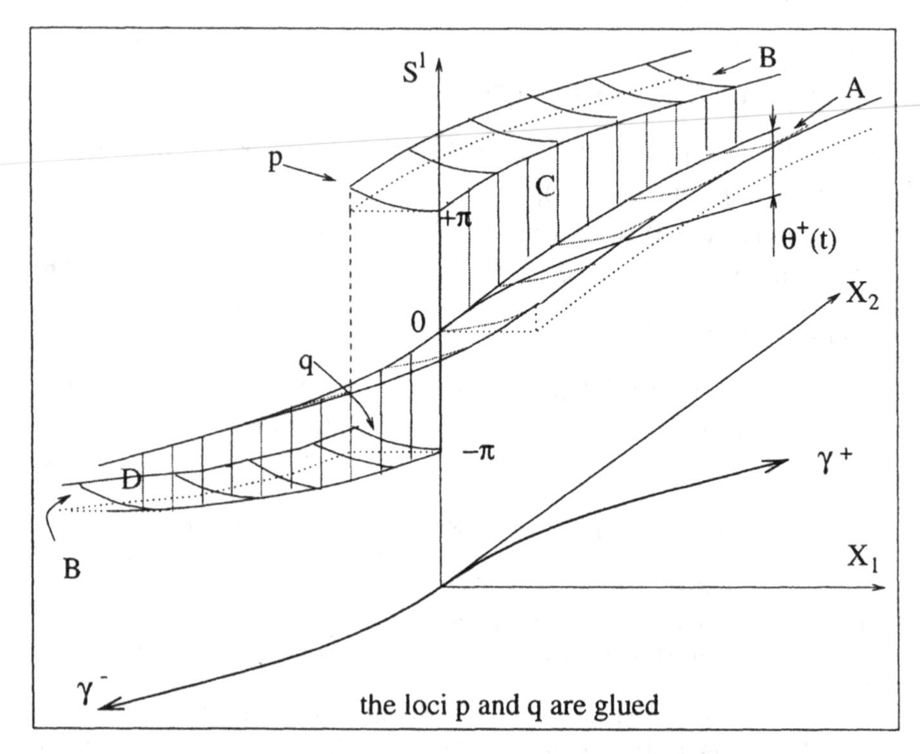

Fig. 4.3. N_0 in a neighborhood of the origin.

4.1.5 Number of pre–Images

If we project the extremal synthesis on $\mathcal{R}(\tau)$, it may happen that subsets of $\mathcal{R}(\tau)$ are covered by more than one trajectory. In fact the borders of these subsets correspond to singularities of the projection from N and they are treated in the following (and in Chapter 5) in more detail.

Fix $\bar{x} \in \mathcal{R}(\tau)$ and an extremal trajectory $\bar{\gamma} : [0, \bar{a}] \to \mathcal{R}(\tau)$ such that $\bar{\gamma}(0) = 0$ and $\bar{\gamma}(\bar{a}) = \bar{x}$, and define the function:

$$K_\varepsilon^{\bar{x},\bar{\gamma}}(x) := \sharp\{\text{extremal trajectories } \gamma : \ \gamma(0) = 0,\ \gamma(a) = x,\ |\gamma(t) - \bar{\gamma}(t)| < \varepsilon \\ \forall\, t \in [0, \min(a, \bar{a})],\ |a - \bar{a}| < \varepsilon\}. \tag{4.10}$$

We refer to Figure 4.5 where for simplicity we drop the dependence on $\bar{x}$, $\bar{\gamma}$ and ε.

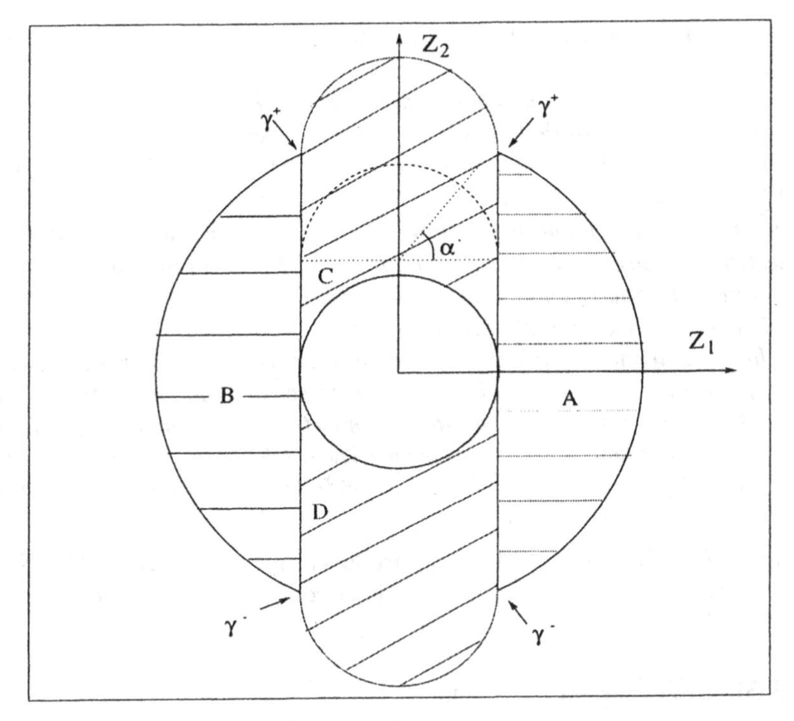

Fig. 4.4. The set R.

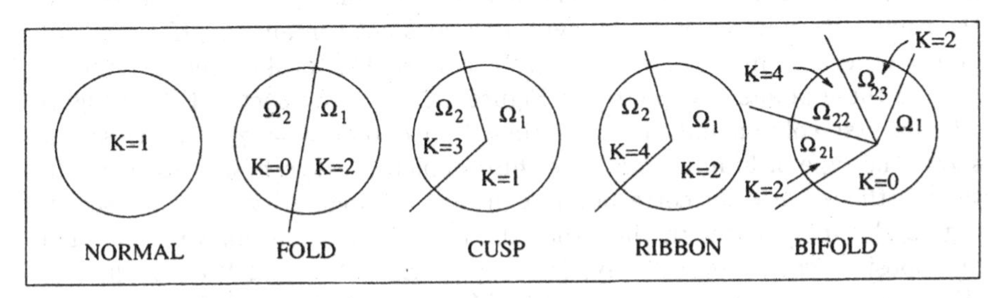

Fig. 4.5. Number of pre-images

Definition 55 *Fix $\bar{x} \in \mathcal{R}(\tau)$ and an extremal trajectory $\bar{\gamma} : [0, \bar{a}] \to \mathcal{R}(\tau)$ such that $\bar{\gamma}(0) = 0$ and $\bar{\gamma}(\bar{a}) = \bar{x}$;*

- *we say that $\bar{x}$ is a normal point along $\bar{\gamma}$ if for ε sufficiently small there exists a neighborhood U of $\bar{x}$ such that $K_{\varepsilon}^{\bar{x},\bar{\gamma}}(U) = 1$;*
- *we say that $\bar{x}$ is a fold point along $\bar{\gamma}$ if there exists a one dimensional piecewise-C^1 manifold l, with $\bar{x} \in l$, satisfying the following. For ε sufficiently small there exists a neighborhood U of $\bar{x}$ divided by l into two*

connected components Ω_1, Ω_2 *such that* $K_\varepsilon^{\bar{x},\bar{\gamma}}(\Omega_1) = 2$, $K_\varepsilon^{\bar{x},\bar{\gamma}}(\Omega_2) = 0$, $K_\varepsilon^{\bar{x},\bar{\gamma}}(l) = 1$;

- *if* l, $U(\varepsilon)$, Ω_1, Ω_2 *are as above, then if* $K_\varepsilon^{\bar{x},\bar{\gamma}}(\Omega_1) = 1$, $K_\varepsilon^{\bar{x},\bar{\gamma}}(\Omega_2) = 3$ *and* $K_\varepsilon^{\bar{x},\bar{\gamma}}(l) = 2$, *we say that* $\bar{x}$ *is a cusp point along* $\bar{\gamma}$;
- *if* l, $U(\varepsilon)$, Ω_1, Ω_2 *are as above then if* $K_\varepsilon^{\bar{x},\bar{\gamma}}(\Omega_1) = 2$, $K_\varepsilon^{\bar{x},\bar{\gamma}}(\Omega_2) = 4$ *and* $K_\varepsilon^{\bar{x},\bar{\gamma}}(l) = 3$, *we say that* $\bar{x}$ *is a ribbon point along* $\bar{\gamma}$;
- *we say that* $\bar{x}$ *is a bifold point along* $\bar{\gamma}$ *if there exists a one dimensional connected piecewise-*C^1 *embedded manifold* l *and two connected* C^1 *embedded manifolds* l_1 *and* l_2 *satisfying the following:*
  - $l \cap \partial l_i = \{x\}$ $(i = 1, 2)$; $\partial l_1 \cap \partial l_2 = \{x\}$;
  - *for* ε *sufficiently small there exists a neighborhood* U *of* x *satisfying the following. The set* $U \setminus l$ *has two connected components* Ω_1, Ω_2 *and the set* $\Omega_2 \setminus \{l_1 \cap l_2\}$ *has three connected components* Ω_{21}, Ω_{22}, *and* Ω_{23} *(the name are chosen as in Figure 4.5) such that* $K_\varepsilon^{\bar{x},\bar{\gamma}}(\Omega_1) = 0$, $K_\varepsilon^{\bar{x},\bar{\gamma}}(\Omega_{21}) = 2$, $K_\varepsilon^{\bar{x},\bar{\gamma}}(\Omega_{22}) = 4$, $K_\varepsilon^{\bar{x},\bar{\gamma}}(\Omega_{23}) = 2$, $K_\varepsilon^{\bar{x},\bar{\gamma}}(l) = 1$ *and* $K_\varepsilon^{\bar{x},\bar{\gamma}}(l_i) = 3$ $(i = 1, 2)$.

Example of ribbon and bifold points are given in Section 4.3, p. 171. Notice that the manifolds l, l_1, l_2 where $K_\varepsilon^{\bar{x},\bar{\gamma}}$ is discontinuous correspond to FCs of kind $\bar{C}$ and W.

4.1.6 Singularities Along $\gamma^+ \cup \gamma^-$

The FPs of the optimal synthesis along $\gamma_{op}^+ \cup \gamma_{op}^-$ are described in Chapter 2 (see in particular Figure 2.9, p. 61). For the extremal synthesis case we have the same FPs, but the extremal synthesis does not contain any K curve and, on the other side may present $\bar{C}$, γ_0, W curves. The local extremal synthesis around these points is depicted in Figure 4.6. Let x be such a FP. In Figure 4.6 the first column indicates the time t such that $x = \gamma^+(t)$, the second shows the type of Frame Point, the third depicts the synthesis near x, finally the fourth indicates the corresponding projection singularity. The superscripts "tg" and "t-o" denotes the fact that the two FCs are tangent with the same or opposite versus, respectively. The superscripts "C" and "D" on the FCs W are explained in Section 4.3, p. 171. Moreover at the points labeled with $*$ (e.g. $(Y\bar{C})_1^{t-o}$) the trajectories self intersect (but their lifts in the cotangent bundle do not!). The situation is studied in [37]. The singularities along γ^- are similar.

4.1.7 Singularities Along Singular Trajectories

After the singularities along $Supp(\gamma^+) \cup Supp(\gamma^-)$ the first singularities we need to study are the end points of turnpikes. These singularities were studied in Chapter 2 (see Example 8,14 Section 2.6.4) and are called $(C,S)_2$ and $(S,K)_3$. For the extremal synthesis the latter singularity does not contain a K curve and the new singularity is called (S,S) singularity, the name being clear from the following:

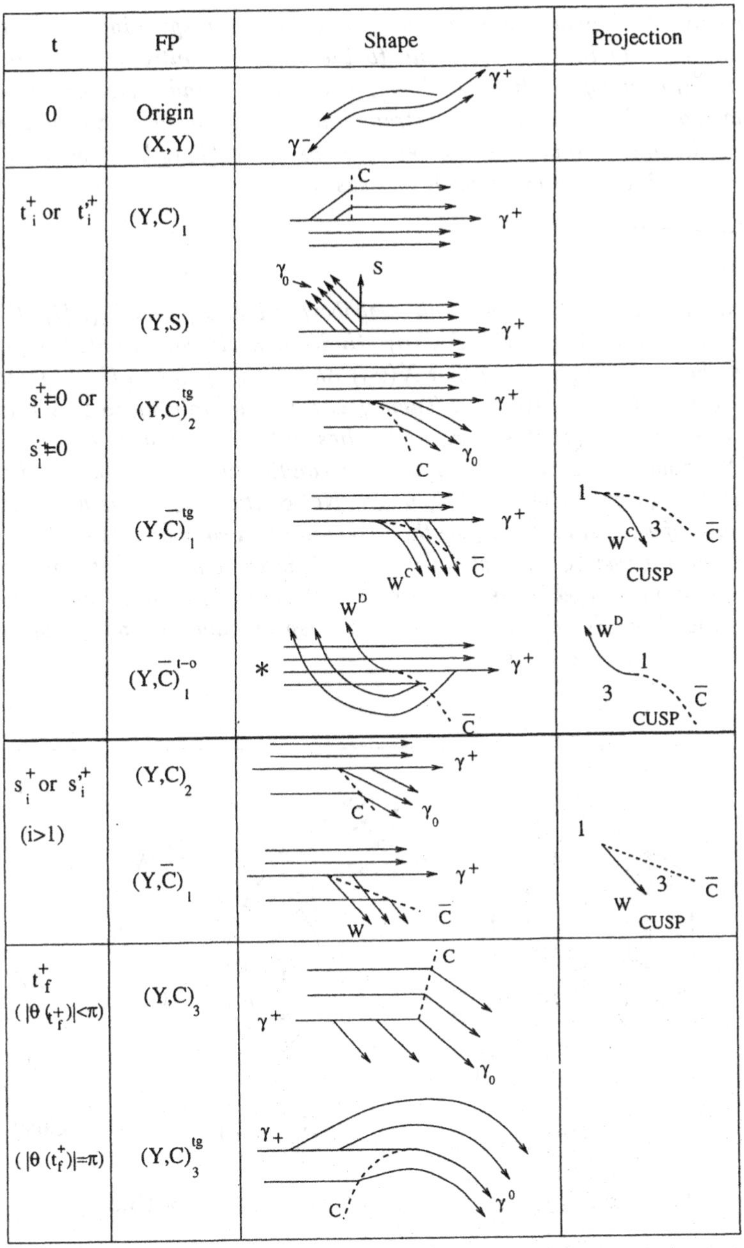

t	FP	Shape	Projection
0	Origin (X,Y)	γ^+, γ^-	
t_i^+ or $t_i'^+$	$(Y,C)_1$	C, γ^+	
	(Y,S)	γ_0, S, γ^+	
$s_1^+ \neq 0$ or $s_1'^+ \neq 0$	$(Y,C)_2^{tg}$	γ^+, C, γ_0	
	$(Y,\bar{C})_1^{tg}$	γ^+, W^C, $\bar{C}$	1, W^C, 3, $\bar{C}$, CUSP
	$(Y,\bar{C})_1^{t-o}$	W^D, $*$, γ^+, $\bar{C}$	W^D, 1, 3, $\bar{C}$, CUSP
s_i^+ or $s_i'^+$ $(i>1)$	$(Y,C)_2$	γ^+, C, γ_0	
	$(Y,\bar{C})_1$	γ^+, W, $\bar{C}$	1, W, 3, $\bar{C}$, CUSP
t_f^+ $(\lvert\theta(t_f^+)\rvert<\pi)$	$(Y,C)_3$	C, γ^+, γ_0	
$(\lvert\theta(t_f^+)\rvert=\pi)$	$(Y,C)_3^{tg}$	γ_+, C, γ^0	

Fig. 4.6. FPs along $\gamma^+ \cup \gamma^-$. In the point labeled with $*$ the trajectories self intersect.

Theorem 32 *Consider an extremal trajectory γ corresponding to a constant control (say +1) in $]t_0 - \varepsilon, t_0]$ and to the singular control φ in $]t_0, t_0 + \varepsilon[$. Then the trajectory $\tilde{\gamma}$ satisfying $\tilde{\gamma}|_{[0,t_0+\varepsilon[} = \gamma|_{[0,t_0+\varepsilon[}$ and corresponding to the singular control φ after $t_0 + \varepsilon$ is extremal at least up to the first time in which X or Y becomes tangent to the set of zeros of $\Delta_B^{-1}(0)$. Let $\bar{t}$ be such time, define $x_0 := \tilde{\gamma}(\bar{t})$, and distinguish two cases:*

a) $\Delta_A(x_0) \neq 0$;
b) $\Delta_A(x_0) = 0$.

If **a)** *holds then let U be a small neighborhood of x_0 and U_A, U_B the two connected components of $U \setminus \Delta_B^{-1}(0)$. Under generic conditions, U_A and U_B can be chosen in such a way that $X(x_0)$ and $Y(x_0)$ point in U_A. Assume for instance $Y(x_0) \cdot \nabla \Delta_B^{-1}(0) = 0$. Then $\tilde{\gamma}$ is extremal only up to $\bar{t}$, a switching curve of kind C originates at x_0 and it lies in U_A (see Figure 4.7 case A).*

If **b)** *holds true, then, under generic conditions, we are in the situation of condition* **(P6)** *of Section 2.4, p. 48. Notice that $\Delta_B^{-1}(0)$ is a turnpike on both sides of x_0 and let S_1, S_2 be the two connected components of $\Delta_B^{-1}(0) \setminus x_0$ ordered by increasing time along $\tilde{\gamma}$. Then $\tilde{\gamma}$ is extremal after $\bar{t}$ and from S_1 and S_2 extremal trajectories generate on both sides. Moreover, the trajectories generating from S_1, crosses S_2, x_0 is a cusp singularity and we are in the situation of Figure 4.7 case B.*

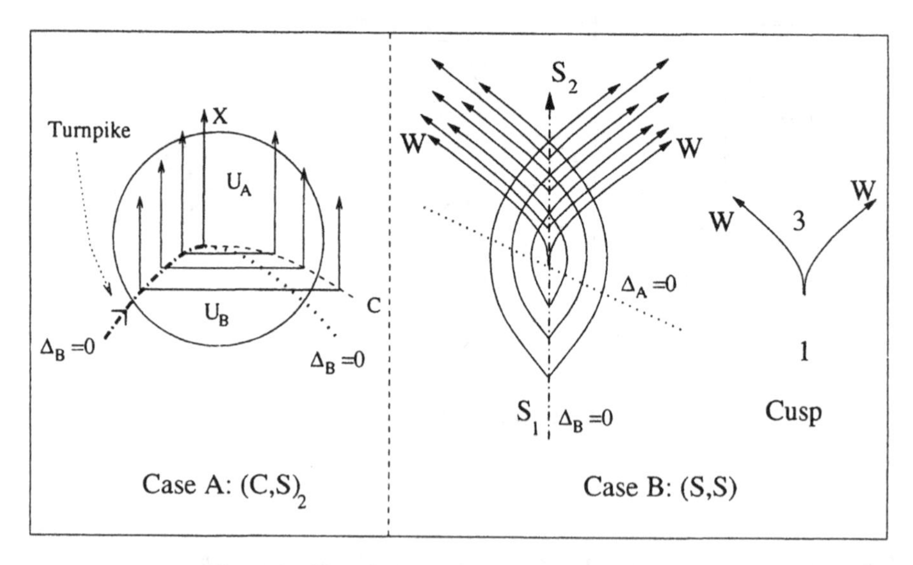

Fig. 4.7. Singularities along singular trajectories

4.2 Extremal Strips and the Algorithm

Now we give the key definition of extremal strip. An extremal strip is essentially a one parameter continuous family of extremal bang–bang trajectories having the same switching strategy.

Definition 56 *Let a, b be two real numbers such that $0 \leq a < b \leq \tau$ and $x \in \mathcal{R}(\tau)$. A set of trajectories $\mathcal{S}^{a,b,x} = \{\gamma_\alpha : \ \alpha \in [a,b],\ \gamma_a(a) = x\}$ is called a extremal strip if:*

i) $\forall\ \alpha \in [a,b],\ \ \gamma_\alpha : [0, \tau(\alpha)] \to \mathbb{R}^2$, $\tau(\alpha) > \alpha$ is an extremal trajectory for the control problem (4.1). Moreover there exists $\varepsilon > 0$ s. t. $\gamma_{[\alpha,\alpha+\varepsilon]}$ corresponds to a constant control ± 1;
ii) $\forall\ \alpha \in]a,b[,\ \ \gamma_\alpha$ does not switch on $\Delta_A^{-1}(0) \cup \Delta_B^{-1}(0)$ after time α;
iii) The set $\mathcal{B}^{a,b,x} = \{y \in \mathcal{R}(\tau) :\ \exists\ \alpha \in\ ]a,b[$ and $t \in]\alpha, \tau(\alpha)[$ s. t. $y = \gamma_\alpha(t)$, t is a switching time for $\gamma_\alpha\}$ is never tangent to X or Y;
iv) The map $\eta : \ \alpha \in\ [a,b] \mapsto \gamma_\alpha(\alpha) \in \mathbb{R}^2$ is a bang or singular arc, $\gamma_\alpha|_{[0,\alpha']} = \gamma_{\alpha'}|_{[0,\alpha']}$ if $a \leq \alpha' \leq \alpha \leq b$ and $G(\eta(\alpha)) \neq 0$ for every $\alpha \in]a,b[$.

The function $\eta : [a,b] \to \mathbb{R}^2$ is called the base of the extremal strip, $\overset{\circ}{\mathcal{S}}{}^{a,b,x} := \{\gamma_\alpha : \ \ \alpha \in]a,b[\}$ is called an open extremal strip and $\partial\mathcal{S}^{a,b,x} := \{\gamma_a,\ \gamma_b\}$ is called extremal strip border (see Figure 4.8).

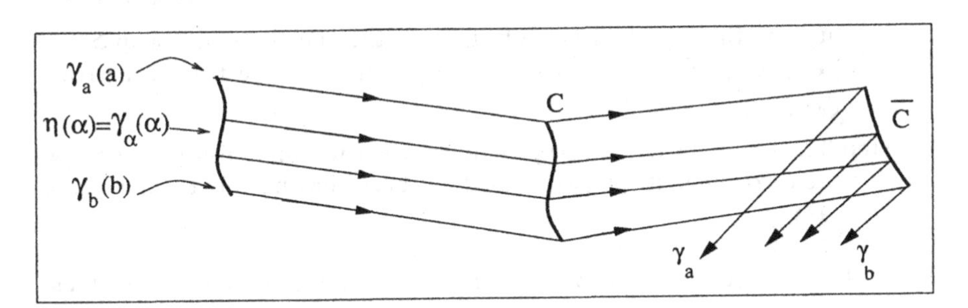

Fig. 4.8. Extremal Strip

Since trajectories generate from both side of a turnpike, one gets immediately:

Proposition 13 *If $\mathcal{S}^{a,b,x}$ is an extremal strip and its base η is a singular arc then there exists a extremal strip $\mathcal{S}' \neq \mathcal{S}^{a,b,x}$ with the same base (the two extremal strips start from the two sides of the turnpike).*

Now we define an algorithm that constructs all the extremal trajectories up to time τ.

ALGORITHM

STEP 1 We construct all the open extremal strips $\overset{\circ}{\mathcal{S}}^{a,b,x}$ with x belonging to $Supp(\gamma^+) \cup Supp(\gamma^-)$ and maximal base. In particular these strips are the strips which bases are $\gamma^+|_{]s_i^+,t_i^+[}$, $\gamma^+|_{]s_i'^+,t_i'^+[}$, $\gamma^+|_{]s_i^-,t_i^-[}$, $\gamma^+|_{]s_i'^-,t_i'^-[}$ and the strips obtained in the following way. Suppose that at time t_i^+ (resp. $t_i'^+$, t_i^-, $t_i'^-$) a turnpike starts and let γ be the trajectory corresponding to the constant control +1 (resp. +1, -1, -1) in the interval $[0,t_i^+]$ (resp. $[0,t_i'^+]$, $[0,t_i^-]$, $[0,t_i'^-]$) and to the singular control φ in $]t_i^+,b]$ (resp. $]t_i'^+,b]$, $]t_i^-,b]$, $]t_i'^-,b]$) where $b=\tau$ (if the turnpike does not reach a $(C,S)_2$ or a (S,S) singular point before τ) or b is the time when the turnpike meets a $(C,S)_2$ or a (S,S) singular point (see Theorem 32, p. 166). In this case we have also the two extremal strips having $\gamma|_{]t_i^+,b[}$ (resp. $\gamma|_{]t_i'^+,b[}$, $\gamma|_{]t_i^-,b[}$, $\gamma|_{]t_i'^-,b[}$) as a base (cfr. Proposition 13, p. 167). Notice that there is a finite number of strips built in this way.

STEP 2 We extend every trajectory of all the extremal strips built in the previous step(s) up to the maximal time at which **ii)** and **iii)** of Definition 56, p. 167 are satisfied.

STEP 3 If all the trajectories built in the previous steps satisfy $\tau(\alpha)=\tau$ we jump to step 4. Otherwise:

step 3.1 let γ_α be a trajectory built in the previous steps that satisfies $\tau(\alpha)<\tau$ (that means that **ii)** or **iii)** are violated at $\gamma_\alpha(\tau(\alpha))$), and suppose that this trajectory belongs to the open extremal strip $\overset{\circ}{\mathcal{S}}^{a,b,x}$ ($\alpha \in]a,b[$). Then we substitute this strip with the two open strips $\overset{\circ}{\mathcal{S}}^{a,\alpha,x}$ and $\overset{\circ}{\mathcal{S}}^{\alpha,b,x}$;

step 3.2 if at $\gamma_\alpha(\tau(\alpha))$ a turnpike starts then we build also the two new open extremal strips (with maximal base) having this turnpike as a base;

step 3.3 jump to step 2.

STEP 4 We construct all the borders of the strips built in the previous steps.

Notice that a border never corresponds to the singular control φ. Moreover, from the conditions **(P1)–(P7)**, **(GA1)–(GA9)**, **(GAτ)** and from the analysis of next Sections it follows that the Algorithm stops after a finite number of steps. Indeed from Chapter 2, we have that under the generic conditions **(P1)–(P7)**, all extremals are finite concatenations (with an uniform bound on the number of arcs) of bang and singular arcs. The number of extremal strips is also finite, because under generic conditions, the assumptions **ii)** and **iii)** of Definition 56, p. 167 are violated in a finite number of points. In order to construct a stable extremal synthesis we assume various generic conditions on FCs and FPs during the execution of the algorithm. Some of these conditions are explicitly stated in Chapter 2 and others are given in the following Sections, namely conditions **(GA10)–(GA16)**. After the end of the algorithm, we need some more generic stability conditions. In Chapter 2, Definition 33,

p. 59 it was defined an equivalence among Frame Points, we introduce here a similar one:

Definition 57 *If x_1, x_2 are two FPs, we say that they are equivalent if there exist some points $y_0 = x_1$, $y_2, ..., y_n = x_2$ such that the following holds. Each y_i belongs to a frame curve D_i. For every y_i, $i = 1, ..., n-1$, there exists an extremal trajectory γ_i and a_i, $b_i \in Dom(\gamma_i)$, satisfying $\gamma_i(a_i) = y_i$, $\gamma_i(b_i) = y_{i+1}$, where $\gamma_i|_{[a_i,b_i]}$ is an X or Y trajectory and $\gamma_i|_{[a_i,b_i]}$ is not part of $\gamma^{\pm}$, γ^0 or W FCs. That is there exists a curve connecting x_1 with x_2 formed by X and Y arcs of extremal trajectories and no one of these arcs is part of $\gamma^{\pm}$, γ^0, or W FCs.*

To understand the meaning of this equivalence see the corresponding definition given in Chapter 2. Moreover it is possible to give natural stability conditions for each FP. We assume the following:

($\mathcal{A}$) all FPs are stable. If x_1, x_2 are two equivalent FPs then $x_1 = x_2$.

Compare with the condition ($\mathcal{A}1$) and ($\mathcal{A}2$) of Chapter 2, Section 2.8.2, p. 89. To ensure stability, we need also the following:

($\mathcal{F}$) the inequality $\phi(t) \geq 0$ is verified in strict sense in the interior of each interval where an extremal trajectory corresponds to a constant bang control.

This is equivalent to condition ($\mathcal{F}1$) of Chapter 2, Section 2.8.2, p. 89. Finally, notice that the stability assumption ($\mathcal{F}2$) of Chapter 2, Section 2.8.2, p. 89, is a consequence of **(GA$_T$)**.

Definition 58 *We denote by Γ the set of trajectories built by the Algorithm. The set of trajectories built before step 4 are denoted by $\mathring{\Gamma}$ and the set of the borders by $\partial\Gamma$.*

Notice that Γ is the canonical projection of Γ^* (see Section 4.1.6, p. 164) from $\mathbb{R}^2 \times (\mathbb{R}^2)_*$ to $\mathbb{R}^2$. Moreover, with the above Definitions: **i)** all the FPs are contained in the base or borders of the extremal strips; **ii)** in the interior of a strip we can have only C and $\bar{C}$ one dimensional singularities. This is due to the following:

Theorem 33 *Let $S^{a,b,x}$ be an extremal strip built by the Algorithm, $\gamma_\alpha \in \mathring{S}^{a,b,x}$, $\gamma_\alpha : [0,\tau] \to \mathcal{R}(\tau)$ and $t_0 \in \;]\alpha,\tau[$. Then there exist ε_1, ε_2, δ ($\varepsilon_1 > \delta > 0$, $\varepsilon_2 > 0$) such that for each $\beta \in]\alpha-\varepsilon_2, \alpha+\varepsilon_2[$ all the trajectories γ_β have the same constant control, say +1, in the interval $]t_0-\varepsilon_1, t_0-\varepsilon_1+\delta[$ and one of the following possibilities occur:*

1) for each $\beta \in]\alpha-\varepsilon_2, \alpha+\varepsilon_2[$, γ_β does not switch in $]t_0-\varepsilon_1, t_0+\varepsilon_1[$;
2) for each $\beta \in]\alpha-\varepsilon_2, \alpha+\varepsilon_2[$, γ_β switches from Y to X (and there are not other switchings) in $]t_0-\varepsilon_1, t_0+\varepsilon_1[$, and the switching locus is of kind C;

3) for each $\beta \in]\alpha - \varepsilon_2, \alpha + \varepsilon_2[$, γ_β switches from Y to X (and there are not other switchings) in $]t_0 - \varepsilon_1, t_0 + \varepsilon_1[$, and the switching locus is of kind $\bar{C}$.

Proof. From **ii)** of Definition 56, p. 167 we have only the following possibilities:

1) t_0 is not a switching time for γ_α,
2) t_0 is a switching time from Y to X for γ_α.

By the generic assumptions **(P1)–(P7)**, **(GA1)–(GA9)**, **(GA$_\mathcal{T}$)** there exists $\varepsilon_1 > 0$ such that:

- γ_α does not switch in $]t_0 - \varepsilon_1, t_0 + \varepsilon_1[$, in the first case;
- γ_α switches only once in $]t_0 - \varepsilon_1, t_0 + \varepsilon_1[$ with a YX switching, in the second case.

By continuity $\Delta_A^{-1}(0)$ and $\Delta_B^{-1}(0)$ never vanish in a neighborhood of $\gamma_\alpha(t_0)$, hence there exists ε_2 such that for each $\beta \in]\alpha - \varepsilon_2, \alpha + \varepsilon_2[$

- γ_β does not switch in $]t_0 - \varepsilon_1, t_0 + \varepsilon_1[$ in the first case;
- γ_β switches only once in $]t_0 - \varepsilon_1, t_0 + \varepsilon_1[$ (with a YX switching) in the second case.

To conclude the proof it remains to check that the switching locus is entirely of kind C or $\bar{C}$, and this is a consequence of **iii)** of Definition 56, p. 167. ■

The singularities of the bases were already studied in Section 4.1.6, p. 164 except for the starting points of the turnpikes if they do not belong to $Supp(\gamma^+) \cup Supp(\gamma^-)$. To enter a turnpike we need to switch on $\Delta_B^{-1}(0)$, so **ii)** of Definition 56, p. 167 implies that these singularities happen only on borders of extremal strips.

4.2.1 Extremal Strip Borders: the FCs of Kind γ_0 and W

We clearly have:

Proposition 14 *$\partial\Gamma$ is a finite set.*

Moreover, by construction, every border belongs to two different strips and adjacent strips correspond locally to the same control. More precisely it holds:

Proposition 15 *Let $\gamma \in \partial\Gamma$, S^1 and S^2 be the two strips such that $\{\gamma\} = S^1 \cap S^2$ and $\bar{t}$ be a switching time for γ. Then the switching loci of S^1 and S^2 passing trough $\gamma(\bar{t})$ correspond both to switchings from Y to X or both to switchings from X to Y..*

Definition 59 *Let $\gamma \in \partial\Gamma$ and suppose that it corresponds to a constant control in the interval $]b, c[$ $(0 < b < c \leq \tau)$. We say that in $]b, c[$ γ is an extremal strip border of kind W if for every $t \in]b, c[$, $\gamma(t)$ is a fold point. Viceversa if for every $t \in]b, c[$, $\gamma(t)$ is a normal point, we say that in $]b, c[$ γ is an extremal strip border of kind γ_0.*

By construction we get:

Proposition 16 *If $\gamma \in \partial \Gamma$, γ is of kind W in $]a,b[$ and of kind γ_0 in $]b,c[$ $(0 \le a < b < c \le \tau)$ or viceversa then b is a switching time for γ.*

In Section 4.3 we study the singularities of the borders of extremal strips that are abnormal extremals. In Section 4.4, p. 181 we study all the singularities of the other types of borders.

4.3 Abnormal Extremals

4.3.1 Generic Properties of Abnormal Extremals

In this section we state and prove the main results about the switching strategies of abnormal extremals.

We already know that optimal abnormal extremals are finite concatenations of bang arcs with switchings happening on the set of zeroes of the function Δ_A (that is where the two vector fields F and G are collinear), see Proposition 2, p. 49. We now prove the same result for abnormal extremals not necessarily optimal:

Proposition 17 *Under generic conditions, it holds:*

- *up to normalization of the covector, there exist two maximal abnormal extremals;*
- *the two maximal abnormal extremals are bang–bang and do not switch on $G^{-1}(0)$.*

Proof. From the definition of abnormal extremal we have $0 = H(0, \lambda(0)) = \lambda(0) \cdot F(0)$, hence up to normalization there are two choices for $\lambda(0)$ and the first claim is proved.

By condition **(P5)** (see Section 2.4, p. 48), generically an abnormal extremal γ does not cross the set $\Delta_A^{-1}(0) \cup \Delta_B^{-1}(0)$. Hence any non regular time of γ is isolated and it is a switching time. Therefore γ is bang–bang. ■

From now on we assume to be in the generic situation of Proposition 17, so all abnormal extremals are bang-bang.

Definition 60 *Let $\gamma : [0,\tau] \to \mathbb{R}^2$ be (the first component of) an abnormal extremal for the control problem (4.1) such that it switches at least once and let t_1 be its first switching time. We call the couple (γ, t_1) a Non Trivial Abnormal Extremal (in the following NTAE). By definition a NTAE is maximal if it is defined on $[0,\tau]$.*

Definition 61 *Let $\gamma : [0,\tau] \to \mathbb{R}^2$ be a NTAE, and $t_1 < t_2 < ... < t_{n(\gamma)-1} < t_{n(\gamma)} := \tau$ the sequence of switching times. We set $AA(i) = Supp\,(\gamma|_{[t_i,t_{i+1}]})$ $(i = 1, ..., n(\gamma) - 1)$ and we call it an abnormal arc.*

From Proposition 17 we get:

Proposition 18 *Let $\gamma : [0, \tau] \to \mathbb{R}^2$ be an extremal trajectory for the control problem (4.1) such that it switches at least once at a time t_1 with $G(\gamma(t_1)) \neq 0$. Let $\lambda : [0, \tau] \to (\mathbb{R}^2)_*$ be the corresponding covector and assume that $t_1 < t_2 < ... < t_{n(\gamma)-1} < t_{n(\gamma)} := \tau$ is the sequence of switching times such that $G(\gamma(t_i)) \neq 0$, $i = 1, ..., n$. Then, under generic conditions, it holds:*

A) *$\lambda(\cdot)$ is unique (up to the multiplication by a positive constant);*

B) *the following conditions are equivalent:*

(a) (γ, t_1) is a NTAE;
(b) $\gamma(t_i) \in \Delta_A^{-1}(0)$ for some $i \in \{1, ..., n(\gamma) - 1\}$;
(c) $\gamma(t_i) \in \Delta_A^{-1}(0)$ for each $i \in \{1, ..., n(\gamma) - 1\}$;
(d) $\gamma(\bar{t}) \in \Delta_A^{-1}(0)$ $(\bar{t} \in Dom(\gamma),$ $G(\gamma(\bar{t}) \neq 0)$ iff $\bar{t} = t_i$ for some $i \in \{1, ...n(\gamma) - 1\}$;

Proof. For the proofs that **(a)** $\Rightarrow$ **(c)**, and **(b)** $\Rightarrow$ **(a)**, see Proposition 1, p. 45. The implications **(c)**$\Rightarrow$**(b)**, **(d)**$\Rightarrow$**(b)** are obvious.
Proof of **A)**. The covector associated to an extremal trajectory is completely determined after the first switching. From $n(\gamma) \geq 2$ it follows that $\lambda(\cdot)$ is unique up to a positive constant.
Proof that **(a)** implies **(d)**. Fix $\bar{t}$ such that $\gamma(\bar{t}) \in \Delta_A^{-1}(0)$. We have $F(\gamma(\bar{t})) = \beta G(\gamma(\bar{t}))$ (by genericity we may assume $\beta \neq 0, \pm 1$) and there exists a sequence $t'_m \nearrow \bar{t}$ such that $|u(t'_m)| = 1$ and

$$\mathcal{H}(\gamma(t'_m), \lambda(t'_m), u(t'_m)) = \lambda(t'_m) \cdot (F + u(t'_m)G)(\gamma(t'_m)) = 0.$$

Hence $(1 + u(t'_m)\beta)\lambda(t'_m) \cdot G(t'_m) \to 0$ and being $\lim_{m\to\infty} u(t'_m) = \pm 1$ we have $\lambda(\bar{t}) \cdot G(\bar{t}) = 0$. Under generic assumptions, $\Delta_B(\gamma(\bar{t})) \neq 0$ thus $\dot{\theta}(\bar{t}) \neq 0$ and $\bar{t}$ is a switching time. Viceversa, since **a)** implies **c)**, we get that $\Delta_A(\gamma(t_i)) = 0$ for each i.
This concludes the proof. ■

From Proposition 17 and Definition 56, p. 167 it follows:

Lemma 20 *Let (γ, t_1) be a NTAE then $\gamma \in \partial\Gamma$.*

Hence it is natural to define $\partial\Gamma_A = \{\gamma \in \partial\Gamma, \gamma \text{ is an abnormal extremal}\}$.

Definition 62 *Let $x \in \Delta_A^{-1}(0)$ be a switching point for a NTAE, then clearly $X(x) \neq 0$, $Y(x) \neq 0$ and $(F + G)(x) = \alpha(F - G)(x)$ for some $\alpha \neq 0$. If $\alpha > 0$ (resp. $\alpha < 0$) we say that at x, $\Delta_A^{-1}(0)$ is direct (resp. inverse).*

From Lemma 20, p. 172 it follows that any $AA(i)$ is an extremal strip border of kind γ_0 or W, but for abnormal extremals a more precise definition for the strip borders of kind W is necessary.

Definition 63 *We refer to Figure 4.9. Let $\gamma \in \partial\Gamma_A$, and suppose that it corresponds to a constant control (say +1) in the interval $]b,c[$ $(0 < b < c \leq \tau)$. Let $\mathcal{S}^1$ and $\mathcal{S}^2$ be the two strips such that $\{\gamma\} = \mathcal{S}^1 \cap \mathcal{S}^2$ and suppose that in the interval $]b,c[$ γ is an extremal strip border of kind W.*

- *We say that in $]b,c[$ γ is a strip border of kind W^C if $\mathcal{S}^1$ and $\mathcal{S}^2$ both lie on the right (resp. on the left) of $\gamma|_{]b,c[}$ and X points to the right (resp. to the left) of $\gamma|_{]b,c[}$ at every point of $Supp(\gamma|_{]b,c[})$.*
- *We say that in $]b,c[$ γ is a strip border of kind W^D if $\mathcal{S}^1$ and $\mathcal{S}^2$ both lie on the right (resp. on the left) of $\gamma|_{]b,c[}$ and X points to the left (resp. to the right) of $\gamma|_{]b,c[}$ at every points of $Supp(\gamma|_{]b,c[})$.*

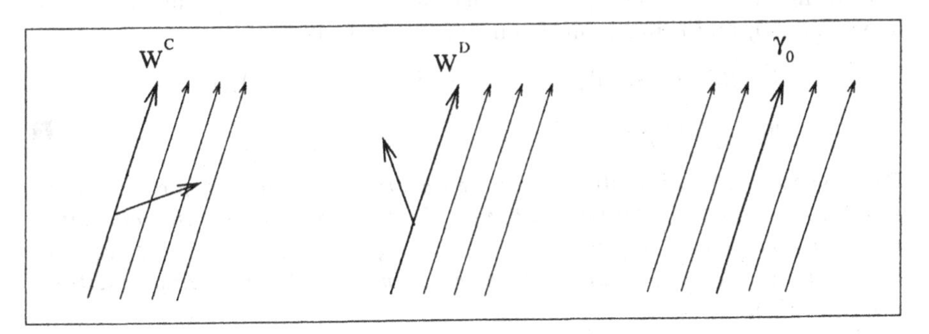

Fig. 4.9. The abnormal extremals of type W^C, W^D and γ_0.

From Proposition 18, p. 172 we have the following:

Lemma 21 *Let $A_1, A_2 \in \{W^C, W^D, \gamma_0\}$ and $\gamma \in \partial\Gamma_A$. If γ is of kind A_1 in $]a,b[$ and of kind A_2 in $]b,c[$ $(A_1 \neq A_2,\ 0 < a < b < c \leq \tau)$ then b is a switching time for γ.*

Now it is clear the meaning of this more fine definition in the case of abnormal extremal strip borders. An abnormal extremal can be of kind W^C (resp. W^D) on $[t-\varepsilon, t]$, $\varepsilon > 0$ and of kind W^D (resp. W^C) on $[t, t+\varepsilon]$, only if t is a switching time. On the contrary in the case of strip borders that are not abnormal extremals the change from W^C to W^D or viceversa can occur without switching and so the difference between W^C and W^D is not crucial.

Proposition 19 *Let γ be a NTAE, $t_1 < t_2 < ... < t_{n(\gamma)-1} < t_{n(\gamma)} := \tau$ be the sequence of its switching times and set $t_0 = 0$. From Proposition 17, p. 171 generically $G(\gamma(t_i)) \neq 0$. If $F(\gamma(t_i)) = \beta_i G(\gamma(t_i))$, then clearly $\beta_i \neq \pm 1$ (otherwise $X(\gamma(t_i)) = 0$ or $Y(\gamma(t_i)) = 0$.) For all $i = 0, .., n(\gamma) - 2$ it holds:*

$$\begin{cases} v(G(\gamma(t_{i+1})), t_{i+1}; 0) = (\beta_i + 1)/(\beta_{i+1} + 1) v(G(\gamma(t_i)), t_i; 0) \\ \qquad \textit{if } \gamma \textit{ corresponds to constant control +1 in } [t_i, t_{i+1}] \\ \\ v(G(\gamma(t_{i+1})), t_{i+1}; 0) = (\beta_i - 1)/(\beta_{i+1} - 1) v(G(\gamma(t_i)), t_i; 0) \\ \qquad \textit{if } \gamma \textit{ corresponds to constant control -1 in } [t_i, t_{i+1}] \end{cases} \tag{4.11}$$

Proof. Fix i and suppose that γ corresponds to constant control +1 in $[t_i, t_{i+1}]$, the opposite case being similar. From $F(\gamma(t_{i+1})) = \beta_{i+1} G(\gamma(t_{i+1}))$, recalling Lemma 7, p. 42 we have

$$\begin{aligned} F(\gamma(t_i)) + G(\gamma(t_i)) &= v(F(\gamma(t_{i+1})) + G(\gamma(t_{i+1})), t_{i+1}; t_i) \\ &= (1 + \beta_{i+1}) v(G(\gamma(t_{i+1})), t_{i+1}; t_i). \end{aligned}$$

Now from $F(\gamma(t_i)) = \beta_i G(\gamma(t_i))$ (notice that in case $i = 0$ we have $F(0) = 0$, hence $\beta_0 = 0$) and using again Lemma 7, p. 42 we have

$$(1 + \beta_i) v(G(\gamma(t_i)), t_i; 0) = (1 + \beta_{i+1}) v(G(\gamma(t_{i+1})), t_{i+1}; 0)$$

that concludes the proof. ■

Proposition 20 *Let $\gamma : [0, \tau] \to \mathbb{R}^2$ be an extremal trajectory for the control problem (4.1) that switches at least once, θ^γ the corresponding function defined in (2.5) (Chapter 2) and $t_1 < t_2 < ... < t_{n(\gamma)-1} < t_{n(\gamma)} := \tau$ the sequence of switching times. Then under generic assumptions the following conditions are equivalent:*

(a) (γ, t_1) is a NTAE;
(b) $\theta^\gamma(t_i) \in \{0, \pm\pi\}$ for some $i \in \{1, ..., n(\gamma) - 1\}$;
(c) $\theta^\gamma(t_i) \in \{0, \pm\pi\}$ for each $i \in \{1, ..., n(\gamma) - 1\}$;
(d) $\theta^\gamma(\bar{t}) \in \{0, \pm\pi\}$, $(\bar{t} \in Dom(\gamma))$ iff $\bar{t} = t_i$ for some $i \in \{1, ...n(\gamma) - 1\}$.

Proof. Proof that **(a)** implies **(c)**. By definition $\theta^\gamma(0) = 0$. Proposition 19 implies that the vectors $v(G(\gamma(t_{i+1}), t_{i+1}; 0)$ and $v(G(\gamma(t_i), t_i; 0)$ are parallel. Since $\theta^\gamma(t_{i+1})$ and $\theta^\gamma(t_i)$ measure precisely the angle between these vectors and $G(0)$, we have $\theta^\gamma(t_{i+1}) = \theta^\gamma(t_i) \pm \pi$.
Proof that **(c)** implies **(a)**. From **c)** we have $\theta^\gamma(t_1) \in \{0, \pm\pi\}$, then for some $b \in \mathbb{R}$ (that by genericity we may assume different from 0 and 1) it holds $v^\gamma(G(\gamma(t_1)), t_1; 0) = bG(\gamma(0))$. Now if we suppose that γ corresponds to constant control +1 in the interval $[0, t_1]$ (the opposite case being similar), we have $bG(\gamma(0)) = b(F + G)(\gamma(0)) = v^\gamma(b(F + G)(\gamma(t_1)), t_1; 0)$. From the injectivity of the map $v_0 \to v^\gamma(v_0, t_0; t_1)$ we obtain $\gamma(t_1) \in \Delta_A^{-1}(0)$. By Proposition 18, p. 172 it follows **(a)**.
Proof that **(b)** implies **(c)** and viceversa. Clearly **(b)** follows from **(c)**, let us prove the opposite. Let λ be the covector associated to γ. From **(b)** we have that $\lambda(0)$ is orthogonal to $G(0)$, hence γ switches iff $\theta^\gamma \in \{0, \pm\pi\}$. Thus **(c)** follows.
Proof that **(a)** implies **(d)**. It is a consequence of Proposition 18, p. 172
The implication **(d)**$\Rightarrow$**(c)** is obvious. This concludes the proof. ■

Proposition 21 *Let (γ, t_1) be a NTAE and $0 =: t_0 < t_1 < t_2 < ... < t_{n(\gamma)-1} < t_{n(\gamma)} := \tau$ the sequence of switching times. Suppose that for some $i \in \{0, 1, ..., n(\gamma) - 2\}$, $\Delta_A^{-1}(0)$ is inverse at the points $\gamma(t_i)$, $\gamma(t_{i+1})$. Then $\theta^\gamma(t_{i+1}) = \theta^\gamma(t_i)$.*

Proof. Set $(F + G)(\gamma(t_i)) = \alpha_i(F - G)(\gamma(t_i))$ and $F(\gamma(t_i)) = \beta_i G(\gamma(t_i))$. Under generic assumptions, α_i and β_i are well defined for each $i = 1, ...n(\gamma)-1$ and it holds:

$$\alpha_0 = -1, \quad \beta_0 = 0 \text{ and } \alpha_i, \beta_i \notin \{0, \pm 1\}, \quad \beta_i = \frac{1 + \alpha_i}{1 - \alpha_i}, \quad i \in \{1, ..., n(\gamma) - 1\}.$$

Now if $\Delta_A^{-1}(0)$ is inverse at both points $\gamma(t_i), \gamma(t_{i+1})$ $(i \in \{0, ..., n(\gamma) - 2\}$ then $\alpha_i, \alpha_{i+1} < 0$ and we have $\beta_i, \beta_{i+1} \in]-1, 1[$ (see Figure 4.10). Recalling the definition of θ^γ, from Proposition 19, p. 173 it follows the conclusion. ■

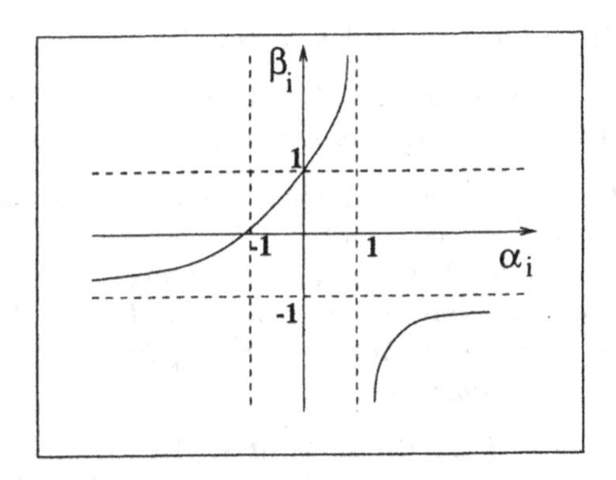

Fig. 4.10. Proof of Proposition 21

From Propositions 17, p. 171 and 20, p. 174, using the definitions of the times s_1^+, $s_1'^+$, t_f^+, s_1^-, $s_1'^-$, t_f^-, it follows:

Corollary 2 *Let γ be an extremal trajectory exiting the origin with control $+1$, then its first switching can occur on $\Delta_A^{-1}(0)$ only if $s_1 \neq 0$ (cond. A) or $s_1' \neq 0$ (cond. B) or $|\theta^+(t_f^+)| = \pi$ (cond. C). Moreover, at most one of the conditions A, B, C holds and the corresponding time is the first switching time of γ and the first time at which γ^+ intersect $\Delta_A^{-1}(0)$. A similar result holds for γ^- and for the times s_1^-, $s_1'^-$, t_f^-.*

Referring to conditions 1, 2, 3 of Remark 35, p. 90, conditions A and B correspond both to case 2 or case 1 (with $s_1^+ \neq 0$ or $s_1'^+ \neq 0$) and condition C corresponds to case 3. Moreover it is clear that for a NTAE (γ, t_1), t_1 is the first time at which γ reaches $\Delta_A^{-1}(0)$. In particular, if the trajectory exits the origin with control $+1$, we have $t_1 = s_1^+$ or $t_1 = s_1'^+$ or $t_1 = t_f^+$. In the case

$s_1^+ = 0$ and $s_1'^+$ is not defined or viceversa and $|\theta^+(t_f^+)| < \pi$ (that implies $t_f^+ = \tau$) there are not NTAE exiting the origin with control +1 (an abnormal extremal exists but it never switches). This case corresponds to case 1 of Remark 35, p. 90 with $s_1^+ = 0$ and s'^+ is not defined or viceversa. These observations are collected in the following:

Corollary 3 *There are at most two maximal NTAE. Moreover:*

(♠) one exits the origin with control $+1$ *and its first switching is at*
- s_1^+ *iff* $s_1^+ \neq 0$;
- $s_1'^+$ *iff* $s_1^+ \neq 0$;
- t_f^+ *iff* $|\theta^+(t_f^+)| = \pi$;

(♣) the other exits the origin with control -1 *and its first switching is at*
- s_1^- *iff* $s_1^- \neq 0$;
- $s_1'^-$ *iff* $s_1'^- \neq 0$;
- t_f^- *iff* $|\theta^-(t_f^-)| = \pi$.

Finally if $|\theta^\pm(t_f^\pm)| = \pi$ *then* $\Delta_A^{-1}(0)$ *is direct at* $\gamma(t_f^\pm)$.

The following two Propositions describe the position of the switching curves of the extremal strips whose borders are abnormal extremals.

Proposition 22 *Let* (γ, t_1) *be a NTAE,* t_i *and* t_{i+1} *two consecutive switching times,* S *an extremal strip such that* $\gamma \in \partial S$ *and* U^i, U^{i+1} *two sufficiently small neighborhoods of* $\gamma(t_i)$ *and* $\gamma(t_{i+1})$. *Moreover let* U^i_{in} *and* U^i_{out} *(resp.* U^{i+1}_{in}, U^{i+1}_{out}*) be the two connected components of* $U^i \setminus \Delta_A^{-1}(0)$ *(resp.* $U^{i+1} \setminus \Delta_A^{-1}(0)$*) chosen in such a way that* γ *enters* U^i_{in} *(resp.* U^{i+1}_{in}*). Under generic conditions we have the following cases:*

(1) $\theta^\gamma(t_i) = \theta^\gamma(t_{i+1})$ *and* $\Delta_A^{-1}(0)$ *direct at* $\gamma(t_i)$.
In this case if the switching locus of S *passing through* $\gamma(t_i)$ *lies in* U^i_{in} *(resp.* U^i_{out}*) then the switching locus of* S *passing through* $\gamma(t_{i+1})$ *lies in* U^{i+1}_{out} *(resp.* U^{i+1}_{in}*).*

(2) $\theta^\gamma(t_i) = \theta^\gamma(t_{i+1})$ *and* $\Delta_A^{-1}(0)$ *inverse at* $\gamma(t_i)$.
In this case if the switching locus of S *passing through* $\gamma(t_i)$ *lies in* U^i_{in} *(resp.* U^i_{out}*) then the switching locus of* S *passing through* $\gamma(t_{i+1})$ *lies in* U^{i+1}_{in} *(resp.* U^{i+1}_{out}*).*

(3) $\theta^\gamma(t_i) = \theta^\gamma(t_{i+1}) \pm \pi$ *and* $\Delta_A^{-1}(0)$ *direct at* $\gamma(t_i)$.
In this case we have the same conclusion as in case **(2)**.

(4) $\theta^\gamma(t_i) = \theta^\gamma(t_{i+1}) \pm \pi$ *and* $\Delta_A^{-1}(0)$ *inverse at* $\gamma(t_i)$.
In this case we have the same conclusion as in case **(1)**.

Proof. Let f_i (resp. A_i) be the sign of $-\Delta_B/\Delta_A$ (resp. Δ_A) on U^i_{in} and B_i the sign of Δ_B on U^i. By hypothesis, taking U^i sufficiently small, these quantities are well defined and we have $f_i = -A_i B_i$. Moreover, set $\theta_i = +1$ if $\theta^\gamma(t_i) = \theta^\gamma(t_{i+1}) \pm \pi$ and $\theta_i = -1$ if $\theta^\gamma(t_i) = \theta^\gamma(t_{i+1})$.

Claim $B_{i+1} = \theta_i B_i$

Proof of the Claim From $sgn(\dot{\theta}^\gamma(t)) = sgn(\Delta_B(\gamma(t)))$ we have that $\bar{v}^\gamma(t)$ (see formula (2.5)) is a vector rotating counterclockwise in U^i (resp. U^{i+1}) iff $B_i > 0$ (resp. $B_{i+1} > 0$). Recalling that $\theta_i, \theta_{i+1} \in \{0, \pm\pi\}$ it is clear that B_i and B_{i+1} have the same sign iff $\theta_{i+1} = \theta_i \pm \pi$.

Case 1 First suppose that Δ_A is direct at $\gamma(t_i)$. In this case from Proposition 18, p. 172 we clearly have $A_{i+1} = -A_i$. Now if the switching loci of $\mathcal{S}$ lie one in U^i_{in} and the other in U^{i+1}_{in} (resp. U^i_{out} and U^{i+1}_{out}), then from Lemma 11 we have $f_i = -f_{i+1}$. This occurs iff $-A_iB_i = +A_{i+1}B_{i+1} = -A_iB_i\theta_i$, from which it follows $\theta_i = +1$.

On the other hand, if the switching loci of $\mathcal{S}$ lie one in U^i_{in} and the other in U^{i+1}_{out} (resp. U^i_{out} and U^{i+1}_{in}), then $f_i = +f_{i+1}$ (that occurs iff $\theta_i = -1$).

Case 2 If Δ_A is inverse at $\gamma(t_i)$ we have $A_{i+1} = +A_i$. Now if the switching loci of $\mathcal{S}$ lie one in U^i_{in} and the other in U^{i+1}_{in} (resp. U^i_{out} and U^{i+1}_{out}) we have $f_i = -f_{i+1}$. This occurs iff $\theta_i = -1$.

On the other hand, if the switching loci of $\mathcal{S}$ lie one in U^i_{in} and the other in U^{i+1}_{out} (resp. U^i_{out} and U^{i+1}_{in}) then $f_i = +f_{i+1}$ (that occurs iff $\theta_i = +1$). ■

Proposition 23 *Let (γ, t_1) be a NTAE and let $\mathcal{S}^1$ and $\mathcal{S}^2$ be two extremal strips such that $\{\gamma\} = \mathcal{S}^1 \cap \mathcal{S}^2$. Let $\bar{t}$ be a switching time for γ and U a small neighborhood of $\gamma(\bar{t})$ such that $U \setminus \Delta_A^{-1}(0)$ has two connected components U_{in} and U_{out}, chosen in such a way that γ enters U from U_{in}. Then, under generic conditions, the switching loci of $\mathcal{S}^1$ and $\mathcal{S}^2$ passing through $\gamma(\bar{t})$ satisfy the following:*

(a) they both lie in U_{in} or in U_{out};
(b) they are tangent to $Supp(\gamma)$ in $\gamma(\bar{t})$.

Proof. of (a). By the analysis of the singularities at the first switching time (see Figure 4.11) we know that **(a)** is true in the special case $\bar{t} = t_1$. Using Proposition 22, p. 176 and by induction it follows the thesis.

Proof. of (b). See Section 3.2.7, p. 148 , Chapter 3. ■

4.3.2 Singularities

In this Section we describe all possible Frame Points occurring along a NTAE.

We start to describe the first singularity for the NTAE exiting the origin with control +1, the opposite case being similar. We refer to Figure 4.11 where all extremal trajectories are depicted in a neighborhood of the Frame Points. Following Corollary 3, a NTAE generates at time s_1^+ (iff $s_1^+ \neq 0$), or at time

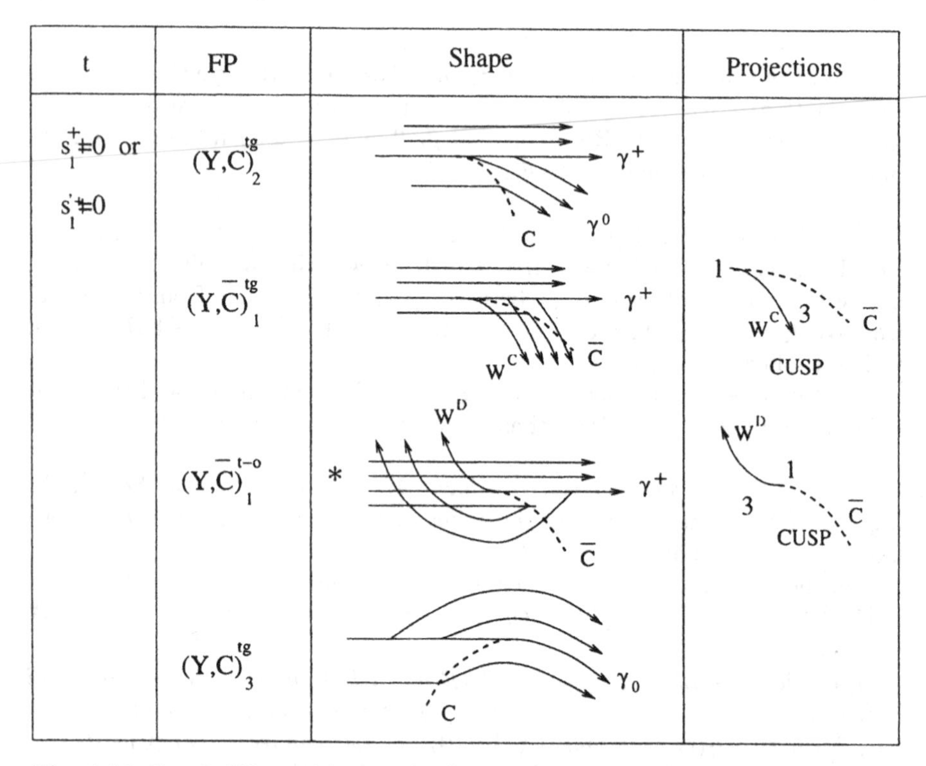

Fig. 4.11. Proof of Proposition 23, the first switching time of an abnormal extremal

$s_1'^+$ (iff $s_1'^+ \neq 0$), or at time t_f^+ (iff $|\theta^+(t_f^+)| = \pi$). Assume $s_1^+ \neq 0$, then a switching curve tangent to $Supp(\gamma^+)$ bifurcates from $\gamma^+(s_1^+)$. Recall formula (3.17) of Chapter 3. Using the definition of θ^+ and s_1^+, and reasoning as in Chapter 2, one gets that $b_1 > 0$, $d_1 < 0$ (we would have $b_1 < 0$, $d_1 > 0$, if $s_1^+ \neq 0$) and $c_0 < 1$. Hence the switching curve is bifurcating to the right of $Supp(\gamma^+)$. Moreover it follows that m, defined in formula (2.72), is bigger than 1 and being $b_1 > 0$ we have (in formula (3.17)) $x_1 > s_1^+$. If the switching curve is of kind C we call the singularity $(Y,C)_2^{tg}$. If the switching curve is of kind $\bar{C}$ and $\Delta_A^{-1}(0)$ is direct at $\gamma^+(s_1^+)$ we call the singularity $(Y,\bar{C})_1^{tg}$. Finally if the switching curve is of kind $\bar{C}$ and $\Delta_A^{-1}(0)$ is inverse at $\gamma^+(s_1^+)$ we call the singularity $(Y,\bar{C})_1^{t-o}$. These names are chosen in accordance with Chapter 2. The case $s_1'^+ \neq 0$ is entirely similar.

The case when the NTAE starts at $\gamma(t_f^+)$ (that happens iff $|\theta^+(t_f^+)| = \pi$), is again described by formula (3.17) with s_1^+ replaced by t_f^+. In this case, reasoning as in Chapter 2, Section 2.8.2, we get $c_0 > 1$ and $b_1 d_1 < 0$. It follows that the switching curve bifurcates to the right, $m < 1$, and in formula (3.17) we have $x_1 < t_f^+$. Moreover, at t_f^+, γ^+ stops to be extremal and $\Delta_A^{-1}(0)$ is direct at $\gamma(t_f^+)$ (see Proposition 21, p. 175). We call this singularity $(Y,C)_3^{tg}$.

To classify the other generic singularities involving a NTAE we consider at the Frame Points:

- if $\Delta_A^{-1}(0)$ is direct or inverse;
- if the switching happens in U_{in} or U_{out}, according to Proposition 22;
- all the essentially different directions of the exiting abnormal trajectory.

We obtain 24 types of singularities. The singularities with entering abnormal extremals of kind γ_0 are showed in Figure 4.12, while in Figure 4.13 all the possible singularities for a NTAE of the kind W^C, and W^D are listed. In these figures we also indicate the labels *fold, cusp, bifold* or *ribbon* in accordance with Definition 55, p. 163.

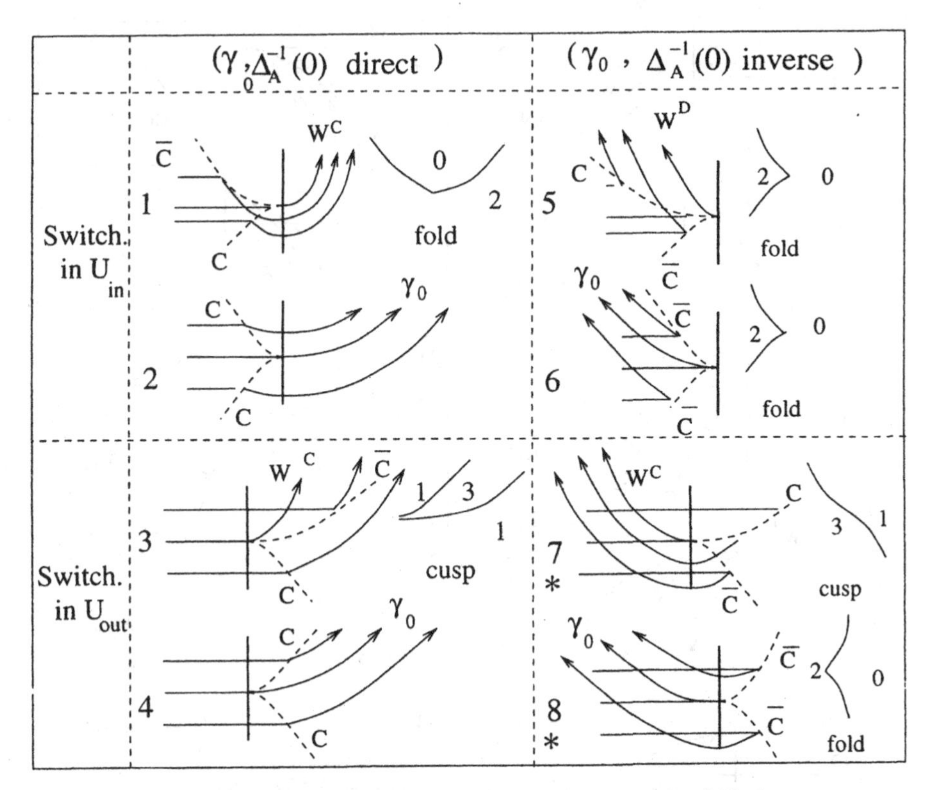

Fig. 4.12. Singularities along abnormal extremals of kind γ_0

In particular call $\bar{x} := \gamma(\bar{t})$ the singular point and let $I \in \{X, Y\}$ be the vector field such that γ is a I–trajectory in $[\bar{t} - \varepsilon, \bar{t}]$ for some $\varepsilon > 0$. Moreover choose a local system of coordinates in a small neighborhood U of $\bar{x}$ such that $\bar{x} = (0,0)$, $I = (1,0)$ and $\Delta_A^{-1}(0) \cap U = \{(x,y) \in \mathbb{R}^2 \cap U : \; x = 0\}$.

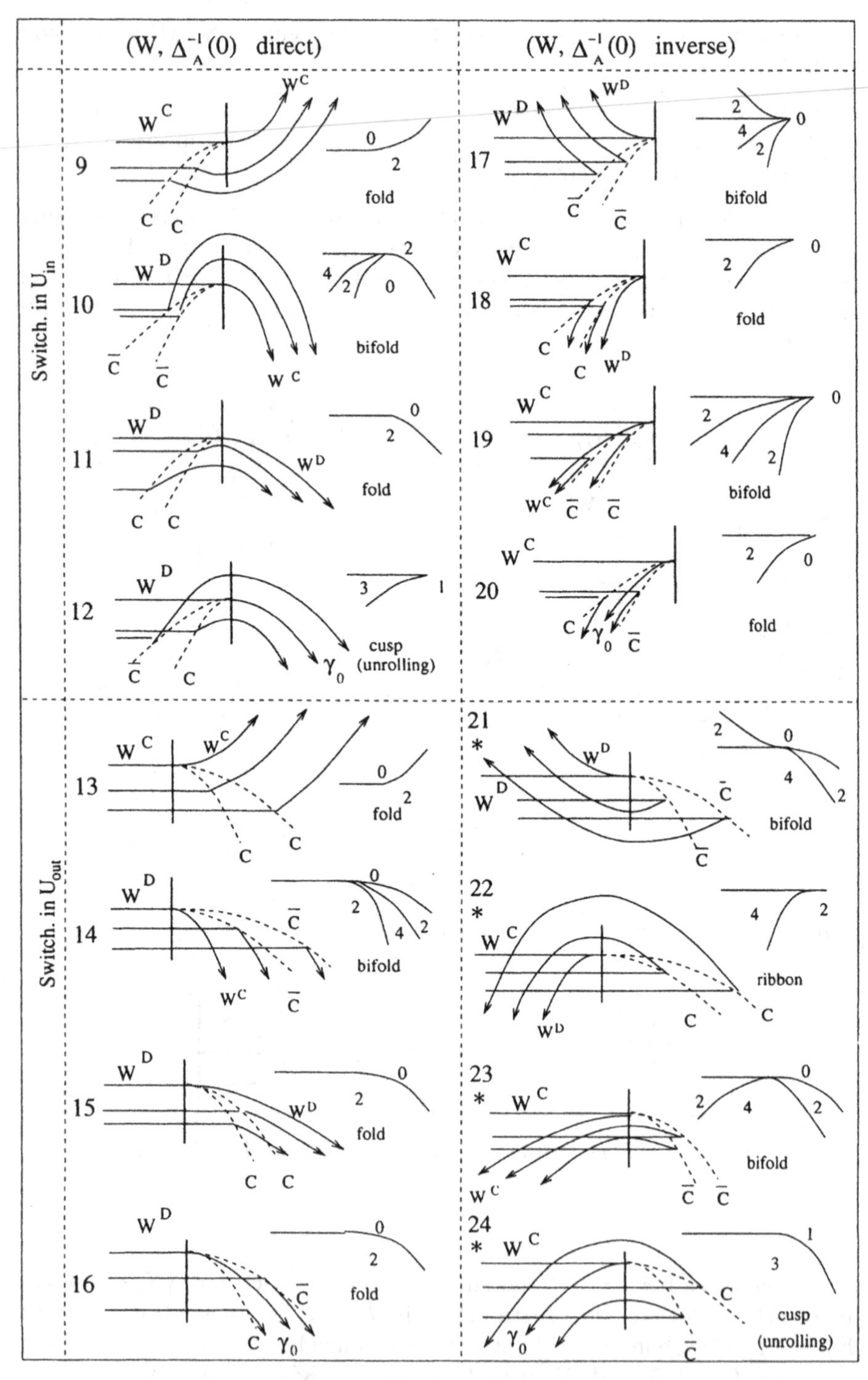

Fig. 4.13. Singularities along abnormal extremals of kind W^C and W^D

In this way (referring to Theorem 22, p. 176) we have:

$$U_{in} = \{(x,y) \in \mathbb{R}^2 \cap U : \quad x \leq 0\}$$
$$U_{out} = \{(x,y) \in \mathbb{R}^2 \cap U : \quad x \geq 0\}.$$

First suppose that γ is a strip border of kind γ_0, $\Delta_A^{-1}(0)$ is direct at $\bar{x}$ and the switching loci of the two strips $\mathcal{S}^1$ and $\mathcal{S}^2$ such that $\{\gamma\} = \mathcal{S}^1 \cap \mathcal{S}^2$ lie in U_{in}. Under these hypotheses $Supp(\gamma) \cap U_{out}$ has equation:

$$x_2 = ax_1^2 + O(x_1^3) \quad \text{for } x_1 > 0$$

and the switching loci of $\mathcal{S}^1$ and $\mathcal{S}^2$ are represented by the following equation.

$$x_2 = bx_1^2 + O(x_1^3)$$
$$x_2 = cx_1^2 + O(x_1^3)$$
$$b > 0, c > 0, x < 0.$$

We assume:

(GA10) a, b, c are distinct and different from zero.

Suppose $a > 0$, the other case being similar. If $a > b$ then for some $\varepsilon > 0$ the trajectory $\gamma|_{[\bar{t},\bar{t}+\varepsilon]}$ is an extremal strip border of kind W^C. In this case we say that the singularity is of type 1 (see figure 2.6). On the other hand, if $a < b$, then for some $\varepsilon > 0$ the trajectory $\gamma|_{[\bar{t},\bar{t}+\varepsilon]}$ is an extremal strip border of kind γ_0. In this case we say that the singularity is of type 2 (see figure 2.6). The other cases can be treated similarly under similar generic assumptions and these singularities include the singularities at t_1. From the above assumptions we have the following:

Proposition 24 *Let (γ, t_1) be a NTAE and $t_1 < t_2 < \dots < t_{n(\gamma)-1}$ the sequence of its switching times. Set $t_0 = 0$ and $t_{n(\gamma)} = \tau$. Under generic conditions, for every $i = 1, \dots, n(\gamma) - 1$, we have the following. If $\gamma|_{[t_{i-1},t_i]}$ is an extremal strip border of type W^C, W^D or γ_0, then $\gamma|_{[t_i,t_{i+1}]}$ is also a strip border of type W^C, W^D or γ_0 (but not necessarily of the same type!).*

By induction one gets:

Proposition 25 *Under generic conditions, the singularities of Figure 4.11, 4.12, 4.13 include all the possible singularities along a NTAE.*

The proof of the fact that all these singularities are in fact realized, is in the next Chapter.

4.4 Extremal Strips Evolution and Frame Points

In this Section we complete the study of strips evolution under generic assumptions and classify all possible Frame Points that are generated as well as the new Frame Curves.

To study extremal strip borders we have to consider the evolution of γ_0 and W Frame Curves, while for the interior of extremal strips the presence of turnpikes and the violation of condition **iii)** (of Definition 56, p. 167) generate new Frame Curves and cause the division of strips in smaller strips.

As a byproduct we obtain a complete classification of all possible stable Frame Points of the extremal synthesis.

4.4.1 Frame Curves and Frame Points Along C and $\bar{C}$

We first prove a theorem used in the following.

Theorem 34 *Consider an extremal trajectory $\bar{\gamma}$ that switches on $\Delta_B^{-1}(0)$ at time $\bar{t}$. Assume $\Delta_A(\bar{\gamma}(\bar{t})) \neq 0$ and that $X(\bar{\gamma}(\bar{t})) \cdot \Delta_B^{-1}(0)(\bar{\gamma}(\bar{t}))$ and $Y(\bar{\gamma}(\bar{t})) \cdot \Delta_B^{-1}(0)(\bar{\gamma}(\bar{t}))$ are both positive or both negative. Under generic assumptions there exists $\varepsilon > 0$ such that every extremal trajectory γ satisfying $\|\gamma - \bar{\gamma}\|_{C^0} < \varepsilon$ switches before crossing $\Delta_B^{-1}(0)$ iff it switches after crossing $\Delta_B^{-1}(0)$.*

Proof. From the assumptions of the statement we have that the set $\Delta_B^{-1}(0)$ is not a turnpike (indeed it is a barrier, see [126]) in a neighborhood of $\bar{\gamma}(\bar{t})$). Moreover, if ε is sufficiently small then every γ satisfying $\|\gamma - \bar{\gamma}\|_{C^0} < \varepsilon$ crosses the set $\Delta_B^{-1}(0)$ near $\bar{\gamma}(\bar{t})$ and we can define the function $\theta_\gamma(t) = arg(G(\gamma(\bar{t})), v(G(\gamma(t)), t; \bar{t}))$. Notice that θ^γ is the same as θ_γ for $\bar{t} = 0$. By definition $\theta_{\bar{\gamma}}(\bar{t}) = 0$ and from Lemma 8, p. 43, we have that there exists $\delta > 0$ such that $\theta_{\bar{\gamma}}$ is monotone increasing (decreasing), in $[\bar{t} - \delta, \bar{t}]$ and monotone decreasing (increasing) in $[\bar{t}, \bar{t} + \delta]$. For ε sufficiently small θ_γ has the same behavior, i.e. there exists t_γ such that $|t_\gamma - \bar{t}| < (\delta/4)$, θ_γ is monotonically increasing (decreasing) in $[\bar{t} - (3\delta/4), t_\gamma]$ and monotone decreasing (increasing) in $[t_\gamma, \bar{t} + (3\delta/4)]$, and $|\theta_\gamma(\bar{t} \pm (3\delta/4)) - \theta_{\bar{\gamma}}(\bar{t} \pm (3\delta/4))| < (|\theta_{\bar{\gamma}}(\bar{t} \pm (3\delta/4))|/4)$. (Notice that t_γ is precisely the time at which γ crosses the set $\Delta_B^{-1}(0)$.) Assume now that γ switches at a point s_γ near $\bar{t}$. Then $\lambda_\gamma(s_\gamma) \cdot G(\gamma(s_\gamma)) = 0$ (here λ_γ denotes the covector associated to γ). Generically, $s(\gamma) \neq t(\gamma)$ and if ε is sufficiently small then $|\theta_\gamma(s_\gamma)| < \min\{|\theta_\gamma(\bar{t} - (3\delta/4))|, |\theta_\gamma(\bar{t} + (3\delta/4))|\}$. Thus there exists s'_γ such that $\theta_\gamma(s'_\gamma) = \theta_\gamma(s_\gamma)$. This implies that s'_γ is another switching time for γ. Indeed $\lambda_\gamma(t) \cdot G(\gamma(t)) = \lambda_\gamma(\bar{t}) \cdot v(G(\gamma(t)), t; \bar{t})$ and the vectors $v(G(\gamma(s_\gamma)), s_\gamma; \bar{t})$ and $v(G(\gamma(s'_\gamma)), s'_\gamma; \bar{t})$ are parallel, hence from $\lambda_\gamma(s_\gamma) \cdot G(\gamma(s_\gamma)) = 0$ we deduce $\lambda_\gamma(s'_\gamma) \cdot G(\gamma(s'_\gamma)) = 0$. Moreover, from $s'_\gamma \neq t_\gamma$ we conclude $\frac{d}{ds}\lambda_\gamma(s) \cdot G(\gamma(s))|_{s=s'_\gamma} \neq 0$. ■

Now we describe all possible Frame Points occurring in an extremal strip. We start analyzing Frame Points on C and $\bar{C}$ curves. To this purpose, let x be a point of a Frame Curve F of kind C or $\bar{C}$.

Proposition 26 *Consider a Frame Curve F of C or $\bar{C}$ type, a point $x \in F$ and denote by $\dot{F}$ the tangent unit vector to the curve F. If $\Delta_A(x) \cdot \Delta_B(x) \neq 0$, $X(x) \wedge \dot{F}(x) \neq 0$ and $Y(x) \wedge \dot{F}(x) \neq 0$ then x is not a Frame Point.*

Proof. For each point y in a neighborhood of x we indicate by (γ_y, λ_y) the extremal pair reaching y at time t_y. If $\theta_y(t) = arg(G(y), v(G(\gamma_y(t)), t; t_y)$ then $sgn(\dot{\theta}_y(t_y)) = sgn(\Delta_B(y))$. Hence, by continuity $sgn(\dot{\theta}_y(t)) \neq 0$ for (y,t) in a neighborhood of (x, t_x) and all trajectories γ_y switch. Finally, the nontangency assumptions for the vector fields X and Y ensure that Frame Curve F does not change type (from C to $\bar{C}$ or viceversa) and thus x is not a Frame Point. ■

Since abnormal extremals have been already considered in the previous Section, we assume $\Delta_A(x) \neq 0$. Hence if x is a Frame Point on the curve F, then we have one of the following cases:

i) $\Delta_B(x) = 0$,
ii) $X(x) \wedge \dot{F}(x) = 0$ or $Y(x) \wedge \dot{F}(x) = 0$,

under the generic assumption:

(GA11) i) and **ii)** are mutually exclusive.

Case i) Assume i) happens and denote by T_B the unit tangent vector to the set $\Delta_B^{-1}(0)$. We choose a system of local coordinates so that $x = (0,0)$, $\dot{F} = (0,1)$ and $F \subset \{(x,y) : y \leq 0\}$. Let us denote by $I \in \{X,Y\}$ and $O \in \{X,Y\}$ the vector fields such that the extremal trajectories enter F as I-trajectories and exit F as O-trajectories.

If F is of type C then without loss of generality we assume that $I_1(x) \geq 0$ and $O_1(x) \geq 0$. We have two generic cases:

a) $I_2(x) > O_2(x)$,
b) $O_2(x) > I_2(x)$

under the generic assumption:

(GA12) $I_2(x) \neq O_2(x)$.

Let $K =: \{aX(x) + bY(x) : a, b \in \mathbb{R}\}$ be the cone generated by $X(x)$ and $Y(x)$ (i.e. by $I(x)$ and $O(x)$). For $T_B(x)$ we have three possibilities: 1) $\pm T_B(x) \in K$; 2) $\pm T_B(x) \notin K$, $T_B(x)$ or $-T_B(x)$ is in the first orthant; 3) $\pm T_B(x) \notin K$, $T_B(x)$ or $-T_B(x)$ is in the second orthant, and the generic assumption:

(GA13) $\pm T_B \notin \partial K, \quad \pm T_B \cdot e_1 \neq 0, \quad \pm T_B \cdot e_2 \neq 0.$

In case a1) a turnpike generates from x and x is a $(C,S)_1$ Frame Point. In cases a2) and a3) applying Theorem 34 we obtain that the extremal trajectories must switch before (case 2)) or after (case 3)) crossing F. Thus a second switching points curve generates at x and x is a $(C,C)_1$ point. Moreover, in both cases the I-trajectory trough x is a γ_0 Frame Curve.

The case b1) is impossible (indeed the extremal trajectories would not switch). In cases b2) and b3) applying again Theorem 34 we obtain that the extremal trajectories must switch before (case 2)) or after (case 3)) crossing F. Thus a second switching points curve generates at x. This new Frame

Curve can be of type C or $\bar{C}$. In case b2), with a new C curve generated, we have that the point x is of type $(C,C)_1$ and the I-trajectory trough x is a γ_0 Frame Curve. While if the new curve is a $\bar{C}$ Frame Curve then we say that x is a $(C,\bar{C})_2$ point, the I-trajectory trough x is a γ_0 Frame Curve before x and a W Frame Curve after x, and x is a fold point. The case b3) with a new curve of type C is entirely similar to case b2) with a new C curve. While if the new curve is of type $\bar{C}$ then we say that the point x is of type $(C,\bar{C})_1$, the I-trajectory trough x is a γ_0 Frame Curve before x and a W Frame Curve after x, and x is a cusp point. All possibilities are depicted in Figure 4.14.
The case in which F is a $\bar{C}$ Frame Curve can be treated similarly. We have four possible cases and Frame Points of type $(C,\bar{C})_2$, $(\bar{C},S)_2$, $(C,\bar{C})_1$ and $(\bar{C},S)_1$ respectively. All cases are shown in Figure 4.15.
Case ii). Necessarily the vector $O(x)$ is tangent to F at x. We have two possible subcases under the generic condition:

(GA14) there exists a neighborhood U of x such that the forward O-trajectory from x is contained either in the connected component of $U \setminus F$ where $I(x)$ points into or in the other connected component of $U \setminus F$.

In the former case of **(GA14)** we say that x is a $(C,\bar{C})_3$ Frame Point, the extremal trajectory through x is a γ_0 Frame Curve before x and a W Frame Curve after, and x is a fold point. In the latter case we call x a $(C,\bar{C})_4$ Frame Point, again the extremal trajectory through x is a γ_0 Frame Curve before x and a W Frame Curve after, while x is a cusp point. In Figure 4.16 are listed the possibilities.

4.4.2 Frame Curves and Points Along γ_0

A Frame Curve F of type γ_0 that is an extremal strip border generically does not enter a turnpike (even it may cross the set $\Delta_B^{-1}(0)$). Then if the trajectory γ_0 does not switch, there is no Frame Point. Assume now that x is a switching point for γ_0. Then on each side of F a C curve is generated. Let us call them C_1 and C_2. These two curves curves intersect at x but generically we have:

(GA15) $\dot{C}^1(x)$ and $\dot{C}^2(x)$ are not parallel at x.

Let us denote by $I \in \{X,Y\}$ and $O \in \{X,Y\}$ the vector fields such that the extremal trajectory corresponding to F reaches x as I-trajectory and leaves from x as O-trajectories. We choose a local system of coordinates so that $x = (0,0)$, $C^1 = \{(x,y) : y = 0, x \leq 0\}$, $I(x) = (0,1)$ and $C^2 \subset \{(x,y) : x \geq 0\}$. We distinguish two cases:

i) $\dot{C}^2$ or $-\dot{C}^2$ is in the second orthant,
ii) $\dot{C}^2$ or $-\dot{C}^2$ is in the first orthant.

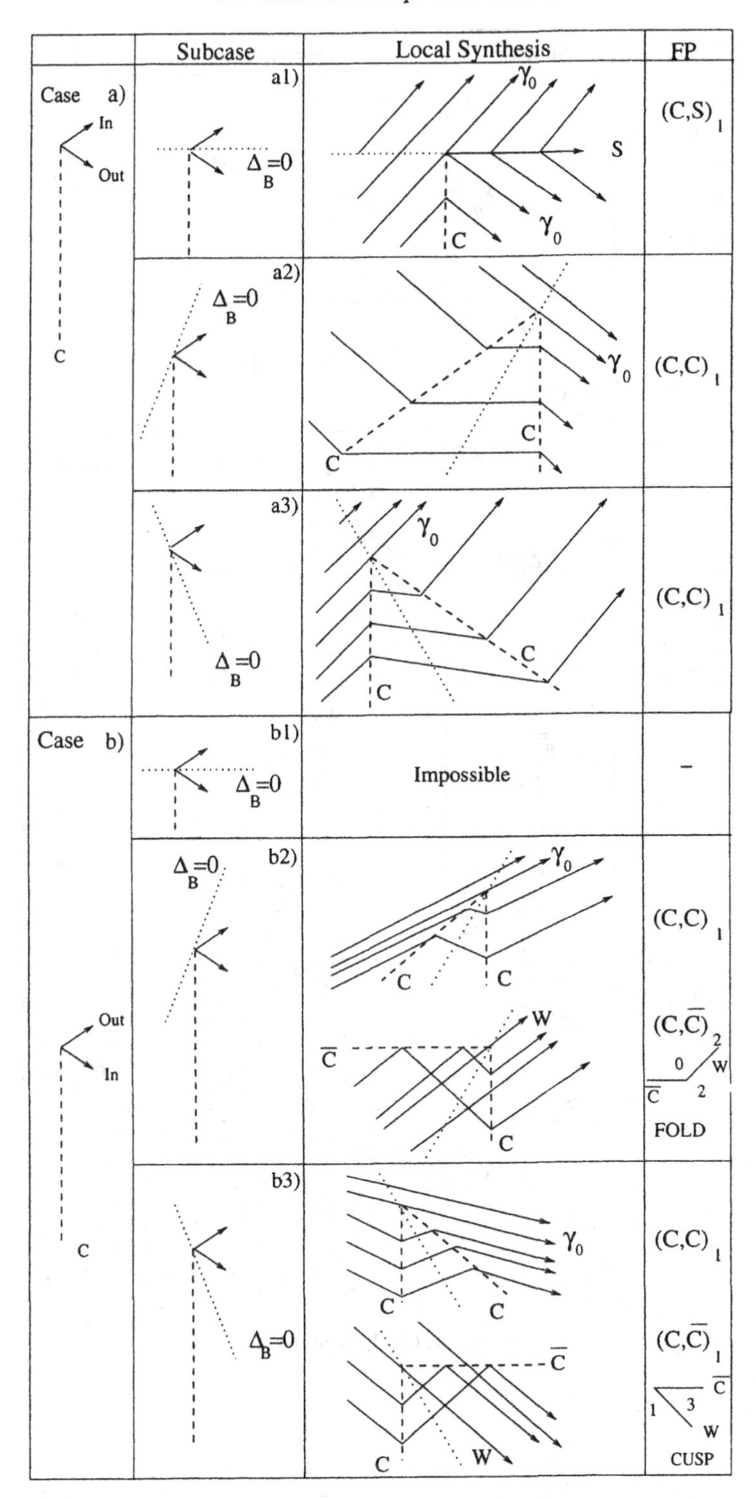

Fig. 4.14. FPs along C

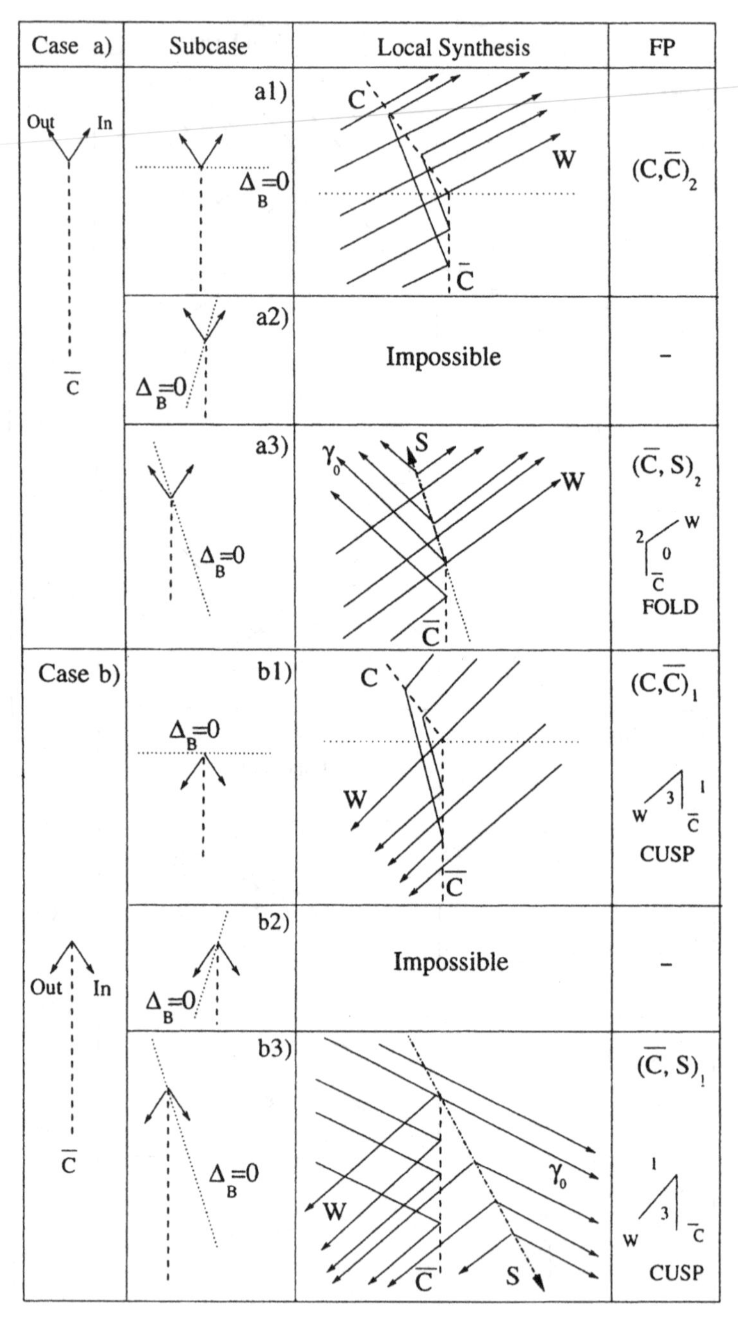

Fig. 4.15. FPs along $\bar{C}$

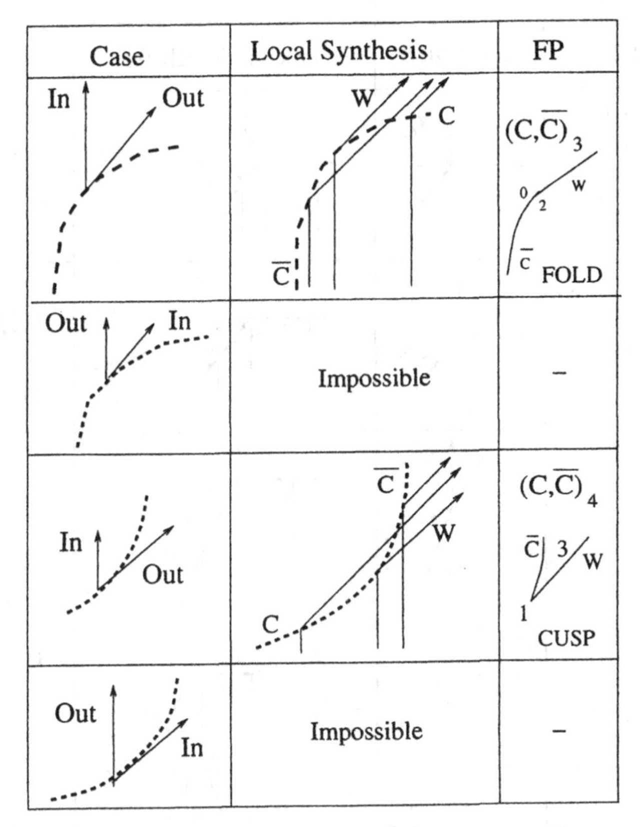

Fig. 4.16. FPs along C and $\bar{C}$ (tangent cases).

Then there exists a small neighborhood U of x such that if r_1 and r_2 are the lines through the origin with tangent vectors $\dot{C}_1(x)$ and $\dot{C}_2(x)$ respectively, then $U \setminus (r_1 \cup r_2)$ has four connected components $U_1, \ldots, U_4$. There are four generic subcases for $O(x)$ lying in one of the regions $U_1, \ldots, U_4$. All the possibilities are shown in Figure 4.17.

4.4.3 Frame Curves and Points Along W

Let us now consider the Frame Points along a Frame Curve F of type W. Also in this case we have a Frame Point only when the trajectory corresponding to W switches. Then assume that x is such a switching point. There exists two switching curves C_1 and C_2 generating at x with generically distinct tangent vectors $\dot{C}^1(x)$ and $\dot{C}^2(x)$ at x. In this case the two curves lie on the same side of W. Again, let us denote by $I \in \{X, Y\}$ and $O \in \{X, Y\}$ the vector fields such that the extremal trajectory corresponding to F reaches x as I-

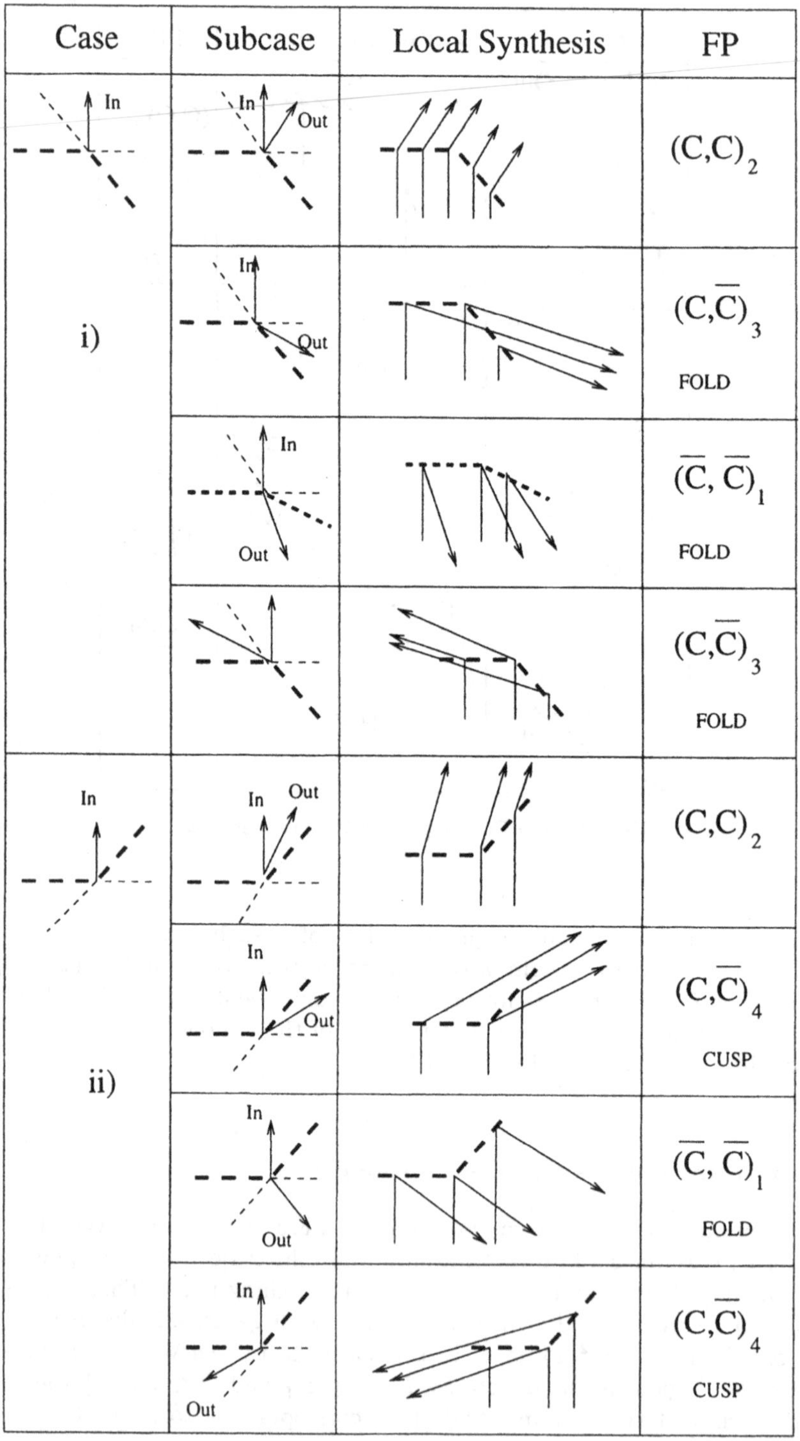

Case	Subcase	Local Synthesis	FP
In	In Out		$(C,C)_2$
i)	In Out		$(C,\overline{C})_3$ FOLD
	In Out		$(\overline{C},\overline{C})_1$ FOLD
			$(C,\overline{C})_3$ FOLD
In	In Out		$(C,C)_2$
ii)	In Out		$(C,\overline{C})_4$ CUSP
	In Out		$(\overline{C},\overline{C})_1$ FOLD
	In Out		$(C,\overline{C})_4$ CUSP

Fig. 4.17. FPs along γ_0.

trajectory and leaves from x as O-trajectories. We choose a local system of coordinates so that $x = (0,0)$, $I(x) = (1,0)$ and the extremal trajectories near F are contained in the set $\{(x,y) : y \geq 0\}$. We have two generic cases:

i) $O_2(x) > 0$,
ii) $O_2(x) < 0$,

under the generic condition:

(GA16) $O_2(x) \neq 0$.

There exists a small neighborhood U of x such that if r is the line through the origin parallel to $O(x)$ then the set $(U \cap \{(x,y) : y \geq 0\}) \setminus r$ in case i) and $(U \cap \{(x,y) : y \leq 0\}) \setminus r$ in case ii) has two connected components U_1 and U_2. Therefore, we have three subcases according to the fact that the vectors tangent to the curves C_1 and C_2 (or their opposite) both point into U_1, both point into U_2 or point to different connected components. All possible Frame Points are illustrated in Figure 4.18.

4.5 Existence of an Extremal Synthesis

In this Section we prove that the algorithm of Section 4.2, p. 167 gives rise to an extremal synthesis. First we describe the extremal locus near the origin and $\gamma^{\pm}$. Then we consider the possible intersections of lift of extremal strips in the cotangent bundle. Finally we prove Theorem 30, p. 155.

Observe that for every point $x \in Supp(\gamma^+) \cup Supp(\gamma^-)$ there are more than one p's such that $(x,p) \in N$.

For every $t > 0$ we have $\{(x,p) \in N : x = \gamma^+(t)\} = [\theta_1(t), \theta_2(t)]$, where the segment is taken in counterclockwise direction on S^1. We define two curves in $\mathbb{R}^2 \times S^1$ both projecting on γ^+ by $\gamma^{1+}(t) := (\gamma^+(t), \theta_1(t))$, $\gamma^{2+}(t) := (\gamma^+(t), \theta_2(t))$. We do the same for γ^-. The FP (X,Y) on the plane (see Figure 2.9, p. 61, Chapter 2) corresponds to a couple of FPs in the cotangent bundle called $(\gamma^{1+}, \gamma^{2-})$, $(\gamma^{2+}, \gamma^{1-})$. This can be easily understood from Figure 4.3 and 4.4 that show N in a neighborhood of the origin and a subset of $\mathbb{R}^2$ to which it is diffeomorphic. The others FPs on the curve γ^+ on the plane correspond to some FPs of N on the curves γ^{1+} or γ^{1-} according to the direction toward which the corresponding FC bifurcates from γ^+.

Definition 64 *Let Bad_N be the set of points $(x,p) \in N$ (we recall that $N = \pi_*(\mathcal{N})$) such that there exist two times $t_1, t_2 \leq \tau$, $d \in [0, \min(t_1,t_2)[$, $\varepsilon > 0$ and two extremal pairs (γ_1, λ_1), (γ_2, λ_2) such that:*

(1) $\gamma_1(0) = \gamma_2(0) = 0$, $\gamma_1(t_1) = \gamma_2(t_2) = x$, $\lambda_1(t_1) = \lambda_2(t_2) = p \in S^1 \subset \mathbb{R}^2$

(2) *γ_1 and γ_2 correspond to the same control in respectively $[t_1 - d, t_1]$, and $[t_2 - d, t_2]$;*

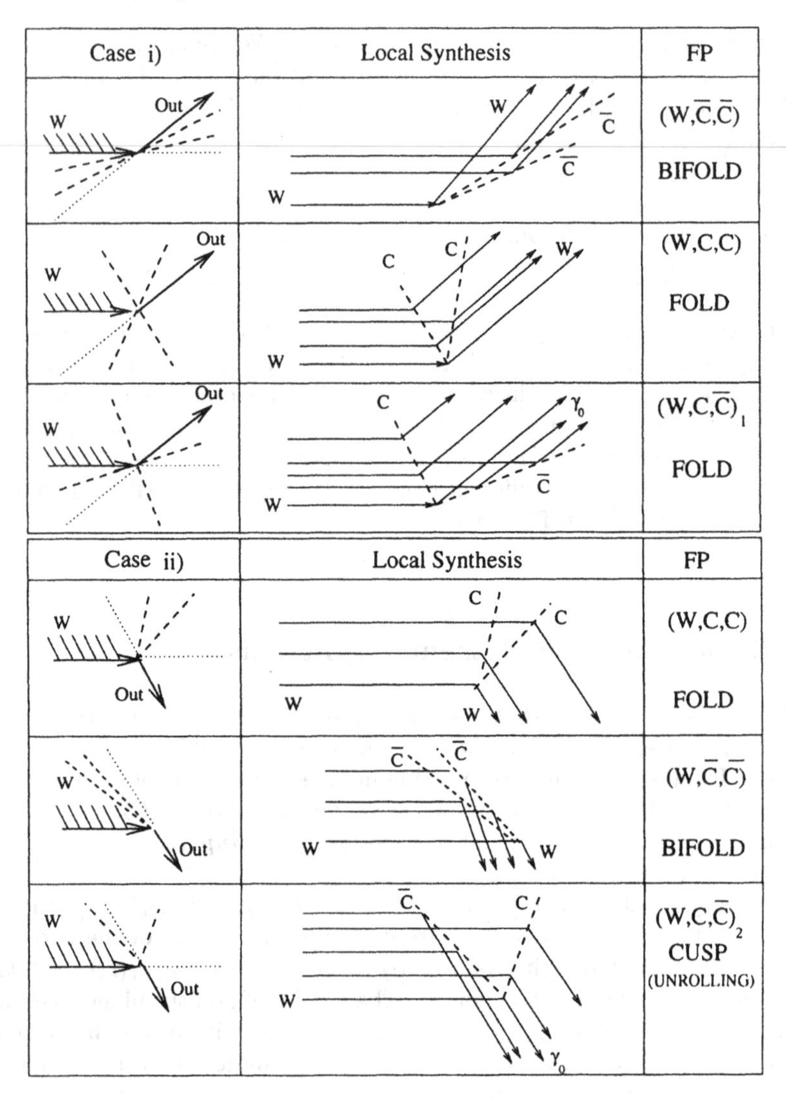

Fig. 4.18. FPs along W (first and second case).

(3) γ_1 *corresponds to the singular control* φ *in* $[t_1 - d - \varepsilon, t_1 - d]$;
γ_2 *corresponds to the constant control +1 or to the constant control -1 in* $[t_1 - d - \varepsilon, t_1 - d]\}$;

The projection of Bad_N on $\mathcal{R}(\tau)$ is showed in Figure 4.19. The solid lines and the dashed lines describe two different strips. Figure 4.20 shows an example of fold entering a turnpike in two different points, hence creating a Bad_N region.

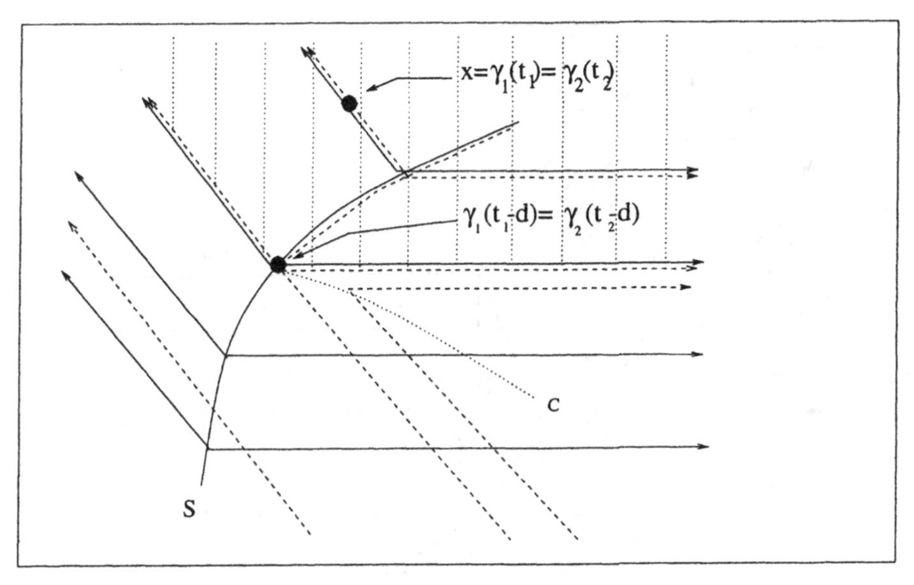

Fig. 4.19. The projection of Bad_N is the dotted region.

Theorem 35 *Let (x,p) belong to N, then generically there exists a unique extremal pair $(\gamma, \lambda) : [0,t] \to \mathbb{R}^4$ such that $\gamma(0) = 0,\ \gamma(t) = x,\ \lambda(t) = p \in S^1$ for some $t \leq \tau$ iff $(x,p) \notin Bad_N$.*

Proof. By definition, the uniqueness implies $(x,p) \notin Bad_N$.
Let us now prove that generically $(x, \lambda) \notin Bad_N$ implies the uniqueness of (γ, λ). Assume by contradiction that $(x,p) \notin Bad_N$ and suppose that there exist two different extremal pairs (γ_1, λ_1), (γ_2, λ_2) such that $\gamma_1(0) = \gamma_2(0) = 0$, $\gamma_1(t_1) = \gamma_2(t_2) = x$, $\lambda_1(t_1) = \lambda_2(t_2) = p$ for some $t_1,\ t_2 \leq \tau$. Let $A := \{t \geq 0 : \ \ \gamma_1(t_1 - t) = \gamma_2(t_2 - t),\ \lambda_1(t_1 - t) = \lambda_2(t_2 - t)\}$ and A_c the connected component of A containing 0. Notice that A_c is closed and let $d := \max A_c$. If $d = \min(t_1, t_2)$ (that implies $t_1 = t_2$) then $(\gamma_1, \lambda_1) \equiv (\gamma_2, \lambda_2)$ and there is nothing to prove. Otherwise set:

$$\bar{t}_1 = t_1 - d, \quad \bar{t}_2 = t_2 - d,$$
$$\bar{x} = \gamma_1(\bar{t}_1) = \gamma_2(\bar{t}_2),$$
$$\bar{\lambda} = \lambda_1(\bar{t}_1) = \lambda_2(\bar{t}_2).$$

Claim: $\bar{\lambda} \cdot G(\bar{x}) = 0.$

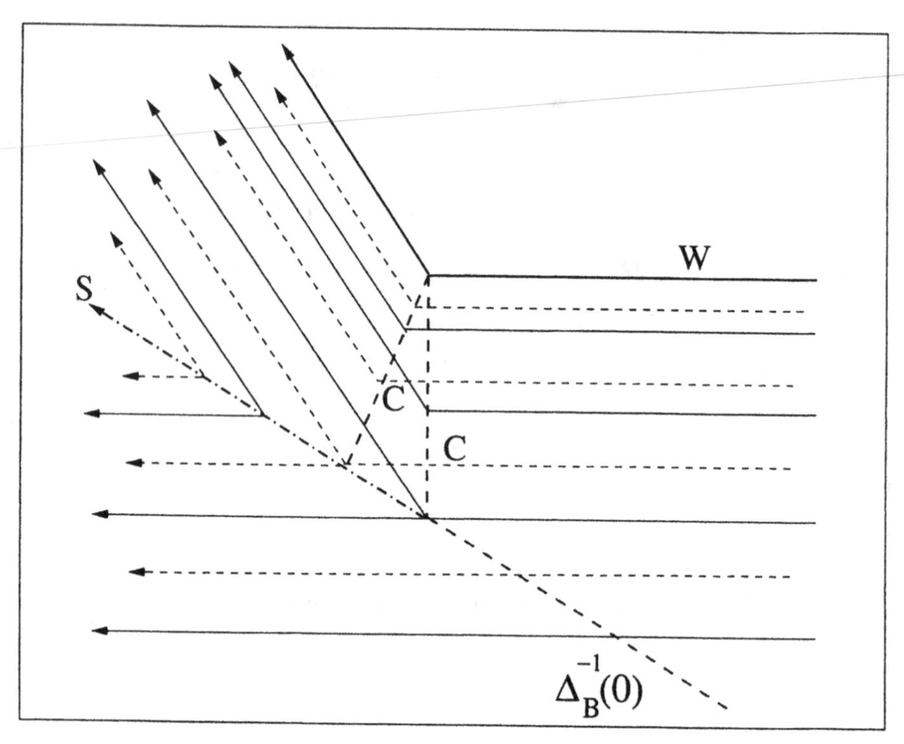

Fig. 4.20. A local example of extremal synthesis creating a Bad_N region. A fold is switching and entering a turnpike in two different points.

Proof of the Claim let u_1 and u_2 be the controls corresponding to γ_1 and γ_2. There exists $\varepsilon > 0$ such that one of the following cases occurs:

(a) $u_1 = \varphi$ on $[\bar{t}_1 - \varepsilon, \bar{t}_1]$, $u_2 = +1$ or $u_2 = -1$ on $[\bar{t}_2 - \varepsilon, \bar{t}_2]$ or viceversa;
(b) $u_1 = +1$ on $[\bar{t}_1 - \varepsilon, \bar{t}_1]$, $u_2 = -1$ on $[\bar{t}_2 - \varepsilon, \bar{t}_2]$, or viceversa.

In case **(a)** from $u_1 = \varphi$ in $[\bar{t}_1 - \varepsilon, \bar{t}_1]$ we have $\bar{\lambda} \cdot G(\bar{x}) = 0$. In case **(b)** (say with $u_1 = +1$) from the PMP we have:

$$\lambda_1(t) \cdot (F + G)(\gamma_1(t)) = \max_{|w| \le 1} \lambda_1(t) \cdot (F + wG)(\gamma_1(t)) \text{ for a.e. } t \in [\bar{t}_1 - \varepsilon, \bar{t}_1],$$
$$\lambda_2(t) \cdot (F - G)(\gamma_2(t)) = \max_{|w| \le 1} \lambda_2(t) \cdot (F + wG)(\gamma_2(t)) \text{ for a.e. } t \in [\bar{t}_2 - \varepsilon, \bar{t}_2],$$

so we have:

$$\lambda_1(t) \cdot G(\gamma_1(t)) \ge 0 \text{ for a.e. } t \in [\bar{t}_1 - \varepsilon, \bar{t}_1],$$
$$\lambda_2(t) \cdot G(\gamma_2(t)) \le 0 \text{ for a.e. } t \in [\bar{t}_2 - \varepsilon, \bar{t}_2].$$

Since all the functions are continuous, we get $\bar{\lambda} \cdot G(\bar{x}) = 0$ that proves the Claim.

Now since $(x,p) \notin Bad_N$ then, condition (3) of Definition 64 must fail. This means that we are in situation **(b)** of the proof of the Claim. Hence $(\gamma_1, \lambda_1)|_{[0,\bar{t}_1]}$ and $(\gamma_2, \lambda_2)|_{[0,\bar{t}_2]}$ are two different extremal pairs reaching $\bar{x}$, say with controls +1 and -1 respectively, and satisfying:

$$\lambda_1(\bar{t}_1) \cdot G(\gamma_1(\bar{t}_1)) = 0, \tag{4.12}$$

$$\lambda_2(\bar{t}_2) \cdot G(\gamma_2(\bar{t}_2)) = 0. \tag{4.13}$$

Generically $\Delta_A(x) \neq 0$ otherwise γ_1 and γ_2 would be two NTAE, see Section 4.3, p. 171. We have the following cases:

Case 1 $\bar{x} \notin \Delta_B^{-1}(0)$. In this case $\bar{t}_1$ is a switching from Y to X for γ_1 and $\bar{t}_2$ is a switching from X to Y for γ_2 (hence generically $d = 0$). Define $v^i(t) := v^{\gamma_i}(G(\gamma_i(t)), t; \bar{t}_i)$ $(i = 1, 2)$ (cfr. Definition 14, p. 37) and the functions:

$$\alpha^1(t) := arg(G(\bar{x}), v^1(t))$$
$$\alpha^2(t) := arg(G(\bar{x}), v^2(t)).$$

Now $\alpha^i(t) = \theta^{\gamma_i}(t) + A^i$ $(i = 1, 2)$ where A^1, A^2 are real numbers. Since $\Delta_B(\bar{x}) \neq 0$, by Proposition 8 (Chapter 2) there exists $\varepsilon > 0$ such that $sgn(\dot{\alpha}^1(\bar{t}_1 + h)) = sgn(\dot{\alpha}^2(\bar{t}_2 + h))$ for each $h \in [-\varepsilon, \varepsilon]$. Moreover we can choose ε small enough such that:

C1) for each $t \in [\bar{t}_1 - \varepsilon, \bar{t}_1]$, $\lambda_1(t) \cdot G(\gamma_1(t)) = \lambda_1(\bar{t}_1) \cdot v^1(t) \geq 0$ where the equality holds iff $t = \bar{t}_1$;

C2) for each $t \in [\bar{t}_2 - \varepsilon, \bar{t}_2]$, $\lambda_2(t) \cdot G(\gamma_2(t)) = \lambda_2(\bar{t}_2) \cdot v^2(t) \leq 0$ where the equality holds iff $t = \bar{t}_2$.

Now $\lambda_1(\bar{t}_1) = \lambda_2(\bar{t}_2)$ and v^1, v^2 are two vector functions satisfying $v^1(\bar{t}_1) = v^2(\bar{t}_2)$ and rotating in the same direction. This contradicts **C1)** or **C2)**.

Case 2 $\bar{x} \in \Delta_B^{-1}(0)$ and X, Y point to the same side of $\Delta_B^{-1}(0)$. In this case with an entirely similar argument used for Case 1 we reach a contradiction.

Case 3 $\bar{x} \in \Delta_B^{-1}(0)$ and X, Y point to opposite side of $\Delta_B^{-1}(0)$. In this case from Proposition 12, p. 47 $\Delta_B^{-1}(0)$ is a turnpike near x, hence γ_1 and γ_2 may enter the turnpike at time respectively $\bar{t}_1$ and $\bar{t}_2$. This case is not generic. ■

Let us now introduce the concept of equivalent extremal syntheses and stable extremal synthesis in finite time.

Definition 65 *Let Γ_1^* and Γ_2^* be two extremal syntheses in finite time corresponding to $(F_1, G_1), (F_2, G_2) \in \Xi$ respectively. We say that Γ_1^* and Γ_2^* are equivalent and we write $\Gamma_1^* \sim \Gamma_2^*$ if there exists an homeomorphism $\Psi : \mathbb{R}^4 \to \mathbb{R}^4$ such that:*

- $\Psi(\mathcal{N}_1) = \Psi(\mathcal{N}_2)$;

- *Ψ maps arcs of extremal trajectories corresponding to control ± 1 (resp. φ) to arcs of extremal trajectories corresponding to ± 1 (resp. φ);*
- *$\pi_* \circ \Psi \circ \pi_*^{-1}$ maps FCs (resp. FPs) of N_1 to FCs (resp. FPs) of N_2 of the same kind.*

Definition 66 *An extremal synthesis Γ_τ^* in time $\tau > 0$ corresponding to $(F,G) \in \Xi$ is stable if the following holds. There exists ε such that if $\|(F',G')-(F,G)\| < \varepsilon$ and ${\Gamma^*}'_\tau$ is the extremal synthesis at time τ for (F',G'), then $\Gamma_\tau^* \sim {\Gamma^*}'_\tau$.*

Proof of Theorem 30. By definition, Γ_τ^* is the set of extremal pairs constructed by the Algorithm in time $\tau > 0$. The Algorithm of Section 4.2, p. 167 constructs all the extremal pairs up to time τ. This guarantees 1. of Definition 54, p. 155 at time τ. Under the generic conditions **(P1)–(P7)**, **(GA1)–(GA16)**, **(GAτ)**, **(GAφ)** the Algorithm stops after a finite number of steps generating a finite number of extremal strips, FCs and FPs. This ensures 2. of Definition 54. Moreover, there exists $m \in \mathbb{N}$ such that for every trajectory γ built by the Algorithm, we have $n(\gamma) \leq m$, that is γ is a concatenation of at most m arcs and each arc corresponds to a constant control ± 1 or to the singular control φ. Hence we obtain 3. of Definition 54, p. 155. The lift to $\mathbb{R}^2 \times S^1$ of each extremal strip $\mathcal{S}$ is a piecewise–$\mathcal{C}^1$ manifold and there is a natural stratification setting FCs be the one dimensional strata and FPs the zero dimensional strata. Now for each couple of (lift of) extremal strips we define:

$$\mathcal{S}^1 \cap \mathcal{S}^2 = \big\{(x,p) = (\gamma_1,\lambda_1)(t_1) = (\gamma_2,\lambda_2)(t_2) : \ (\gamma_i,\lambda_i) \in \mathcal{S}^i, \\ t_i \in Dom(\gamma_i), \ i \in \{1,2\}\big\},$$

and:

$$\mathcal{S}^1 \cup \mathcal{S}^2 = \big\{(x,p) = (\gamma_i,\lambda_i)(t_i) : \ (\gamma_i,\lambda_i) \in \mathcal{S}^i, \ t_i \in Dom(\gamma_i), \ i \in \{1,2\}\big\}.$$

From Theorem 35, p. 191 we get a description of the set $\mathcal{S}^1 \cap \mathcal{S}^2$. More precisely, if two extremal strips intersect each other then $\mathcal{S}^1 \cap \mathcal{S}^2 = Bad_N(\mathcal{S}^1,\mathcal{S}^2)$ where $Bad_N(\mathcal{S}^1,\mathcal{S}^2) = Bad_N \cap (\mathcal{S}^1 \cup \mathcal{S}^2)$. It follows that $\mathcal{S}^1 \cup \mathcal{S}^2$ is a stratified set with FCs and FPs as strata, indeed $\partial Bad_N(\mathcal{S}^1,\mathcal{S}^2) \subset FCs$. Since the number of extremal strips is finite we conclude that N is a stratified set in time τ. Now $\mathcal{N} = \pi_*^{-1}(N)$, therefore $\mathcal{N}$ is a locally closed locally finite in $\mathbb{R}^4$ regular Whitney stratified subset of $\mathbb{R}^4$ of dimension three. Thus 4. of Definition 54, p. 155 is verified. The set of discontinuities of $\sharp(\Gamma^*(x,p))$ on $\mathcal{N}$ is precisely the union of the sets $\pi_*^{-1}\left(\partial Bad_N(\mathcal{S}^1,\mathcal{S}^2)\right)$. Hence we obtain 5. of Definition 54, p. 155.

The generic conditions **(P1)–(P7)**, **(GA1)–(GA16)**, **(GAφ)**, **(GAτ)**, $(\mathcal{A})$, and $(\mathcal{F})$, imply that all FCs and FPs are stable. It is easy to check that Γ_τ^* is stable and this concludes the proof. ■

4.6 Classification of the Singularities in $\mathbb{R}^2 \times S^1$

In this Section we topologically classify all generic FCs and FPs in the normalized cotangent bundle $\mathbb{R}^2 \times S^1$.

Definition 67 *Let l_1 and l_2 be two FCs of $N = \pi_*(\mathcal{N})$. We say that l_1 and l_2 are topologically equivalent if for every $(x_1, p_1) \in l_1$ and for every $(x_2, p_2) \in l_2$ there exist a neighborhood U_1 of (x_1, p_1) in N, a neighborhood U_2 of (x_2, p_2) in N and a homeomorphism $\Psi : U_1 \to U_2$ such that:*

1) $\Psi(U_1) = U_2$;
2) $\Psi(l_1 \cap U_1) = l_2 \cap U_2$;
3) $\Psi(x_1, p_1) = (x_2, p_2)$;
4) *Ψ send extremal arcs of trajectory–covector pairs onto extremal arcs of trajectory–covector pairs.*

Definition 68 *Let (x, p_1), (x_2, p_2) be two FPs of N. We say that they are topologically equivalent if there exist a neighborhood U_1 of (x_1, p_1) in N, a neighborhood U_2 of (x_2, p_2) in N and a homeomorphism $\Psi : U_1 \to U_2$ satisfying 1, 3, 4 of Definition 67 and:*

5) if l_1 is a FC that intersects U_1, then $\Psi(l_1 \cap U_1) = l_2 \cap U_2$ where l_2 is a FC equivalent to l_1.

By the previous analysis and from Definition 67, p. 195 we have the following equivalence:

- $\gamma^{1\pm}|_{[s_i^\pm, t_i^\pm]} \sim \gamma^{2\pm}|_{[s_i'^\pm, t_i'^\pm]}$. We call this class TY (topologically equivalent to Y–curve).
- $\gamma^{1\pm}|_{[t_i^\pm, s_{i+1}^\pm]} \sim \gamma^{2\pm}|_{[t_i'^\pm, s_{i+1}'^\pm]} \sim \gamma^0 \sim W \sim W^C \sim W^D$ (for every γ^0, W, W^C, W^D that are not in $\partial Bad_N(\mathcal{S}^1, \mathcal{S}^2)$ for every $\mathcal{S}^1, \mathcal{S}^2$ extremal strips). We call this class $T\gamma^0$;
- all C and $\bar{C}$ are equivalent. We call this class TC;
- all the turnpikes are equivalent. We call this class TS;
- the two connected components of $\pi^{-1}(0) \setminus \{(\gamma^{1+}, \gamma^{2-}), (\gamma^{2+}, \gamma^{1-})\}$ are equivalent. We call this class TB;
- all the $\partial Bad_N(\mathcal{S}^1, \mathcal{S}^2)$, for every $\mathcal{S}^1, \mathcal{S}^2$ extremal strips, are equivalent. We call this class $T\partial Bad_N$.

It follows:

Theorem 36 *The equivalence classes of FCs in $\mathbb{R}^2 \times S^1$ are the following:*

TY $T\gamma^0$ TC TS TB $T\partial Bad_N$

Similarly from Definition 68, p. 195 we have the equivalences:

- $(\gamma^{1+},\gamma^{2-}) \sim (\gamma^{2+},\gamma^{1-})$. We call this class $T(X,Y)$.
- All the FPs of kind $(Y,C)_1$ are equivalent. We call this class $T(Y,C)_1$.
- All the FPs of kind $(Y,C)_2$, $(Y,C)_2^{tg}$, $(Y,\bar{C})_1$, $(Y,\bar{C})_1^{tg}$, $(Y,\bar{C})_1^{t-0}$ are equivalent. We call this class $T(Y,C)_2$.
- All the FPs of kind $(Y,C)_3$, $(Y,C)_3^{tg}$ are equivalent. We call this class $T(Y,C)_3$.
- All the FPs of kind $(C,C)_1$, $(C,\bar{C})_2$, $(C,\bar{C})_1$ are equivalent. We call this class $T(C,C)_1$.
- All the FPs of kind $(C,C)_2$, $(\bar{C},\bar{C})_1$, $(C,\bar{C})_3$, $(C,\bar{C})_4$, $1_{ab}, \ldots, 24_{ab}$, (W,C,C), $(W,\bar{C},\bar{C})$, $(W,C,\bar{C})_1$, $(W,C,\bar{C})_2$ (where $1_{ab} \ldots 24_{ab}$ are the points of figure 5.2, 5.3, 5.4, 5.5) are equivalent. We call this class $T(C,C)_2$.
- All the FPs of kind $(C,S)_1$, $(\bar{C},S)_1$, $(\bar{C},S)_2$ are equivalent. We call this class $T(C,S)$.
- All the FPs of kind (Y,S) are equivalent. We call this class $T(Y,S)$.
- All the FPs of kind $(C,S)_2$ are equivalent. We call this class $T(C,S)_2$.
- All the FPs of kind (S,S) are equivalent. We call this class $T(S,S)$.

It follows:

Theorem 37 *The equivalence classes of FPs in* $\mathbb{R}^2 \times S^1$ *are the following:* $T(X,Y)$, $T(Y,C)_1$, $T(Y,C)_2$, $T(Y,C)_3$, $T(C,C)_1$, $T(C,C)_2$, $T(C,S)$, $T(Y,S)$, $T(C,S)_2$, $T(S,S)$.

5

Projection Singularities

To understand better the link between the singularities of the optimal synthesis, of the extremal synthesis and of the minimum time function, we study the set of pairs (*extremal, cost*) (that live in a five dimensional space) $T^*M \times \mathbb{R}$ and its projections on the various subspaces. For simplicity we restrict to the case $M = \mathbb{R}^2$. Since the covector λ never vanishes (because the Lagrangian cost is constantly equal to 1, see Remark 19, p. 36), we normalize it in such a way its norm is 1. This corresponds to consider extremals on the projectivized cotangent bundle $PT^*\mathbb{R}^2 = \mathbb{R}^2 \times S^1$ of pairs (x, p) where $x \in \mathbb{R}^2$ and $p \in T_x^*\mathbb{R}^2$ with $|p| = 1$. We thus study the set:

$$\begin{aligned}\mathbf{Q} := \{(x,p,t) \in PT^*\mathbb{R}^2 \times \mathbb{R} = \mathbb{R}^2 \times S^1 \times \mathbb{R}^+ : 0 \le t \le \tau, \exists \text{ an} \\ \text{extremal pair } (\gamma,\lambda) \text{ s.t. } \gamma(0)=0, \gamma(t)=x, \lambda(t)=p\}.\end{aligned}$$

We prove that $\mathbf{Q}$ is a two dimensional piecewise $\mathcal{C}^1$ submanifold of $PT^*\mathbb{R}^2 \times \mathbb{R}$ with boundary. Then we consider the projections on (x, p), (x, t) and x spaces, namely on the projectivized cotangent space, on the space containing the graph of the minimum time function and on the base space. We have the situation depicted in Figure 5.1.

Projecting $\mathbf{Q}$ onto the (x, p) space through Π_1, it may happen that two regions glue together as in Figure 5.2. This is the reason why the set of extremals is a Whitney stratified set, but not a manifold, and the set of points where it fails to be a manifold is described by Theorem 35, p. 191 of Chapter 4. Notice that this is the only topological type of projection singularity for Π_1 that happens under generic assumptions.

Let us recall the definition of N and give that of $\tilde{N}$:

$$\begin{aligned}N &= \{(x,p) \in \mathbb{R}^2 \times S^1 : \exists\, t \le \tau, \exists \text{ extremal pair} \\ &\quad (\gamma,\lambda) : [0,t] \to T^*\mathbb{R}^2 \text{ s.t. } \gamma(0)=0, \gamma(t)=x, \lambda(t)=p\}, \\ \tilde{N} &= \{(x,t) \in \mathbb{R}^2 \times \mathbb{R}^+ : \exists \text{ extremal trajectory} \\ &\quad \gamma : [0,t] \to \mathbb{R}^2 \text{ s.t. } \gamma(0)=0, \gamma(t)=x\},\end{aligned}$$

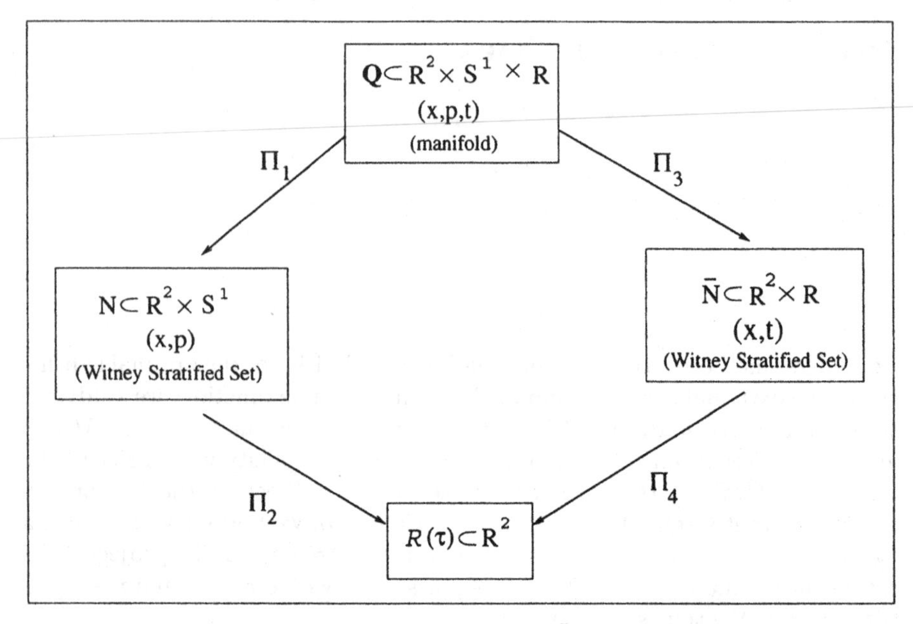

Fig. 5.1. The spaces $\mathbf{Q}$, N, $\tilde{N}$ and $\mathcal{R}(\tau)$

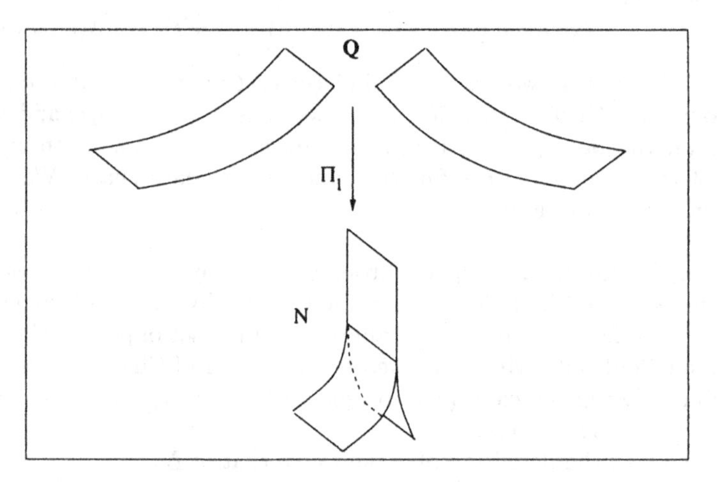

Fig. 5.2. The projection Π_1: N is not a manifold

that is $N = \Pi_1(\mathbf{Q})$ and $\tilde{N} = \Pi_3(\mathbf{Q})$. We study the stable singularities of the projection of N onto $\mathbb{R}^2$ through Π_2. At points where N is a manifold, we are in the situation of the projection of a two dimensional manifold on the plane, hence we may expect to find the classical Whitney singularities of smooth maps between two dimensional manifolds, namely folds and cusps. Apart from some special singularities (called vertical) due to the fact that the target is a point (for the case of a general target see Appendix B), the stable projection singularities encompass, beside folds and cusps, two new singularities called bifold and ribbon, see Figure 5.3. The former appears in relation with not optimal trajectories along fold points, while the second appears only along abnormal extremals. Since N is only a stratified set and not a smooth manifold, it is necessary to point out that the classification of stable projection singularities:

- is done only for points at which N is a manifold,
- has to to be intended in topological sense.

It is nontrivial to prove that all singularities appear under generic conditions. In fact, some singularities are related to the history of extremal trajectories and show up only if some global conditions are verified by the control system. In particular, this is true for the ribbon projection singularity, that can be found only along abnormal extremals. The set of possible singularities along abnormal extremals is formed of 28 (equivalence classes of) singular points, but not all sequences of singularities can be realized. We recall that a set of words Ω, from a given finite alphabet, is recognizable if there exists an automaton generating exactly the words of Ω. An automaton is roughly speaking a directed graph, with labelled edges, and one constructs words considering all the paths starting from a given set of points and ending to another fixed set of points. We are able to prove that the generic sequences of singularities, along abnormal extremals, can be classified by a recognizable set of words. As a byproduct, we obtain the existence of systems presenting projection singularities of ribbon type.

In Section 5.2, we consider also the projection Π_3 on the (x,t) space and Π_4 on $\mathbb{R}^2$. The manifold $\mathbf{Q}$ is two dimensional, thus the projection Π_3 can be seen as the immersion of a two dimensional manifold into a three dimensional space. The only stable singularity, for the smooth case, is the Whitney Umbrella and the same happens in our case (in topological sense). To every cusp in $\mathbb{R}^2 \times S^1$, it corresponds a Whitney Umbrella in the (x,t) space and this is due to the presence of an overlap curve in the optimal synthesis. For smooth classification, the singularity is precisely a Swallowtail, see Figure 5.4. To every ribbon, it corresponds a ribbon point (not a singularity for Π_3 but only for Π_4) and to every bifold it corresponds either a bifold (again not a singularity for Π_3 but only for Π_4) or a Whitney Umbrella.

The analysis of the projections Π_3 and Π_4 is made by studying the evolution of the time front along extremals. It is interesting to notice that the projection of

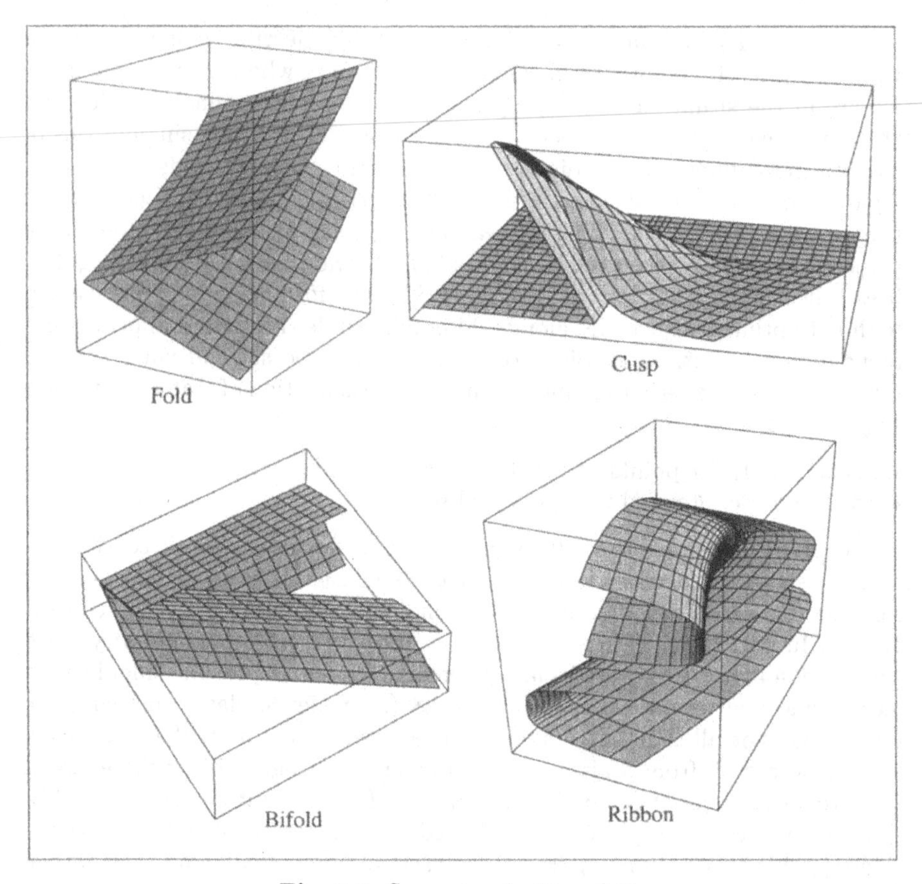

Fig. 5.3. Some singularities of Π_2.

the extremal time front on $\mathbb{R}^2$, generically develops only standard singularities as wave front on a two dimensional manifold, see [16, 17].

Theorem 38 (Q is a manifold) *Under generic conditions, on F and G, $\mathbf{Q}$ is a piecewise C^1 manifold with boundary.*

Proof. In Theorem 35, p. 191, it is proved, under generic conditions, that the set $\{(x,p) : \exists t \text{ s.t. } (x,p,t) \in \mathbf{Q}\}$ fails to be a manifold only because of singular trajectories called turnpikes. More precisely if (x,p) is such a point then there exist two distinct trajectories, say γ_1, γ_2 reaching the same turnpike. It is easy to check that, under generic assumptions, γ_1 and γ_2 reach a same point of the turnpike at different times, so that $\mathbf{Q}$ is a manifold at each (x,p,t). ∎

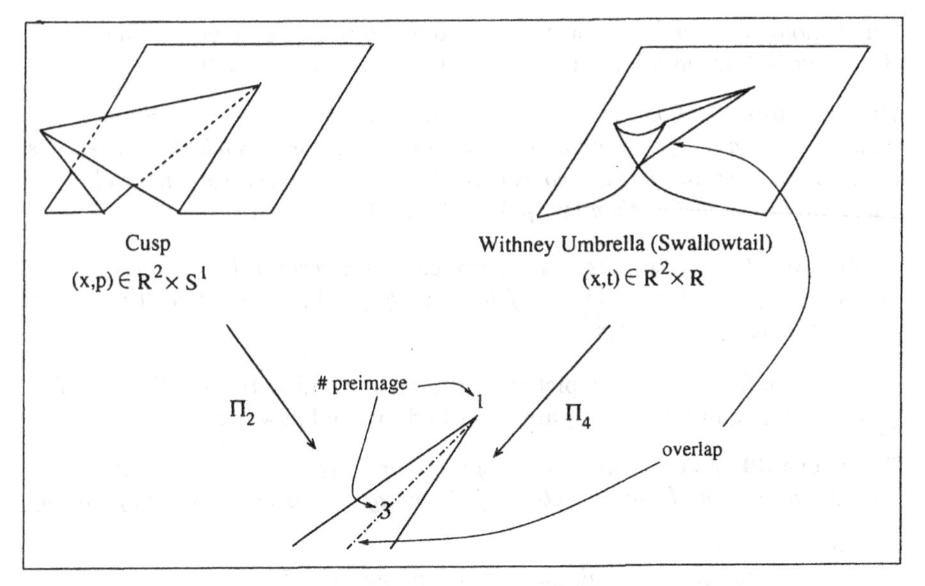

Fig. 5.4. Cusp and Whitney Umbrella.

5.1 Singularities of the Projection Π_2

Notice that Π_2 is a projection from a Whitney stratified set of dimension two, contained in $\mathbb{R}^2 \times S^1$, to the plane. Hence we should expect the classical singularities, namely fold and cusp. It happens that indeed there are other stable singularities. We first need the following:

Definition 69 *A stable topological singularity of the projection Π_2 is a stable topological singularity for the projection of N onto $\mathbb{R}^2$ outside a neighborhood of the set of discontinuities of $\#(\Gamma^*_{(x,p)})$ (where N fails to be a manifold see Definition 54, p. 155).*

5.1.1 Classification of Projection Singularities

We introduce the definition of topologically equivalent singularities:

Definition 70 *Let (x_1, p_1), (x_2, p_2) be two points of N. We say that they are two equivalent topological singularities of the projection of the extremal synthesis, and we we write $(x_1, p_1) \sim_\pi (x_2, p_2)$ if there exists an homeomorphism $\Psi : U_1 \to U_2$, U_i neighborhood of (x_i, p_i) $(i = 1, 2)$ in $\mathbb{R}^2 \times S^1$, such that for every $p, p' \in S^1$ we have:*

$$1) \quad \pi\left(\Psi(x,p)\right) = \pi\left(\Psi(x,p')\right)$$
$$2) \quad \Psi(U_1 \cap N) = U_2 \cap N$$

Our topological classification follows from the analysis given in Chapter 4. Recalling Definition 55, p. 163, of Chapter 4, we have the following:

Proposition 27 *Take $(x_1, p_1), (x_2, p_2)$ in N such that $x_1, x_2 \notin Supp(\gamma^+) \cup Supp(\gamma^-)$, and let γ_1, γ_2 be the two extremal trajectories such that γ_i reaches x_i with covector p_i. If the two points x_1, x_2 of $\mathcal{R}(\tau)$ are both normal, fold, cusp, bifold or ribbon, then $(x_1, p_1) \sim_\pi (x_2, p_2)$.*

Definition 71 *We call the equivalence classes determined by Proposition 27 respectively normal, fold, cusp, bifold and ribbon. Moreover we call vertical a point of $Supp(\gamma^+) \cup Supp(\gamma^-)$.*

In Figure 5.3, p. 200, we depict the projection singularities in $\mathbb{R}^2 \times S^1$. The projection singularities of Π_2 are described by the following:

Theorem 39 *Under generic assumptions on $(F, G) \in \Xi$, each stable topological singularity of the projection of $\mathbf{Q}$ through Π_2 is of one of the following type:*

1. *Vertical (origin, normal, cusp, fold, ending)*
2. *Fold,*
3. *Cusp,*
4. *Bifold,*
5. *Ribbon.*

Only 1.,2.,3. can happen on a small neighborhood of the intersection with the set of optimal pairs, with fold points not being optimally reached. Moreover 5. appears only along abnormal extremals.

Remark 54 Notice that the difference, with respect to the Whitney classical singularities of smooth maps between two dimensional smooth manifolds, (see [130]), is that now we have a stratified set with "edges" and "corners".

Proof. There are projection singularities along $\gamma^+ \cup \gamma^-$ that we call vertical. In particular all points of $\pi^{-1}(0)$ are called vertical origin. All the points of $\pi^{-1}(Supp(\gamma^+) \cup Supp(\gamma^-))$ that are not FPs are equivalent and they are called vertical normal singularities. The lift in N ot a FP of kind $(Y, \bar{C})_1^{tg}$ is called a vertical cusp point (see figure 4.5 and 4.6). It is easy to check the equivalences $(Y, \bar{C})_1^{tg} \sim_\pi (Y, \bar{C})_1^{t-o} \sim_\pi (Y, \bar{C})_1$. Finally, the lifts of the equivalent points $(Y, C)_3 \sim_\pi (Y, C)_3^{tg}$ are called vertical ending. From the analysis of Chapter 4 we conclude. ∎

5.1.2 Classification of Abnormal Extremals

Notice that Theorem 39 describes all possible stable projection singularities of Π_2 on $\mathbf{Q}$. However it is not proved that these singularities actually appear for some system.

The vertical, fold and cusp singularities appear in the examples given in Section 2.6.4, p. 62, Chapter 2, and are described in details in the analysis of FPs along $\gamma^{\pm}$. Bifold singularities may appear along normal extremal and it is not difficult to construct examples. On the other side, the ribbon singularity may appear only along abnormal extremals.

The set of all possible singularities along abnormal extremals is formed of 28 (equivalence classes of) singular points, but not all sequences of singularities can be realized. We aim to provide a classification of all possible sequences of singularities, and prove the realizability of the ribbon singularity.

A good classification is obtained if one can put the possible sequences in bijective correspondence with some algebraic or combinatorial object Ω with simple structure. If all possible sequences of singularities were admitted, then this classification could be done choosing Ω to be the set of all words formed with letters from the alphabet $\{1, \ldots, 28\}$, with the meaning that each number corresponds to a singularity. This is not the case, however we can still have some regular structure, more precisely Ω can be chosen as a set of words recognizable by an automaton and this is the most natural classification for this problem, see rules **R1,R2,R3** of below. The recognizable sets of words share a regular structure used in theoretical computer science, see [60].

Definition 72 *We say that a set A provides a generic classification of abnormal extremals (in time τ) if there exists a generic subset Π of Ξ and a map $\Phi : \Pi \to A$ such that $\Phi((F,G)) = \Phi((F',G'))$, $(F,G),(F',G') \in \Pi$, if and only if the corresponding maximal abnormal extremals present the same finite sequence of generic singularities.*

Since the sequence of singularities completely describes the abnormal extremal, we obtain that abnormal extremals are recognizable by an automaton:

Theorem 40 *The set of abnormal extremals in a generic set of Ξ, can be classified through a set of words recognizable by an automaton.*

Using Corollary 3, p. 176, we have that for each system $(F,G) \in \Xi$ there exists exactly two maximal abnormal extremals $\gamma_A^{\pm} = \gamma_A^{\pm}((F,G),\tau)$ in time τ, exiting the origin respectively with control ± 1.

in bijective correspondence with some algebraic or combinatorial object Ω with simple structure. If all possible sequences of singularities were admitted, then this classification could be done choosing Ω to be the set of all words formed with letters from the alphabet $\{1, \ldots, 28\}$, with the meaning that each number corresponds to a singularity. This is not the case, however we can still have some regular structure, more precisely Ω can be chosen as a set of words recognizable by an automaton and this is the most natural classification for this problem, see rules **R1,R2,R3** of below. The recognizable sets of words share a regular structure used in theoretical computer science, see [60].

Definition 72 *We say that a set A provides a generic classification of abnormal extremals (in time τ) if there exists a generic subset Π of Ξ and a map $\Phi : \Pi \to A$ such that $\Phi((F,G)) = \Phi((F',G'))$, $(F,G),(F',G') \in \Pi$, if and only if the corresponding maximal abnormal extremals present the same finite sequence of generic singularities.*

Since the sequence of singularities completely describes the abnormal extremal, we obtain that abnormal extremals are recognizable by an automaton:

Definition 73 *Let Σ be a finite set and consider the set Σ^* of ordered $n-$tuples $s = (\sigma_1, ..., \sigma_k)$, $\sigma_i \in \Sigma$ $(i = 1, ...k)$, $k \geq 0$. We call Σ the* alphabet, *$\sigma \in \Sigma$ a letter, $s = (\sigma_1, ..., \sigma_k) \in \Sigma^*$ a word of length k and Σ^* the set of words generated by Σ.*

The set of words generated by an alphabet is a set with a simple structure and a classification based on such a set is quite satisfying. Such a kind of classification was given, for example, for the sequence of generic singularities along $\gamma^\pm$ in [108]. For abnormal extremals we have to use a set with a more complicate structure.

Definition 74 *Let Σ be a finite alphabet, an automaton $\mathcal{A}$ over Σ consists of the following:*

- *a finite set $\mathbb{S}$ whose elements are called states;*
- *a set of initial states $\mathbb{I} \subseteq \mathbb{S}$;*
- *a set of terminal states $\mathbb{T} \subseteq \mathbb{S}$;*
- *a set of edges $\mathbb{E} \subseteq \mathbb{S} \times \Sigma \times \mathbb{S}$. An edge is indicated as (S_1, σ, S_2) and we say that it begins at S_1, it ends at S_2 and it carries the label σ.*

Usually an automaton is represented by a set of circles (states) and a set of arrows that connect the circles (the edges). The initial (resp. final) states are labelled by arrows pointing towards (resp. away from) the circle. If there are several edges beginning and ending at the same states they are replaced by a single arrow carrying several labels. An example of automaton is shown in Figure 5.5.

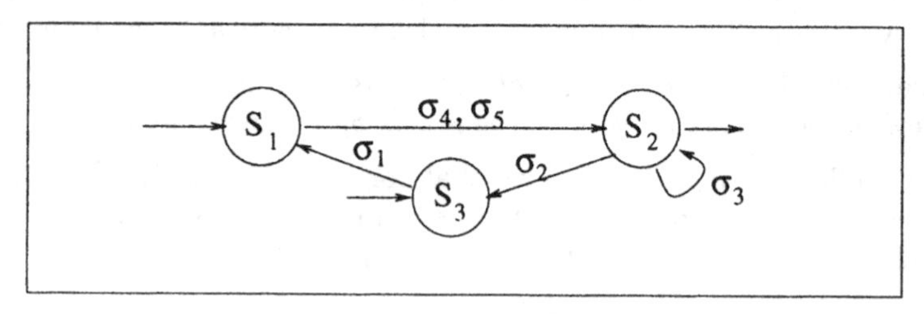

Fig. 5.5. Example of automaton

A path in $\mathcal{A}$ is a finite sequence of edges of the type $(S_1, \sigma_1, S_2)(S_2, \sigma_2, S_3)...(S_k, \sigma_k, S_{k+1})$. If $S_1 \in \mathbb{I}$ and $S_{k+1} \in \mathbb{T}$ we say that the path is successful.

Definition 75 *A set of words $\Omega \subset \Sigma^*$ is said to be recognizable by $\mathcal{A}$ if for every word $(\sigma_1, \sigma_2, ..., \sigma_m) \in \Omega$ of length m there exists $S_1, ..., S_{m+1} \in \mathbb{S}$ such that:*

- $(S_i, \sigma_i, S_{i+1}) \in \mathbb{E}$ *for every* $i = 1, ..., m$;

- $(S_1, \sigma_1, S_2)(S_2, \sigma_2, S_3)...(S_m, \sigma_m, S_{m+1})$ *is a successful path.*

Let us resume all the information on the switching strategy of an abnormal extremal via three rules. Let (γ, t_1) be a NTAE and $t_1 < t_2 ... < t_{n(\gamma)-1} < t_{n(\gamma)} := \tau$ the sequence of its switching times.

R1. Let $\mathcal{S}^1$ and $\mathcal{S}^2$ be the two extremal strips such that $\gamma \in \mathcal{S}^1 \cap \mathcal{S}^2$ and let $\bar{\mathcal{S}} := \mathcal{S}^1 \cup \mathcal{S}^2$. We have the following cases:

(1) $\theta^\gamma(t_i) = \theta^\gamma(t_{i+1})$ and $\Delta_A^{-1}(0)$ direct at $\gamma(t_i)$.
In this case if the switching locus of $\bar{\mathcal{S}}$ passing through $\gamma(t_i)$ lies in U^i_{in} (resp. U^i_{out}) then the switching locus of $\bar{\mathcal{S}}$ passing through $\gamma(t_{i+1})$ lies in U^{i+1}_{out} (resp. U^{i+1}_{in}).

(2) $\theta^\gamma(t_i) = \theta^\gamma(t_{i+1})$ and $\Delta_A^{-1}(0)$ inverse at $\gamma(t_i)$.
In this case if the switching locus of $\bar{\mathcal{S}}$ passing through $\gamma(t_i)$ lies in U^i_{in} (resp. U^i_{out}) then the switching locus of $\bar{\mathcal{S}}$ passing through $\gamma(t_{i+1})$ lies in U^{i+1}_{in} (resp. U^{i+1}_{out}).

(3) $\theta^\gamma(t_i) = \theta^\gamma(t_{i+1}) \pm \pi$ and $\Delta_A^{-1}(0)$ direct at $\gamma(t_i)$.
In this case we have the same conclusion as in case **(2)**.

(4) $\theta^\gamma(t_i) = \theta^\gamma(t_{i+1}) \pm \pi$ and $\Delta_A^{-1}(0)$ inverse at $\gamma(t_i)$.
In this case we have the same conclusion as in case **(1)**.

The rule **R1.** is a direct consequence of Propositions 22, p. 176 and 23, p. 177 **a)**.

R2. If $\Delta_A^{-1}(0)$ is inverse at t_i and t_{i+1} then $\theta^\gamma(t_{i+1}) = \theta^\gamma(t_i)$.

The rule **R2.** follows from Proposition 21, p. 175.

R3. Only the following consecutive singularities are possible:

singularity at t_i	possible singularity at t_{i+1}	AA(i)
$(YC)_2^{tg}$,$(YC)_3^{tg}$,2,4,6,8,12,16,20,24	1,2,3,4,5,6,7,8	γ^0
$(Y\bar{C})_1^{tg}$,1,3,7,9,10,13,14,19,23	9,13,18,19,20,22,23,24	W^C
$(Y\bar{C})_1^{t-o}$,5,11,15,17,18,21,22	10,11,12,14,15,16,17,21	W^D

Clearly if an abnormal arc of kind γ^0 (resp. W^C, W^D) exits from the singularity $\gamma(t_i)$ $(i = 1, ..., n(\gamma) - 1)$, then an abnormal arc of kind γ^0 (resp. W^C, W^D) enters into the singularity $\gamma(t_{i+1})$. Thus **R3.** can be directly checked using Figure 4.1, 4.2 and 4.3.

We now are ready to build an automaton $\mathcal{A}$. For us the set of states is the set of the 28 singularities:

$$\mathbb{S} := \{(YC)_2^{tg}, (Y\bar{C})_1^{tg}, (Y\bar{C})_1^{t-o}, (YC)_3^{tg}, \qquad 1, 2, ..., 23, 24\}$$

and the alphabet is:

$$\Sigma := \{0, \pi\}$$

that is (if we are considering two singularities at t_i and t_{i+1}) the set of values assumed by the function $\Delta\theta_i^\gamma := |\theta^\gamma(t_{i+1}) - \theta^\gamma(t_i)|$. The set of initial states is formed by the singularities $(YC)_2^{tg}$, $(Y\bar{C})_1^{tg}$, $(Y\bar{C})_1^{t-o}$ and $(YC)_3^{tg}$ and the set of terminal states coincides with $\mathbb{S}$. Using rules **R1**$\div$ **R3** we obtain Table A that shows how the edges connect the states, i.e. it describes the set of edges $\mathbb{E}$.

letters→ *states* ↓	0	π
$(YC)_2^{tg}$	1,2,5,6	3,4,7,8
$(Y\bar{C})_1^{tg}$	9,18,19,20	13,22,23,24
$(Y\bar{C})_1^{t-o}$	14,15,16,21	10,11,12
$(YC)_3^{tg}$	1,2,5,6	3,4,7,8
1	13,22,23,24	9,18,19,20
2	3,4,7,8	1,2,5,6
3	9,18,19,20	13,22,23,24
4	1,2,5,6	3,4,7,8
5	10,11,12,17	14,15,16
6	1,2,5,6	3,4
7	13,22,23,24	9
8	3,4,7,8	1,2
9	13,22,23,24	9,18,19,20
10	13,22,23,24	9,18,19,20
11	14,15,16,21	10,11,12,17
12	3,4,7,8	1,2,5,6
13	9,18,19,20	13,22,23,24
14	9,18,19,20	13,22,23,24
15	10,11,12,17	14,15,16,21
16	1,2,5,6	3,4,7,8
17	10,11,12,17	14,15,16
18	10,11,12,17	14,15,16
19	9,18,19,20	13
20	1,2,5,6	3,4
21	14,15,16,21	10,11,12
22	14,15,16,21	10,11,12
23	13,22,23,24	9
24	3,4,7,8	1,2

Table A

For example from the state (singularity) 18, using the letter π we may reach the states 14, 15, 16. This means that the edges of $\mathbb{E}$ with label π, that start at the state 18 are: $(18, \pi, 14), (18, \pi, 15)$ and $(18, \pi, 16)$. It is clear that for this automaton every word of Σ^* is recognizable, but $\mathcal{A}$ does not provide a generic classification because a word corresponds to more than one sequence of singularities. However it describes in a simple way the set of abnormal extremals and, in particular, from Table A we have:

Theorem 41 *All states* $1 \div 24$ *can be reached with at most two edges. More precisely the singularity 17 is the only one that needs more than one edge. Moreover, the singularity number 22 (the ribbon) can be realized with the edge* $((Y\bar{C})_1^{tg}, \pi, 22)$ *and the singularity number 10 (that is a bifold) with the edge* $((Y\bar{C})_1^{t-o}, \pi, 10)$.

Figure 5.6 shows the automaton with all states, but not all the edges. More precisely only the edges of the kind $((YC)_2^{tg}, \ . \ , \ . \ )$, $((Y\bar{C})_1^{tg}, \ . \ , \ . \ )$ and $(15, \ . \ , \ . \ )$ are drawn.

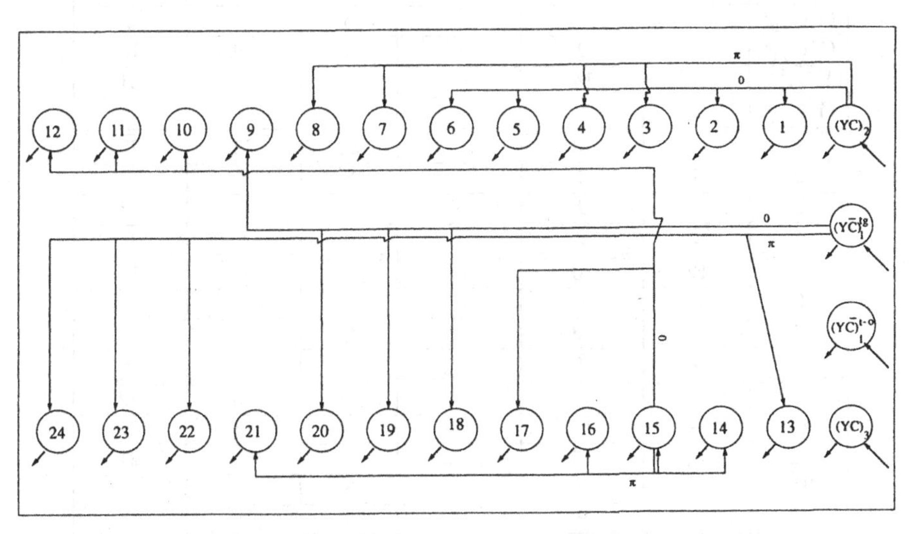

Fig. 5.6. The automaton (some edges are omitted).

To build a new automaton $\mathcal{A}'$ that provides a generic classification, we need to include more information in the alphabet. First of all we assign a label to the entry arrows I_1, I_2, I_3, and I_4, corresponding respectively to the singularities $(YC)_2^{tg}$, $(Y\bar{C})_1^{tg}$, $(Y\bar{C})_1^{t-o}$, and $(YC)_3^{tg}$. Then we introduce more information in the letters, i.e. we use a bigger alphabet. To do this, given a generic singularity on a NTAE, we want to include the following data relatively to the subsequent singularity:

- $\Delta_A^{-1}(0)$ direct or inverse (indicated by D and I respectively),
- the kind of exiting abnormal arc (i.e. γ^0, W^C or W^D).

In this way, the automaton $\mathcal{A}'$ is formed by $\mathbb{S}' = \mathbb{S}$, $\Sigma' = \{0, \pi\} \times \{D, I\} \times \{\gamma^0, W^C, W^D\} \cup \Sigma_1'$ (where $\Sigma_1' = \{I_1, I_2, I_3, I_4\}$) and set of edges $\mathbb{E}'$. Every element of Σ' (that is not in Σ_1') is indicated by a triplet $(\ . \ , \ . \ , \ . \ )$. In Table B the set of edges $\mathbb{E}'$ (with labels not in Σ_1') is completely described. Notice that not all words of $(\Sigma')^*$ are recognizable by $\mathcal{A}'$. For example the word $I_2(\pi, I, W^D)(0, D, W^C)$ is recognizable and it corresponds to the sequence of singularities $(Y\bar{C})_1^{tg} \to 22 \to 14$ while the word $I_1(0, I, W^C)$ is not recognizable.

With these definitions, to every word recognizable by $\mathcal{A}'$ it corresponds one and only one possible sequence of generic singularities along a NTAE. Theorem 40 is therefore proved.

letters→ *states* ↓	$(0, D, \gamma^0)$	$(0, D, W^C)$	$(0, D, W^D)$	$(0, I, \gamma^0)$	$(0, I, W^C)$	$(0, I, W^D)$
$(YC)_2^{tg}$	2	1	–	6	–	5
$(Y\bar{C})_1^{tg}$	–	9	–	20	19	18
$(Y\bar{C})_1^{t-o}$	16	14	15	–	–	21
$(YC)_3^{tg}$	2	1	–	6	–	5
1	–	13	–	24	23	22
2	4	3	–	8	7	–
3	–	9	–	20	19	18
4	2	1	–	6	–	5
5	12	10	11	–	–	17
6	2	1	–	6	–	5
7	–	13	–	24	23	22
8	4	3	–	8	7	–
9	–	13	–	24	23	22
10	–	13	–	24	23	22
11	16	14	15	–	–	21
12	4	3	–	8	7	–
13	–	9	–	20	19	18
14	–	9	–	20	19	18
15	12	10	11	–	–	17
16	2	1	–	6	–	5
17	12	10	11	–	–	17
18	12	10	11	–	–	17
19	–	9	–	20	19	18
20	2	1	–	6	–	5
21	16	14	15	–	–	21
22	16	14	15	–	–	21
23	–	13	–	24	23	22
24	4	3	–	8	7	–

Table B (first part)

letters→ states ↓	(π, D, γ^0)	(π, D, W^C)	(π, D, W^D)	(π, I, γ^0)	(π, I, W^C)	(π, I, W^D)
$(YC)_2^{tg}$	4	3	–	8	–	7
$(Y\bar{C})_1^{tg}$	–	13	–	24	23	22
$(Y\bar{C})_1^{t-o}$	12	10	11	–	–	–
$(YC)_3^{tg}$	4	3	–	8	7	–
1	–	9	–	20	19	18
2	2	1	–	6	–	5
3	–	13	–	24	23	22
4	4	3	–	8	7	–
5	16	14	15	–	–	–
6	4	3	–	–	–	–
7	–	9	–	–	–	–
8	2	1	–	–	–	–
9	–	9	–	20	19	18
10	–	9	–	20	19	18
11	12	10	11	–	–	17
12	2	1	–	6	–	5
13	–	13	–	24	23	22
14	–	13	–	24	23	22
15	16	14	15	–	–	21
16	4	3	–	8	7	–
17	16	14	15	–	–	–
18	16	14	15	–	–	–
19	–	–	–	–	13	–
20	4	3	–	–	–	–
21	12	10	11	–	–	–
22	12	10	11	–	–	–
23	–	9	–	–	–	–
24	2	1	–	–	–	–

Table B (second part)

We refer to Figure 5.7 for a graphic example of synthesis involving a ribbon singularity.

5.2 Projection Singularities for Π_3

Let us introduce some more definitions. Since $\mathbf{Q}$ is a piecewise smooth two dimensional manifold and the codomain of Π_3 is $\mathbb{R}^2 \times S^1$, the singularities of Π_3 can be seen as the singularities of the immersion of a surface in a three dimensional space. For the smooth case, the only generic singularity is called Whitney Umbrella. More precisely,

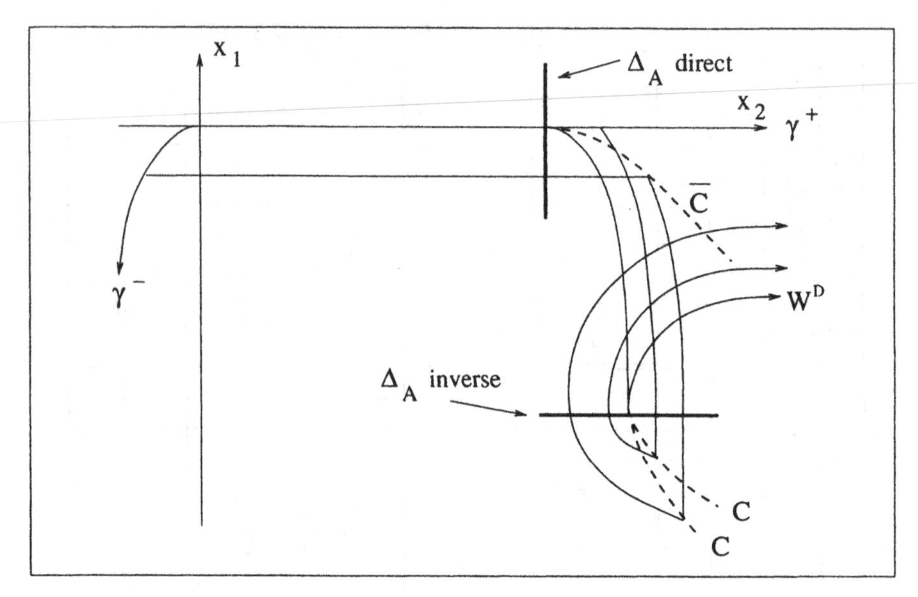

Fig. 5.7. An example of synthesis involving a ribbon.

Definition 76 *We say that a point (x,p,t) is a* Whitney Umbrella *singularity for the projection Π_3 if the following holds. There exist a neighborhood U of (x,p,t) and a one dimensional connected manifold $\ell \subset \mathbf{Q}$ containing (x,p,t), such that:* **i)** *$\ell \setminus \{(x,p,t)\}$ has two connected components ℓ_1 and ℓ_2,* **ii)** *Π_3 is injective on $U \setminus \ell$,* **iii)** *for every $z \in \ell_1$ there exists one and only one $w \in \ell_2$ such that $\Pi_3(z) = \Pi_3(w)$.*

Also in this case it happens that generically only Whitney Umbrella singularities appear. Notice that we can identify N and $\tilde{N}$ with subsets of $\mathbf{Q}$, more precisely $(x,p,t) \in \mathbf{Q}$ implies $(x,p) \in N$ and $(x,t) \in \tilde{N}$. We obtain:

Theorem 42 *Let (x,p,t) belong to $\mathbf{Q}$. Then:*

- *if (x,p) is a normal point in N then $\tilde{N}$ is a manifold in a neighborhood of (x,t) and the projections Π_3, and Π_4, are regular at (x,p,t) and (x,t) respectively;*
- *if (x,p) is a fold point in N then $\tilde{N}$ is a manifold in a neighborhood of (x,t), the projection Π_3 is regular at (x,p,t) and the projection Π_4 has a fold singularity at (x,t);*
- *if (x,p) is a cusp point in N then Π_3 has a Whitney Umbrella singularity at (x,p,t), hence $\tilde{N}$ is not a manifold at (x,t);*
- *if (x,p) is a bifold point in N then either Π_3 has a Whitney Umbrella singularity at (x,p,t) (hence $\tilde{N}$ is not a manifold at (x,t)), or Π_3 is regular at (x,p,t) and Π_4 has a bifold singularity at (x,t);*
- *if (x,p) is a ribbon point in N then Π_3 is regular at (x,p,t) and Π_4 has a ribbon singularity at (x,t).*

To prove Theorem 42, we study the time front along extremal trajectories. This analysis, and in particular that of the projection of the front on the base $\mathbb{R}^2$, is of own interest, hence we organize the obtained results in a subsection.

5.2.1 The Extremal Front

If we consider all extremal trajectories, including the not optimal ones, we can define the extremal front F_T^e as the set of points that are reached by an extremal trajectory at time T. Obviously we have $F_T \subset F_T^e$. Our aim is to study the singularities developed by F_T^e as a one dimensional wave front on a two dimensional manifold.

It is well known that for such wave fronts there is only a generic kind of local singularity called standard singularity, see [16, 17]. We want prove that at Frame Points of the optimal synthesis the extremal front F_T^e develops only standard singularities.

We start by describing the situation at the Frame Point $y^0 = \gamma^+(s_1^+)$ which is of kind $(Y,K)_1$. The extremal synthesis (that is the collection of all extremals in the cotangent bundle) has a cusp projection singularity at $(Y,K)_1$. The borders of the cusp are the X trajectory from y^0 (that is a γ_0 Frame Curve for the extremal synthesis) and a curve of kind $\bar{C}$, that is a curve of switching points reached by not optimal extremal trajectories that reflects "back" after the switching.

Let $x = (x_1, x_2)$ and choose a system of coordinates in such a way that (2.66) and (2.67), p. 94 hold. The fact that a K curve starts at $(s_1^+, 0)$ implies (see Proposition 5, p. 95):

$$\delta := c_0 d_1 - b_1 \left(\frac{1-c_0}{2}\right)^2 > 0. \tag{5.1}$$

For time $t < \mathbf{T}(y^0)$, in a neighborhood of $\mathbf{T}(y^0)$, $F_t^e = F_t$ and the extremal front F_t^e is a one dimensional piecewise smooth manifold. For time $t > \mathbf{T}(y^0)$, in a neighborhood of $\mathbf{T}(y^0)$, the extremal front F_t^e is formed by three branches:

- the first corresponds to trajectories that bifurcate from γ^+ after x_0. Its expression as a function of x_1 is an arc of the function $x_2^A(x_1)$ defined in (3.20);
- the second corresponds to trajectories that switched on the $\bar{C}$ curve. Its expression as a function of x_1 is an arc of the function $x_2^B(x_1)$ defined in (3.22);
- the third corresponds to Y trajectories that did not reach yet the curve $\bar{C}$. Its expression as a function of x_1 is an arc of the function $x_2^C(x_1)$ defined in (3.23).

Let α^A, α^B, and α^C be the coefficients of x_1^2 in the expression of x_2^A, x_2^B, and x_2^C respectively. With our choice of parameters (i.e. $b_1 > 0$, $d_1 < 0$, $c_0 < 1$,

$\delta > 0$) one can check that $\alpha^A < 0, \quad \alpha^B > \alpha^C > 0$. Moreover $x^A(x_1), \; x^B(x_1)$ have a first order tangency at the point:

$$x^{AB} = \frac{(1-c_0)b_1 t + 4d_1 s_1^+}{(1-c_0)b_1 + 4d_1},$$

while $x^B(x_1)$ and $x^C(x_1)$ have a first order tangency between them and with the trajectory exiting y^0 with constant control -1 at the point

$$x^{BC} = s_1^+ + c_0(t - s_1^+).$$

We have $x^{AB} > x^{BC}$ thus the extremal front F_t^e develops a standard singularity. Figure 5.10 shows the extremal synthesis and F_t^e close to the point $(Y, K)_1$, in the two cases in which $c_0 > 0$ and $c_0 < 0$. The case $c_0 = 0$ is not generic.

The cases of Frame Points of type $(C, K)_2$, $(S, K)_1$ and $(S, K)_3$ can be analyzed in an entirely similar way. At the other Frame Points the extremal front F_T^e presents no singularity. We finally obtain:

Theorem 43 *The extremal front F_T^e presents only standard singularities. Moreover, all singularities are developed at Frame Points (namely $(Y, K)_1$, $(C, K)_2$, $(S, K)_1$ and $(S, K)_3$) corresponding to cusp projection singularities of the extremal synthesis.*

In the following we further analyze the properties of F_T^e, in particular along abnormal extremals. First we need a technical Lemma. Recall Definition 12, p. 25, Theorem 8, p. 25 and the Definition of extremal strips 56, p. 167.

Lemma 22 *Let $S^{a,b,x}$ be an extremal strip and $\gamma_\alpha \in Int(S^{a,b,x})$ with $\alpha \in]a, b[$. Assume that $\bar{x} = \gamma_\alpha(\bar{t})$ is not on a frame curve of the extremal strip (that are curves of kind C or $\bar{C}$). Then given $\eta : [0, \varepsilon] \to \mathbb{R}^2$ of class C^1 such that:*

- $\eta(0) = \bar{x}$,
- *for every $s \in [0, \varepsilon]$, $\eta(s) = \gamma_\beta(t)$ for some $\beta \in]a, b[$ and $t \leq \bar{t}$,*

there exists a one-parameter variational family u_ε, $\varepsilon \in [0, \bar{\varepsilon}]$, of controls, generating the vector $A\frac{d}{ds}\eta(s)|_{s=0}$ for some $A > 0$. Hence, if λ_α is the covector of γ_α,

$$\lambda_\alpha(\bar{t}) \cdot \frac{d}{ds}\eta(s)|_{s=0} \leq 0$$

Proof. Inside each extremal strip but not on a frame curve the required one–parameter variational family u_ε can be constructed in the following way.

Assume that the trajectory γ_α, reaching $\bar{x}$, is bang–bang and let u_α be the corresponding control that switches to control (say) $+1$ at α. Then we can define $u_\varepsilon(t) = 0$ for $t \in [0, f_1(\varepsilon)]$, $u_\varepsilon(t) = u_\alpha(t - f_1(\varepsilon))$ for $t \in [f_1(\varepsilon), \alpha + f_1(\varepsilon) - f_2(\varepsilon)]$ and finally $u_\varepsilon(t) = 1$ on $[\alpha + f_1(\varepsilon) - f_2(\varepsilon), \bar{t}]$. Let γ_ε be the

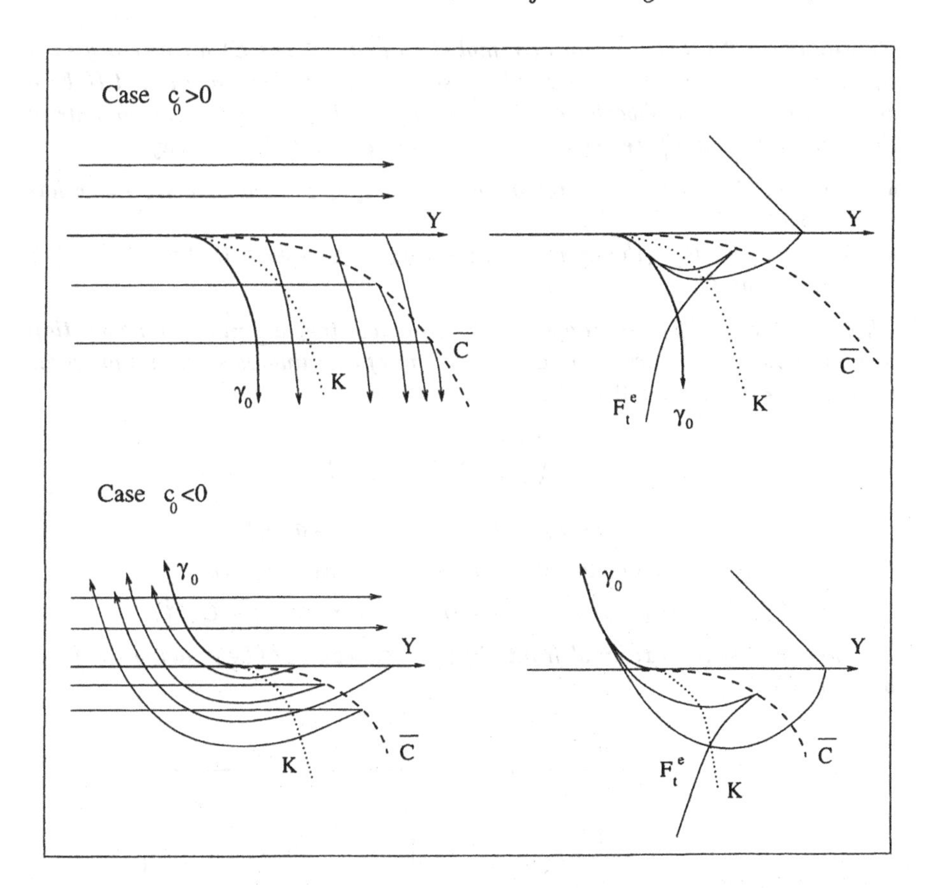

Fig. 5.8. Extremal synthesis and MTF in a neighborhood of the first FP of kind $(Y, K)_1$.

trajectory corresponding to u_ε. If we choose a system of coordinates such that $F + G \equiv (1,0)$ in a neighborhood of $\gamma_\alpha([\alpha, \bar{t}])$ and $(F - G)(\gamma_\alpha(\alpha)) = (0,1)$ then $(d/d\varepsilon)\gamma_\varepsilon(\bar{t})|_{\varepsilon=0} = (\dot{f}_2 - \dot{f}_1, \dot{f}_2)$, where $\dot{f}_i = (d/d\varepsilon)f_i(\varepsilon)|_{\varepsilon=0}$. Choosing properly the two functions f_i we get the required conclusion.

The case in which γ_α is not bang-bang can be treated making suitable variations along the last singular part. We omit details. ■

One of the one dimensional singularities of the extremal synthesis is called $\bar{C}$, see Section 4.1.3, p. 158. This curve is formed by switching points of extremal trajectories that reflect backward. Next Theorem gives information about the position of extremal time fronts near $\bar{C}$.

Proposition 28 *Let S be an extremal strip and $\gamma \in Int(S)$ a trajectory such that $\bar{x} := \gamma(\bar{t})$ is a switching point on some $\bar{C}$ curve for some $\bar{t}$. Let U be a sufficiently small neighborhood of $\bar{x}$ and $(F^e_{\bar{t}})^1$, $(F^e_{\bar{t}})^2$, the two components of the extremal front $F^e_{\bar{t}}$ restricted to U, chosen in the following way:*

- *if $x \in (F^e_{\bar{t}})^1$, then the extremal trajectory $\gamma_x \in S$ such that $\gamma_x(\bar{t}) = x$ has not switched on $\bar{C}$.*
- *if $x \in (F^e_{\bar{t}})^2$, then the extremal trajectory $\gamma_x \in S$ such that $\gamma_x(\bar{t}) = x$ has switched on $\bar{C}$.*

Then $(F^e_{\bar{t}})^1$ and $(F^e_{\bar{t}})^2$ are tangent at $\bar{x}$. Moreover they are in the same position of Figure 5.9, i.e. one can find a local system of coordinates such that for some $a_0 > 0$, $a_1 < 0$ and $b < 0$ we have $\bar{x} = (0,0)$ and:

$$Y = \begin{pmatrix} 1 \\ 0 \end{pmatrix}, X(\bar{x}) = \begin{pmatrix} 0 \\ 1 \end{pmatrix},$$

$$\bar{C} = \{(x_1, x_2) \in U : x_1 \in \mathbb{R}, x_2 = a_0 x_1\},$$

$$(F^e_{\bar{t}})^1 = \{(x_1, x_2) \in U : x_1 \leq 0, x_2 = a_1 x_1 + O(x_1^3)\},$$

$$(F^e_{\bar{t}})^2 = \{(x_1, x_2) \in U : x_1 \leq 0, x_2 = a_1 x_1 + b x_1^2 + O(x_1^3)\}.$$

In other words, the extremal front $(F^e_{\bar{t}})^1$ is "ahead" of $(F^e_{\bar{t}})^2$ along the flows of X and Y.

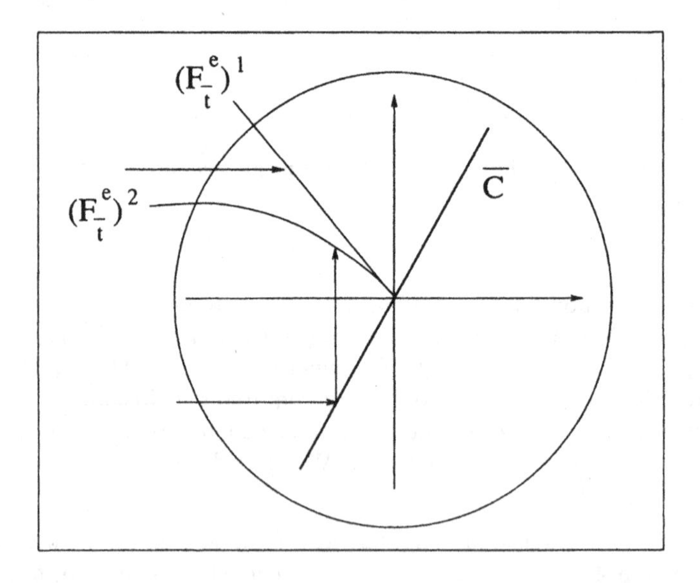

Fig. 5.9. Proposition 28

Proof. The fact that the two extremal fronts are tangent on the $\bar{C}$ curve can be proved with arguments completely similar to the ones used in the analyzes along a C curve (see Section 3.2.5, p. 141 and in particular Remark 50, p. 141).

Now assume by contradiction that we are in the situation of Figure 5.10, that is $(F^e_{\bar{t}})^2$ is ahead of $(F^e_{\bar{t}})^1$. Take $\varepsilon > 0$ small and consider the minimum time problem with the following initial set R. Consider the trajectories of S restricted to $[0, \bar{t} - \varepsilon]$ and do not let them switch on $\bar{C}$. Thus we obtain a set R that locally contains the reachable set in time $\bar{t} - \varepsilon$ denoted by $\mathcal{R}(\bar{t} - \varepsilon)$. Notice that $(F^e_{\bar{t}-\varepsilon})^1 \subset \partial R$.

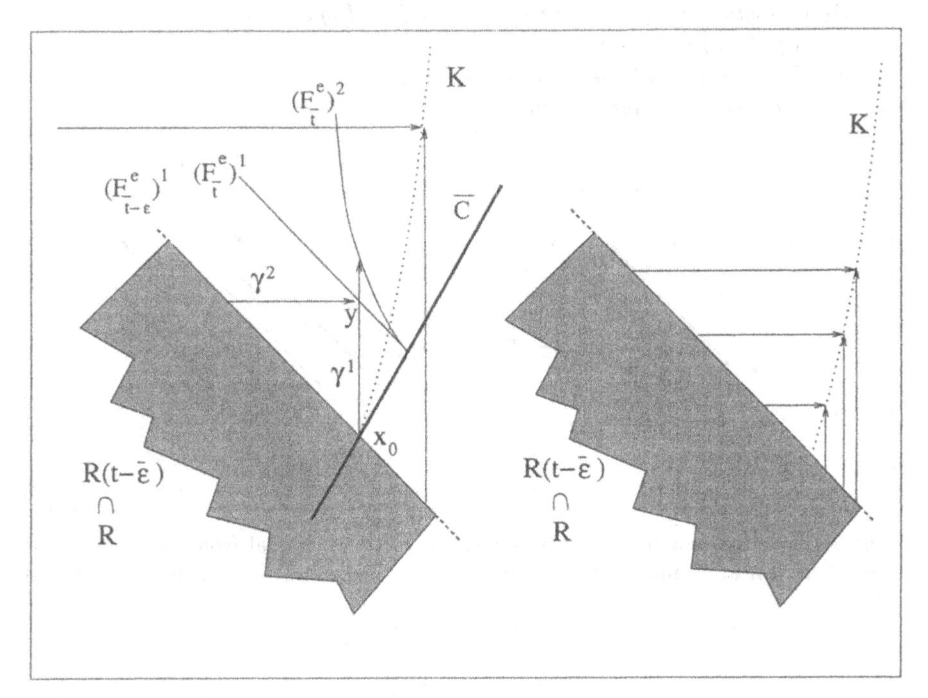

Fig. 5.10. Proof of Proposition 28

The transversality condition of PMP ensures the following. Let $x_0 = \mathcal{R}(\bar{t}-\varepsilon) \cap \bar{C}$. The extremal trajectories start from $\partial R \setminus (F^e_{\bar{t}-\varepsilon})^1$ with control -1 and from $(F^e_{\bar{t}-\varepsilon})^1$ with control $+1$. Reasoning as for $(Y, \bar{C})_1$ (see Section 2.8.2, p. 89), we get that there exists an overlap curve K, starting at x_0, such that the optimal synthesis is formed by X and Y trajectories leaving ∂R and reaching K from opposite sides (see the Figure 5.10). Let γ_1 be the X trajectory starting from x_0. Notice that $\gamma_1(\varepsilon)$ belongs to $(F^e_{\bar{t}})^2$ and $Supp(\gamma_1) \cap (F^e_{\bar{t}})^1 = y = \gamma_1(t)$ with $t \in]0, \varepsilon[$. Let γ_2 be the Y trajectory starting from $F^e_{\bar{t}-\varepsilon}$ and passing through y. We obviously have $y = \gamma_2(\varepsilon)$, which contradicts the optimality of γ_2. ■

The following Proposition describes the position of the extremal front with respect to the position of the vector fields X and Y along abnormal extremals.

Proposition 29 *Let $\mathcal{S}^{a,b}$ be an extremal strip, $\gamma = \gamma_a \in \partial\mathcal{S}$ an abnormal extremal with at least one switching, λ the associated covector and $x = \gamma(t)$, such that $\gamma|_{[t-\varepsilon,t+\varepsilon]}$ corresponds to control $+1$. Let F_t^e be the extremal time front on $\mathcal{S}$ and $v : [0,\varepsilon] \to \mathbb{R}^2$ a smooth local parameterization of F_t^e with $v(0) = x$. Recall that $\dot{v}(0)$ is parallel to $Y(x)$, see Proposition 28, p. 132. Then:*

- $\lambda(t) \cdot Y(x) = 0$;
- *the following conditions are equivalent (see Figure 5.11):*
  i) $\dot{v}(0) \cdot Y(x) < 0$ (> 0),
  ii) $X(x)$ points outside (inside) $\mathcal{S}$,
  iii) $\lambda(t)$ points inside (outside) $\mathcal{S}$.

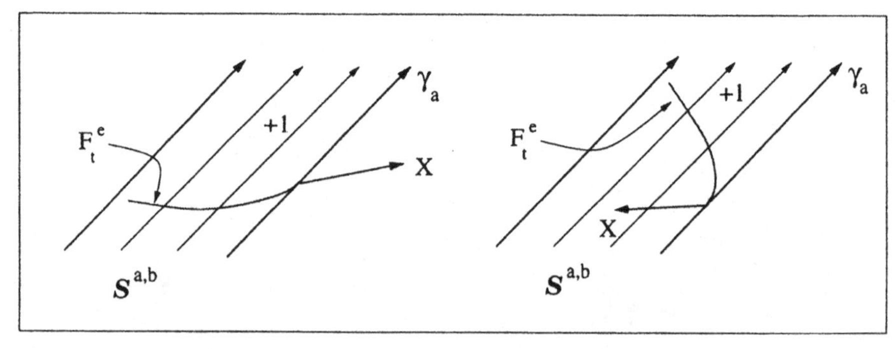

Fig. 5.11. Proposition 29: relative position of the extremal front and the vector field X for an abnormal extremal corresponding to control $+1$, along the extremal strip.

Proof. Since the Hamiltonian of PMP vanishes, we get $\lambda(t) \cdot Y(x) = 0$.

Let us show first that **i)** implies **ii)**. Assume $\dot{v}(0) \cdot Y(x) < 0$ and by contradiction that $X(x)$ points inside $\mathcal{S}$. Consider α sufficiently close to a so that $X(\gamma_\alpha(t))$ points outside the extremal strip restricted to time less than or equal to t. This is possible because γ_a does not switch at x, hence $X(x)$ is not parallel to $Y(x)$. Then for every $\varepsilon \in [0,\alpha]$ define γ^ε to be the trajectory that runs on $\gamma_{\alpha-\varepsilon}$ up to some time t_ε and then reach $X(\gamma_\alpha(t))$ with control -1. One has that $t_\varepsilon \sim t - (\sqrt{\alpha} - \sqrt{\alpha - \varepsilon})$, while the time to reach $X(\gamma_\alpha(t))$ from $\gamma^\varepsilon(t_\varepsilon)$ is of order ε. We thus obtain that the trajectories γ_ε reach $X(\gamma_\alpha(t))$ in time of order $t + \varepsilon - (\sqrt{\alpha} - \sqrt{\alpha-\varepsilon}) = t + \varepsilon(1 - (2\sqrt{\alpha})^{-1}) + o(\varepsilon^2) < t$, for α and ε sufficiently small. Thus, $\gamma_\alpha(t)$ can not be extremal.

Now we prove that **ii)** implies **iii)**. From $\lambda(t) \cdot Y(x) = 0$, $\lambda(t) \cdot G(x) > 0$ (from the maximization condition of PMP), and $X = Y - 2G$ we get $\lambda(t) \cdot X(x) < 0$, hence the conclusion.

Finally, if $\lambda(t)$ points inside S, then from Lemma 22 we obtain that $\dot{v}(0) \cdot Y(x) < 0$. Otherwise we can produce variations having positive scalar product with $\lambda(t)$. This proves that **iii)** implies **i)** and concludes the proof.■

In Section 4.3, p. 171 it is shown that, out of frame points, abnormal extremals correspond to normal or fold points, and (see Section 4.1.3, p. 158) we say that the abnormal extremal is of γ_0 type if it is formed by normal points and of W type if it is formed by fold points. Moreover, we say that it is of W^C type if the field corresponding to the other bang control points towards points with two preimages for Π_2 and we say that it is of W^D type if the opposite happens, see Definition 63, p. 173.
From Proposition 29, we immediately get the following:

Corollary 4 *Consider an abnormal extremal (γ, λ), $x = \gamma(t)$, with at least one switching, such that $\gamma|_{[t-\varepsilon,t+\varepsilon]}$ corresponds to control $+1$, and let $(F_t^e)^1$, $(F_t^e)^2$ be the two fronts of the extremal strips having γ as border. Let $v_i : [0,\varepsilon] \to \mathbb{R}^2$ be a smooth local parameterization of $(F_t^e)^i$ with $v_i(0) = x$. We have the following:*

- *if γ is of type γ_0 then: $\dot{v}_1(0) \cdot Y(x) > 0$ (< 0) iff $\dot{v}_2(0) \cdot Y(x) < 0$ (> 0);*
- *if γ is of type W then: $\dot{v}_1(0) \cdot Y(x) < 0$ (> 0) iff $\dot{v}_2(0) \cdot Y(x) < 0$ (> 0) iff γ is of kind W^D (W^C) iff $\lambda(t)$ point inside (outside) the extremal strip.*

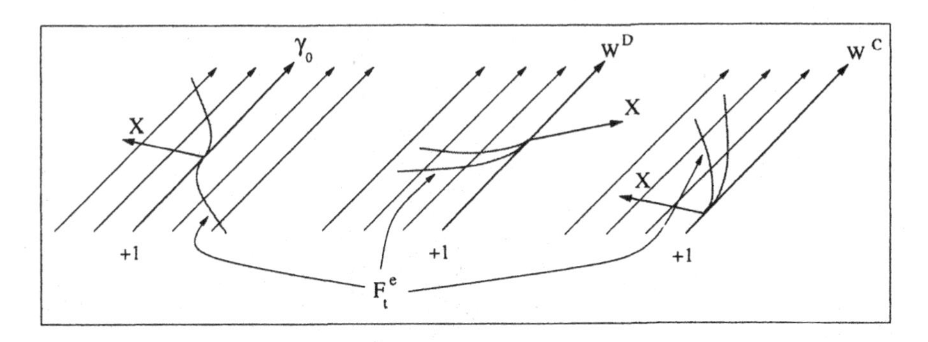

Fig. 5.12. Corollary 4: relative position of the extremal front for an abnormal extremal of kind γ_0, W^D and W^C.

Proposition 30 *Assume that x is a ribbon point, then at x it enters a W^C FC. Moreover, the extremal time fronts never self intersect.*

Proof. The analysis of Section 4.3, p. 171 implies the first assertion. Corollary 4, using the same notation, gives $\dot{v}_i(0) \cdot Y(x) > 0$. Hence the time fronts are ahead of the abnormal extremal and we obtain the conclusion from the shape of the extremal synthesis. ■

We are now ready to prove Theorem 42, p. 210.

Proof of Theorem 42. The first two claims are easy to prove. The third claim follows from the fact that the overlap curve is reached at the same time by the two extremal trajectories, coming from the two sides.

Now, if we have a bifold point then, by Proposition 28, p. 214, the two extremal fronts can either cross each other along a one dimensional manifold or never cross each others, depending on the position of the switching curve of type C or $\bar{C}$. This implies the fourth claim.

Finally, Proposition 30 implies that Π_3 cannot be singular at a ribbon point. ■

A

Some Technical Proofs of Chapter 2

In this Appendix we collect some proofs of Theorems stated in Chapter 2, that can be skipped at first reading.

A.1 Proof of Theorem 15, p. 60

In this section we give a complete description of Frame Curves generated by the algorithm $\mathcal{A}$. We use the notation introduced in Section 2.6.1, p. 58, for the six types of Frame Curves and we refer to the Examples of Section 2.6.4. From now on we consider a fixed $\tau \geq 0$ and a fixed system Σ for which $\mathcal{A}$ succeeds at time τ.

A FC D is simple if $D \setminus \partial D$ does not contain any frame point. Every FC can be divided into a finite number of simple FC's. The classification of simple FC's in connection with the classification of frame points gives a complete classification of FC's. In fact two FC's D_1, D_2 are equivalent if we can divide them in two families $D_1^1, \ldots, D_1^n$ and $D_2^1, \ldots, D_2^n$ such that:

$$D_1^i \equiv D_2^i \qquad\qquad D_1^i \cap D_1^j \equiv D_2^i \cap D_2^j \quad \forall\, i, j \in \{1, \ldots, n\}$$

where we assume, by definition, that $\emptyset \equiv \emptyset$. Therefore, we consider only simple FC's.

Y-curve. Consider a Y-curve D and $x \in D \setminus \partial D$. There exists a neighborhood U of x such that the control $u_{\mathcal{A}}$ is constant in each one of the two connected components U_1, U_2 of $U \setminus D$. If, for example, $u_{\mathcal{A}} = 1$ on U_1 then Y-trajectories leave from D entering U_1. It is clear that there are only two possibilities:

(Y1) $u_{\mathcal{A}} = 1$ on U_1 and $u_{\mathcal{A}} = -1$ on U_2, or viceversa

(Y2) $u_{\mathcal{A}} = -1$ on $U_1 \cup U_2$.

Consider the system Σ_1 of Example 1 at time $\frac{\pi}{2\sqrt{1-\varepsilon}}$. If (Y1) holds true then:

$$\Gamma_{\mathcal{A}}(\Sigma,\tau)|x \equiv \Gamma_{\mathcal{A}}(\Sigma_1,\tau_1)|\gamma^-(t_0)$$

where:

$$0 < t_0 < 1.$$

Hence D is equivalent to $\gamma^-|[0,1]$. In this case we say that D is of type Y_1 or that D is an Y_1-curve.
If (Y2) holds true then:

$$\Gamma_{\mathcal{A}}(\Sigma,\tau)|x \equiv \Gamma_{\mathcal{A}}(\Sigma_1,\tau_1)|\gamma^-(t_0)$$

where:

$$1 < t_0 < 2.$$

Then D is equivalent to $\gamma^-|[1,2]$. As before, we say that D is of type Y_2 or that D is an Y_2-curve.
F-curve. Consider an $F-$curve D and $x \in D \setminus \partial D$. There exists a neighborhood U of x in $\mathcal{R}(\tau)$ such that $u_{\mathcal{A}}$ is constant on $U \setminus Fr(\mathcal{R}(\tau))$. Consider the system Σ_1 at time τ_1 of Example 1. Choose $0 < \varepsilon < 1$ and let x_1 be the point reached by the trajectory corresponding to the control:

$$u_1 = -1 \text{ on } [0,\varepsilon], \qquad u_1 = 1 \text{ on } [\varepsilon,\tau].$$

We have:

$$\Gamma_{\mathcal{A}}(\Sigma,\tau)|x \equiv \Gamma_{\mathcal{A}}(\Sigma_2,\tau_2)|x_1.$$

C-curve. Let D be a $C-$curve and consider a point x of $D \setminus \partial D$. There exists a neighborhood U of x such that the control $u_{\mathcal{A}}$ is constant in each one of the two connected components of $U \setminus D$. From the description of the switching curves it is clear that $u_{\mathcal{A}}$ is equal to 1 on one component and equal to -1 on the other.
Consider the system Σ_2 of Example 2 at time $\tau_2 > \frac{3}{2}\pi$. We have:

$$\Gamma_{\mathcal{A}}(\Sigma,\tau)|x \equiv \Gamma_{\mathcal{A}}(\Sigma_2,\tau_2)|(-3,-1).$$

S-curve. Let x be a point of the relative interior of an $S-$curve D. As for the previous case, there exists a neighborhood U of x such that $u_{\mathcal{A}}$ is constant in each connected component of $U \setminus D$. From the definition of turnpike we have that $u_{\mathcal{A}}$ has different signs on the two components.

Consider the system Σ_1 at time τ_1 of Example 1. The following equivalence holds:

$$\Gamma_{\mathcal{A}}(\Sigma,\tau)|x \equiv \Gamma_{\mathcal{A}}(\Sigma_1,\tau_1)|\left(-1,-\frac{1}{2}\right).$$

K-curve. Consider a $K-$curve D and $x \in D \setminus \partial D$. If U is a suitably small neighborhood of x, then the control $u_{\mathcal{A}}$ is constant in each connected component of $U \setminus D$. As before, $u_{\mathcal{A}}$ has different signs on the two components. Consider the system Σ_1 of Example 1 at time $\tau_1 > 4$, we have the equivalence:

$$\Gamma_{\mathcal{A}}(\Sigma,\tau)|x \equiv \Gamma_{\mathcal{A}}(\Sigma_1,\tau_1)|(-2,0).$$

Thank to this analysis we have proved Theorem 15.

A.2 Proof of Theorem 16, p. 60

In the following we use notations of Section 2.6.2, p. 60 and refer to Examples of Section 2.6.4, p. 62. We consider only stable frame points, therefore, all frame points are intersections of no more than two frame curves. Indeed, an intersection of three or more frame curves can be destroyed by an arbitrary small perturbation of the system.
From now on we consider a fixed $\tau > 0$ and a fixed system Σ for which $\mathcal{A}$ succeeds at time τ. In particular Σ is locally controllable. For each type of frame point there are only a finite number of equivalence classes.

It is easy to check that, by construction, (FP0) can never occur. The case in which one frame curve is of type X, Y, F, S or K is immediate. The case in which D_1, D_2 are both C frame curve is consequence of the following observation. If we assume that D_1, D_2 are not tangent, and this is a generic situation, then there are some curves of zeros of either Δ_A or Δ_B to which x belongs. Indeed, near x, there are trajectories switching from control $+1$ to -1 and viceversa. Moreover, from Theorem 11, p. 44, it follows that the possibility of switching from control $+1$ to -1 and viceversa, depends on the sign of the function f. In all possible cases, we obtain the existence of trajectories having two switchings near such curves. But this is prohibited by Lemma 14, p. 52.

(X,Y)-point. Consider an (X,Y)–point x of $\Gamma_{\mathcal{A}}(\Sigma,\tau)$. If $x = (0,0)$ then it is a stable (X,Y)–frame point. Indeed if Σ' is ε–near to Σ and ε is sufficiently small, then Σ' is locally controllable and $\Gamma_{\mathcal{A}}(\Sigma,\tau)|(0,0) \equiv \Gamma_{\mathcal{A}}(\Sigma',\tau)|(0,0)$. Let Σ_1 be the system of Example 1 at time $\tau_1 > 0$, then:

$$\Gamma_{\mathcal{A}}(\Sigma,\tau)|(0,0) \equiv \Gamma_{\mathcal{A}}(\Sigma_1,\tau_1)|(0,0).$$

Now suppose that $x \neq (0,0)$. It follows that $x = \gamma^-(t^-) = \gamma^+(t^+)$, $t^- > 0$, $t^+ > 0$. We have that $t^- = t^+$ otherwise one of the two trajectories is deleted from the synthesis. Since the condition $t^- = t^+$ can be destroyed by a small perturbation, x is not stable. In fact in this case x belongs to an overlap curve, hence it is the intersection of at least three frame curves.
(Y,F)-point. Let x be an (Y,F)–frame point. The cases (FP1), (FP3) cannot occur because $\partial(Fr(\mathcal{R}(\tau))) = \emptyset$. Therefore we are in the case (FP2). There exists a neighborhood U of x (in $\mathcal{R}(\tau)$) such that $u_{\mathcal{A}}$ is constant in each one of the two connected components U_1, U_2 of $U \setminus (\gamma^- \cup F)$. One of the two following cases holds:

(YF1) $u_{\mathcal{A}} = -1$ on $U_1 \cup U_2$
(YF2) $u_{\mathcal{A}} = 1$ on U_1 and $u_{\mathcal{A}} = -1$ on U_2, or viceversa.

Consider the system Σ_1 of Example 1 (Section 2.6.4, p. 62) at time τ_1, and let $x_1 = \gamma^-(\tau_1)$. If (YF1) holds true and:

$$1 < \tau_1 < 2$$

then:

$$\Gamma_{\mathcal{A}}(\Sigma,\tau)|x \equiv \Gamma_{\mathcal{A}}(\Sigma_1,\tau_1)|x_1.$$

In this case we say that x is a frame point of type $(Y,F)_1$.
If (YF2) holds true then some Y–trajectories arise from γ^- and reach F. Let:

$$\tau_1 > 2$$

then:

$$\Gamma_{\mathcal{A}}(\Sigma,\tau)|x \equiv \Gamma_{\mathcal{A}}(\Sigma_1,\tau_1)|x_1$$

and x is a frame point of type $(Y,F)_2$.

(Y,C)-point.

Assume that (FP1) holds true. There exists a neighborhood U of the (Y,C)–frame point x such that $u_{\mathcal{A}}$ is constant in each one of the three connected components U_1, U_2, U_3 of $U \setminus (\gamma^- \cup C)$. We label U_1, U_2, U_3 in such a way that: U_3 is the connected component of $U \setminus \gamma^-$ that does not contain $C \cap U$; U_1 comes before U_2 along γ^- for the orientation of increasing time. Because of the definition of C–curve we have one of the following:

(YC1) $u_{\mathcal{A}} = 1$ on U_1
(YC2) $u_{\mathcal{A}} = 1$ on U_2.

Consider the system Σ_3 at time τ_3 of Example 3. If (YC1) holds true then $u_{\mathcal{A}} = -1$ on $U_2 \cup U_3$ and the Y–trajectories leaving γ^- reach C. We have:

$$\Gamma_{\mathcal{A}}(\Sigma,\tau)|x \equiv \Gamma_{\mathcal{A}}(\Sigma_3,\tau_3)|\gamma^-\left(\frac{1}{3}\ ln\ \frac{5}{2}\right).$$

In this case we say that x is of type $(Y,C)_1$.

<u>Remark</u> **56** Consider an $(Y,C)_1$ Frame Point. In Example 3, $\gamma^-(\frac{1}{3}\ ln\frac{5}{2})$ belongs to a nonordinary arc that is not a turnpike. This happens for every frame point x of type $(Y,C)_1$. Indeed, assume $x = \gamma^-(t_x)$ and let (γ_r, u_r), $r \in [t_x - \varepsilon, t_x + \varepsilon]$ ($\varepsilon > 0$), be the pair such that $\gamma_r(0) = \gamma^-(r)$ and $u_r \equiv 1$. Let λ_r be the covector field along (γ_r, u_r) satisfying:

$$\lambda_r(0) \cdot G(\gamma_r(0)) = 0 \qquad det\left[\lambda_r(0), G(\gamma_r(0))\right] > 0 \qquad \| \lambda_r(0) \| = 1$$

and consider the function:

$$\psi(r,s) = \lambda_r(s) \cdot G(\gamma_r(s)).$$

The equation $\psi(r,s)=0$ has two branches of solutions in $(t_x,0)$, then we have:

$$0=\left.\frac{\partial\psi}{\partial s}\right|_{(t_x,0)}=\lambda_{t_x}(0)\cdot[F,G](\gamma^-(t_x)).$$

Now $0=\lambda_{t_x}(0)\cdot G(x)=\lambda_{t_x}(0)\cdot[F,G](x)$ and $\lambda_{t_x}(0)\neq 0$ then

$$\Delta_B(x)=det\Big(G(x),[F,G](x)\Big)=0.$$

It follows that if $\nabla(\Delta_B(x))\neq 0$ then x belongs to a non ordinary arc. This non ordinary arc cannot be a regular turnpike otherwise it would have been constructed by the algorithm $\mathcal{A}$.

If (YC2) holds true then $u_{\mathcal{A}}=-1$ on $U_1\cup U_3$. Hence:

$$\Gamma_{\mathcal{A}}(\Sigma,\tau)|x\equiv\Gamma_{\mathcal{A}}(\Sigma_3,\tau_3)|\gamma^-\left(\frac{1}{3}\ ln\ 4\right)$$

and we say that x is of type $(Y,C)_2$.

The case (FP2) is not generic. Indeed if (FP2) holds, then there exists a neighborhood U of x in C such that for each $y\in U$ there exists a trajectory γ_y that switches at $y=\gamma_y(t_y)$. One side of C with respect to x is reached by trajectories γ_y that arise from a FC D_1. The other side is reached by trajectories that originate from a different FC, say D_2. Then at x, two different switching curves meet each other and x is not stable.

Suppose that (FP3) holds true. If C lies on the left (right) of γ^- then $u_{\mathcal{A}}\equiv -1$ to the left (right) of γ^-. Consider the system Σ_2 of Example 2 (of Section 2.6.4, p. 62) at time $\tau_2>\pi$. Then we have that:

$$\Gamma_{\mathcal{A}}(\Sigma,\tau)|x\equiv\Gamma_{\mathcal{A}}(\Sigma_2,\tau_2)|(2,0).$$

In this case x is of type $(Y,C)_3$.

(Y,S)-point. Let x be an (Y,S)–point x and assume that (FP1) holds true. There exists a neighborhood U of x such that $u_{\mathcal{A}}$ is constant on each one of the three connected components U_1,U_2,U_3 of $U\setminus(\gamma^-\cup S)$. We suppose that U_1,U_2,U_3 are labelled in such a way that: U_3 is the connected component of $U\setminus\gamma^-$ that does not contain $S\cap U$; U_1 comes before U_2 along γ^- for the orientation of increasing time. From the definition of turnpike it follows that $u_{\mathcal{A}}=1$ on U_1 and $u_{\mathcal{A}}=-1$ on $U_2\cup U_3$. Consider the system Σ_1 at time τ_1 of the first example (Section 2.6.4, p. 62). The following equivalence holds:

$$\Gamma_{\mathcal{A}}(\Sigma,\tau)|x\equiv\Gamma_{\mathcal{A}}(\Sigma_1,\tau_1)|\left(-1,-\frac{1}{3}\right).$$

The cases (FP2),(FP3) can not occur because, from the definition of turnpike, it follows that γ^- can not terminate at x.

(Y,K)-point. Assume that (FP1) holds true. As before, there exists a neighborhood U of x such that $u_{\mathcal{A}}$ is constant on each one of the three connected components U_1, U_2, U_3 of $U \setminus (\gamma^- \cup K)$. We label U_1, U_2, U_3 in such a way that: U_3 is the connected component of $U \setminus \gamma^-$ that does not contain $K \cap U$; U_1 comes before U_2 along γ^- for increasing time. We have that $u_{\mathcal{A}} = 1$ on U_2 and $u_{\mathcal{A}} = -1$ on $U_1 \cup U_3$. Under generic assumptions, the Y–trajectories arising from γ^- reach K. In fact, if the contrary happens then $X(x)$ and $Y(x)$ are parallel, but this is not generic. Consider again the system Σ_1 at time τ_1 of the first example (of Section 2.6.4, p. 62). In this case we have:

$$\Gamma_{\mathcal{A}}(\Sigma, \tau)|x \equiv \Gamma_{\mathcal{A}}(\Sigma_1, \tau_1)| \left(-2, -\frac{2}{3}\right)$$

and we say that x is of type $(Y, K)_1$.
Now let (FP2) hold. For every sufficiently small neighborhood U of x, we have that $u_{\mathcal{A}}$ is constant on each one of the two connected components U_1, U_2 of $U \setminus K$. If, for example, U_1 contains $\gamma^- \cap U$ then $u_{\mathcal{A}} = -1$ on U_1 and $u_{\mathcal{A}} = 1$ on U_2. Consider the system Σ_4 at time τ_4 of Example 4 and let $\bar{x}$ be the point in which γ^- intersects the overlap curve. We have:

$$\Gamma_{\mathcal{A}}(\Sigma, \tau)|x \equiv \Gamma_{\mathcal{A}}(\Sigma_4, \tau_4)|\bar{x}$$

and we say that x is of type $(Y, K)_2$.
Assume that (FP3) holds true and that Y and K are not tangent. There exists a neighborhood U of x such that $u_{\mathcal{A}}$ is constant in each one of the two connected components U_1, U_2 of $U \setminus (\gamma^- \cup K)$. Suppose that U_1, U_2 are labelled in such a way that the vector $X(x)$ points into U_2. It is clear that $u_{\mathcal{A}} = -1$ on U_1 and $u_{\mathcal{A}} = 1$ on U_2. The Y–trajectories leaving from γ^- do not reach K. Consider the synthesis Γ_5 of Example 5. We have that:

$$\Gamma_{\mathcal{A}}(\Sigma, \tau)|x \equiv \Gamma_5| \left(1, \frac{2}{3}\right)$$

and we say that x is of type $(Y, K)_3$.

(X,X),(Y,Y),(F,F)-point. It is easy to verify that points of these types cannot exist.

(F,C)-point. Consider an (F, C)–point x. Since $\partial(Fr(\mathcal{R}(\tau))) = \emptyset$, the cases (FP2) and (FP3) cannot occur. Then (FP1) holds true. There exists a neighborhood U of x in $\mathcal{R}(\tau)$ such that $u_{\mathcal{A}}$ is constant in each one of the two connected components of $U \setminus (F \cup C)$. It is clear that $u_{\mathcal{A}} = 1$ on one connected component and $u_{\mathcal{A}} = -1$ on the other. The trajectories leaving from C reach F. Consider the system Σ_2 of Example 2 (Section 2.6.4, p. 62) at time π.

We have:

$$\Gamma_{\mathcal{A}}(\Sigma,\tau)|x \equiv \Gamma_{\mathcal{A}}(\Sigma_2,\pi)|(-3,-1).$$

(F,S)-point. As for the previous type, only the case (FP1) can hold. There exists a neighborhood U of x in $\mathcal{R}(\tau)$ such that $u_{\mathcal{A}}$ is constant in each one of the two connected components of $U \setminus (F \cup S)$. Again $u_{\mathcal{A}} = 1$ on one connected component and $u_{\mathcal{A}} = -1$ on the other. Under generic assumptions the trajectories leaving from S reach F. Consider the system Σ_1 at time τ_1 of Example 1 (Section 2.6.4, p. 62). Let x_1 be the point in which the turnpike intersects the frontier of the reachable set, namely

$$x_1 = \big(-1, -\frac{1}{3} - \frac{1}{2}(\tau_1 - 1)\big).$$

It follows:

$$\Gamma_{\mathcal{A}}(\Sigma,\tau)|x \equiv \Gamma_{\mathcal{A}}(\Sigma_1,\tau_1)|x_1.$$

(F,K)-point. Consider an (F,K)–point x. The case (FP1) is the only possible one. There exists a neighborhood U of x in $\mathcal{R}(\tau)$ such that $u_{\mathcal{A}}$ is constant in each one of the two connected components of $U \setminus (F \cup K)$. It is clear that $u_{\mathcal{A}} = 1$ on one connected component and $u_{\mathcal{A}} = -1$ on the other. Consider again the system Σ_1 at time τ_1 of Example 1 (Section 2.6.4, p. 62). Let x_1 be the point in which the overlap curve intersects the frontier of the reachable set, namely:

$$x_1 = \left(-2, \frac{2}{3} + \frac{1}{3}\left(1 + \frac{\tau}{2}\right)^3 - \left(1 + \frac{\tau}{2}\right)^2\right).$$

We have the following:

$$\Gamma_{\mathcal{A}}(\Sigma,\tau)|x \equiv \Gamma_{\mathcal{A}}(\Sigma_1,\tau_1)|x_1.$$

(C,C)-point. Let x be a (C,C)–point. From the definition of switching curve we have that the cases (FP1),(FP2) cannot occur. Therefore (FP3) holds. There exist two switching curves C_1, C_2 verifying $x = C_1 \cap C_2$ and a neighborhood U of x such that $u_{\mathcal{A}}$ is constant in each connected component of $U \setminus (C_1 \cup C_2)$. We have that $u_{\mathcal{A}}$ has different signs on the two connected components. Consider the cases:

(CCa) Trajectories leaving from C_1 reach C_2
(CCb) Trajectories leaving from C_2 reach C_1.

It is easy to show that (CCa),(CCb) cannot hold at the same time, otherwise there is no trajectory reaching x. Hence we have two cases:

(CC1) (CCa) holds and (CCb) does not, or viceversa
(CC2) (CCa) and (CCb) do not hold.

Consider the synthesis Γ_6 of Example 6. If (CC1) holds true then:

$$\Gamma_{\mathcal{A}}(\Sigma,\tau)|x \equiv \Gamma_6|\left(-\frac{1}{3}, -\frac{13}{72} + \frac{4}{9}\, ln\left(\sqrt[3]{\frac{5}{2}}\right)\right)$$

and we say that x is of type $(C,C)_1$.
Consider the system Σ_2 at time $\tau_2 > 3\pi$ of Example 2 (Section 2.6.4, p. 62). If (CC2) holds true then:

$$\Gamma_{\mathcal{A}}(\Sigma,\tau)|x \equiv \Gamma_{\mathcal{A}}(\Sigma_2,\tau_2)|(4,0)$$

and we say that x is of type $(C,C)_2$.

Remark **57** Reasoning as in Remark 56, p. 222, one can prove that if x is a frame point of type $(C,C)_1$, then $\Delta_{\mathcal{B}}(x) = 0$.

The frame points of type $(C,C)_2$ are not effective singular points. Indeed, the optimal synthesis near these points is equivalent to the synthesis near a point x of a simple FC of type C, verifying $x \in C \setminus \partial C$. However the curve C may fail to be smooth at these points, as showed by Example 2.

(C,S)-point. Consider a (C,S)–point x. There exists a neighborhood U of x such that $u_{\mathcal{A}}$ is constant in each connected component of $U \setminus (C \cup S)$. The cases $(FP1), (FP2)$ can not occur because the control $u_{\mathcal{A}}$ changes sign crossing S (or C), moreover it has to be constant along each side of C (or S). Therefore $(FP3)$ holds true. There exists a C^1 parameterization $\alpha : [0,\varepsilon] \mapsto \mathbb{R}^2$, $\varepsilon > 0$, such that $\alpha(t) \in C$, $\alpha(0) = x$. Consider the vectors:

$$C(x) = \lim_{t\to 0} \dot{\alpha}(t) \qquad S(x) = F(x) + \varphi(x)G(x)$$

where φ is the control to stay on S (cfr. Lemma 10, p. 46). Assume that $C(x)$ and $S(x)$ are not parallel. Let U_X, U_Y be the connected component of $U \setminus \{x + tS(x) : t \in \mathbb{R}\}$ labelled in such a way that $X(x), Y(x)$ point into U_X, U_Y respectively. Moreover, let U_1, U_2 be the connected component of $U \setminus (C \cup S)$ labelled in such a way that the angle with vertex x and sides $C(x), S(x)$ contained in U_1 is smaller than that one contained in U_2. Now, if U_1 is contained in U_Y then $u_{\mathcal{A}} = 1$ on U_1 otherwise $u_{\mathcal{A}} = -1$ on U_1. There exists $\gamma_S \in Traj(\Sigma)$ such that $\gamma_S(Dom(\gamma_S)) = S \cap U$. We have two cases:

(CS1) $In(\gamma_S) = x$,
(CS2) $Term(\gamma_S) \doteq x$.

Assume that (CS1) holds. There are two subcases:

(CSa) Some constructed trajectories reach C from U_2
(CSb) Some constructed trajectories reach C from U_1.

If (CSb) holds, then no nontrivial trajectory reaches x, but this is not possible. Hence (CSa) holds true. For the same reason the trajectories originating from S and entering U_2 cannot reach C. Consider the synthesis Γ_7 of Example 7 (Section 2.6.4, p. 62), we have:

$$\Gamma_{\mathcal{A}}(\Sigma,\tau)|x \equiv \Gamma_7| \left(-1, -\frac{1}{3}\right)$$

and we say that x is of type $(C, C)_1$. Suppose that (CS2) holds. We have again the subcases (CSa),(CSb). The case (CSa) cannot hold. Indeed, the trajectories arising from S and entering U_2 cannot reach C and then, from the direction of $X(x)$, $Y(x)$, we have that $In(\gamma_S) = x$ contradicting (CS2). Suppose (CSb) holds. We have that the trajectories leaving from S and entering U_1 reach C. From Theorem 11, p. 44, it follows that Δ_B cannot have constant sign on $V \cap U_1$ for any neighborhood V of x. Hence we have the nongeneric condition $\nabla \Delta_B(x) = 0$.
Consider again the case (CS2) and assume now that $C(x)$ and $S(x)$ are parallel. The trajectories arriving onto C come from S. Consider the system Σ_8 at time τ of Example 8, we have:

$$\Gamma_{\mathcal{A}}(\Sigma, \tau)|x \equiv \Gamma_{\mathcal{A}}(\Sigma_8, \tau)| \left(-1 - \frac{1}{2\sqrt[3]{2}}, -1 - \frac{1}{\sqrt[3]{2}}\right)$$

and we say that x is of type $(C, S)_2$.

(C,K)-point. There exists a neighborhood U of the (C, K)–point x such that $u_{\mathcal{A}}$ is constant in each connected component of $U \setminus (C \cup K)$. The cases $(FP1), (FP2)$ can not occur because the control $u_{\mathcal{A}}$ changes sign when we cross K (or C), but it also has to be constant along each side of C (or K). Therefore $(FP3)$ holds true. There exist two $\mathcal{C}^1$ diffeomorphisms $\alpha_{1,2} : [0, \varepsilon] \mapsto \mathbb{R}^2$, $\varepsilon > 0$, such that $\alpha_1(t) \in C$, $\alpha_2(t) \in K$, $\alpha_{1,2}(0) = x$. Consider the vectors:

$$C(x) = \lim_{t \to 0} \dot{\alpha}_1(t) \qquad K(x) = \lim_{t \to 0} \dot{\alpha}_2(t).$$

Suppose that $C(x)$ and $K(x)$ are not parallel. Let U_X, U_Y be the connected component of $U \setminus \{x + tK(x) : t \in \mathbb{R}\}$ labelled in such a way that $X(x), Y(x)$ point into U_X, U_Y respectively. Let U_1, U_2 be the connected component of $U \setminus (C \cup K)$labelled in such a way that the angle with vertex x and sides $C(x), K(x)$ contained in U_1 is smaller than that one contained in U_2. If U_1 is contained in U_X then $u_{\mathcal{A}} = 1$ on U_1 otherwise $u_{\mathcal{A}} = -1$ on U_1.
We have two cases:

(CK1) Some constructed trajectories reach C from U_1
(CK2) Some constructed trajectories reach C from U_2.

Assume that (CK1) holds. The trajectories originating from C cannot reach K, otherwise we have one of the not generic conditions $Y(x) = 0$, $X(x) = 0$. Consider the synthesis Γ_9 of Example 9, we have:

$$\Gamma_{\mathcal{A}}(\Sigma, \tau)|x \equiv \Gamma_9| \left(\frac{8}{3}, \frac{\sqrt{8}}{3}\right) \tag{A.1}$$

and we say that x is of type $(C, K)_1$.

Assume that (CK2) holds. Consider the synthesis Γ_{10} of Example 10, we have:

$$\Gamma_{\mathcal{A}}(\Sigma,\tau)|x \equiv \Gamma_{10}|\left(-\frac{1}{3},-\frac{13}{72}+\frac{4}{9}\,ln\left(\sqrt[3]{\frac{5}{2}}\right)\right)$$

and we say that x is of type $(C,K)_2$.

Now suppose that $C(x)$ and $K(x)$ are parallel. If the trajectories leaving from C do not reach K then the equivalence (A.1) holds and a not stable tangency between C and K is verified. If the opposite happens, then we have a stable tangency between C and K. Consider the synthesis Γ_{11} of Example 11 and define $(\bar{x}_1,\bar{x}_2)$ in the same way. The following equivalences hold:

$$\Gamma_{\mathcal{A}}(\Sigma,\tau)|x \equiv \Gamma_{11}|(\bar{x}_1,\bar{x}_2)$$

$$\equiv \Gamma_{10}|\left(-\frac{1}{3},-\frac{13}{72}+\frac{4}{9}\,ln\left(\sqrt[3]{\frac{5}{2}}\right)\right).$$

Remark **58** The point $(\bar{x}_1,\bar{x}_2)$ of Example 11 and (2.27) of Example 10 are equivalent but they are in some sense different. In fact, proceeding as in Remark 56, p. 222, one can prove that if $K(x), C(x)$ are linearly independent and (CK2) holds then $\Delta_{\mathcal{B}}(x)=0$. If, instead, $K(x), C(x)$ are parallel we can have that $\Delta_{\mathcal{B}}(x)\neq 0$ as in Example 10.

(S,S)-point. It is easy to verify that these points cannot exist.

(S,K)-point. Consider an (S,K)–point x. The control $u_{\mathcal{A}}$ is constant in each connected component of $U\setminus(S\cup K)$, for every sufficiently small neighborhood U of x. The cases $(FP1), (FP2)$ cannot occur because the control $u_{\mathcal{A}}$ changes sign when we cross K (or S) and is constant along each side of S (or K).Therefore $(FP3)$ holds true. Assume that the trajectories leaving from one side of S reach K and those leaving from the other side do not. There exists $\gamma_S\in Traj(\Sigma)$ such that $\gamma_S(Dom(\gamma_S))=S\cap U$. There are two cases:

(SK1) $In(\gamma_S)=x$
(SK2) $Term(\gamma_S)=x$.

If (SK1) holds true then consider the synthesis Γ_{12} of Example 12. We have:

$$\Gamma_{\mathcal{A}}(\Sigma,\tau)|x \equiv \Gamma_{12}|\left(-1,-\frac{1}{3}\right)$$

and we say that x is of type $(S,K)_1$.
If (SK2) holds, consider the system Σ_{13} at time τ and the curves S_1, K_1 of Example 13. Let $x_1=S_1\cap K_1$. We have:

$$\Gamma_{\mathcal{A}}(\Sigma,\tau)|x \equiv \Gamma_{\mathcal{A}}(\Sigma_{13},\tau)|x_1$$

and we say that x is of type $(S,K)_2$.
Assume now that trajectories arising from both side of S reach K, and consider the system Σ_{13}, then:

$$\Gamma_A(\Sigma,\tau)|x \equiv \Gamma_{14}|(0,0)$$

Remark **59** At a point x where $G(x)=0$ and $\Delta_B^{-1}(0)$ is locally a turnpike, generically, the synthesis either is regular, with a C curve passing through x being transversal to $\Delta_A^{-1}(0)$ and $\Delta_B^{-1}(0)$, or it presents an $(S,K)_3$ point. An example of the first case can be found in the synthesis of Example 3 of the Introduction and it is described in detail in [34]. Assume now that a singular trajectory reaches such a point x. Making a local change of coordinates we can assume $F \equiv (0,1)$ and let:

$$\nabla G(x) = \begin{pmatrix} a & b \\ c & d \end{pmatrix}.$$

To fix the ideas assume that Y points to the right of the turnpike. Since the turnpike is tangent to $F(x)$ (see Lemma 13, p. 51), then $b<0$ and $det(\nabla G(x))>0$. Making a linear change of coordinates using the matrix

$$\begin{pmatrix} 1 & 0 \\ -d/b & 1 \end{pmatrix},$$

we are back to the situation of Example 14 of Section 2.6.4.

(K,K)-point. Consider a (K,K)-point x. From the definition of overlap curve we have that the cases (FP1), (FP2) can not occur, then (FP3) holds. Consider the system Σ_{13} at time τ of Example 13. The overlap curve K_1 is union of two overlap curves K_1', K_1''. The set K_1' is formed by intersections of $Y*X$– and $X*Y$–trajectories, while K_1'' is formed by intersections of $Y*X$– and $X*S*Y$–trajectories. Let $x_1 = K_1' \cap K_1''$. We have:

$$\Gamma_A(\Sigma,\tau)|x \equiv \Gamma_A(\Sigma_{13},\tau)|x_1.$$

From the present analysis we have immediately Theorem 16.

A.3 Proof of Proposition 3, p. 92

The trajectory γ^+ is extremal on some interval $[0,\ t^*]$ iff there exists a nonzero adjoint vector which satisfies

$$\dot\lambda(t) = -\lambda(t)\cdot\nabla Y\big(e^{tY}(0)\big), \qquad \lambda(t)\cdot G\big(e^{tY}(0)\big) \geq 0, \tag{A.2}$$

for all $t \in [0,\ t^*]$. From (A.2) it follows

$$\lambda(t)\cdot G\big(e^{tY}(0)\big) = \lambda(0)\cdot v^+(G(\gamma^+(t),t;0) = \lambda(0)\cdot\big(e^{-tY}\big)_* G\big(e^{tY}(0)\big) \geq 0. \tag{A.3}$$

Observe that a vector $\lambda(0)$ satisfying (A.3) for all $t \in [0,\ t^*]$ can exist if and only if the vector $v^+(G(\gamma^+(t), t; 0)$ ranges within an angle $\leq \pi$. By the definition of t_f^+ and by **(GA2)**, this happens if and only if $t^* \leq t_f^+$. This establishes (I).

To prove (II), call γ the trajectory corresponding to the control u in (2.62). Then γ is extremal iff there exists an adjoint vector λ which satisfies condition i) of PMP (Theorem 10, p. 35, Chapter 2) together with

$$\begin{cases} \lambda(t) \cdot G(\gamma(t)) \geq 0 & \text{if} \quad t \in [0,\ t^*], \\ \lambda(t) \cdot G(\gamma(t)) \leq 0 & \text{if} \quad t \in [t^*\ t^* + \varepsilon]. \end{cases} \tag{A.4}$$

The inequalities in (A.4) are equivalent to

$$\begin{cases} \lambda(0) \cdot v^+(G(\gamma^+(t), t; 0) \geq 0 \\ \text{if} \quad t \in [0,\ t^*], \end{cases} \tag{A.5}$$

$$\begin{cases} \lambda(0) \cdot \left(e^{-t^*Y}\right)_* \left(e^{(t^*-t)X}\right)_* G\left(e^{(t-t^*)X} e^{t^*Y}(0)\right) = \lambda(0) \cdot v^+(G(\gamma^+(t), t; 0) \leq 0 \\ \text{if} \quad t \in [t^*,\ t^* + \varepsilon]. \end{cases} \tag{A.6}$$

From the properties of the angular function θ^+ it follows that, if t^* does not belong to any closed interval $[s_i^+,\ t_i^+]$ or $[s_i'^+,\ t_i'^+]$, there can be no vector $\lambda(0) \neq 0$ which satisfies (A.5) together with

$$\lambda(0) \cdot v^+(G(\gamma^+(t), t; 0) = 0. \tag{A.7}$$

On the other hand, if t^* is contained in one of the open intervals $]s_i^+,\ t_i^+[$ or $]s_i'^+,\ t_i'^+[$, then some vector $\lambda(0)$ does exist, such that (A.5), (A.7) hold. We have to show that there exists $\varepsilon > 0$ sufficiently small so that

$$\lambda(0) \cdot v\big(G(\gamma(t)), t; 0\big) < 0 \qquad \forall t \in \ ]t^*,\ t^* + \varepsilon].$$

The above inequality is a consequence of (A.6), (A.7) if we show that, as t ranges in a suitably small neighborhood of t^*, the vector $v\big(G(\gamma(t)), t; 0\big)$ rotates in a constant direction. This is indeed the case, because, by Lemma 8, p. 43,

$$sgn\left(\frac{d}{dt} \arg\Big(G(0),\ v\big(G(\gamma(t)), t; 0\big)\Big)\right) = sgn\big(\Delta_B(\gamma(t))\big) = sgn\big(\Delta_B(e^{t^*Y}(0))\big),$$

for t sufficiently close to t^*. Now **(GA4)** implies $\dot{\theta}(t^*) \neq 0$, with

$$sgn\big(\Delta_B(e^{t^*Y}(0))\big) = sgn\big(\dot{\theta}^+(t^*)\big) = sgn\big(\theta^+(t^*)\big) = \begin{cases} 1 \text{ if } t^* \in]s_i^+,\ t_i^+[\,, \\ -1 \text{ if } t^* \in]s_i'^+,\ t_i'^+[\,. \end{cases} \tag{A.8}$$

As the control u switches from 1 to -1 at time t^*, the corresponding trajectory bifurcates from the curve γ^+ to the right (i.e., clockwise) or to the left (counterclockwise) depending on whether the determinant $X \wedge Y = 2\, F \wedge G$ at the point $e^{t^* Y}(0)$ is positive or negative. Since the Jacobian matrix $(e^{-tY})_*$ preserves orientation,

$$\begin{aligned} sgn\ F(e^{tY}(0)) \wedge G(e^{tY}(0)) &= sgn\ Y(e^{tY}(0)) \wedge G(e^{tY}(0)) \\ &= sgn\ (e^{-tY})_* Y(e^{tY}(0)) \wedge (e^{-tY})_* G(e^{tY}(0)) \\ &= sgn\ Y(0) \wedge (e^{-tY})_* G(e^{tY}(0)) \qquad (A.9) \\ &= sgn\ G(0) \wedge (e^{-tY})_* G(e^{tY}(0)) \\ &= sgn\ \theta^+(t). \end{aligned}$$

Recalling (A.8), this completes the proof of (II).

To prove (III), fix some $t_i^+ \in]0, t_f^+[$. The analysis for a time $t_i'^+$ is entirely similar. From the relation $\text{sgn}(\dot\theta(t))) = \text{sgn}(\Delta_B(\gamma(t)))$, we have

$$\dot\theta^+(t_i^+) = 0 = \Delta_B(e^{t_i^+ Y}(0)). \qquad (A.10)$$

Thus using:

$$\frac{d}{dt}\{arg(v_0, v(t))\} = \frac{v(t) \wedge \dot v(t)}{\|v(t)\|^2}, \qquad (A.11)$$

and the stability assumption **(GA3)** it follows

$$0 > \ddot\theta^+(t_i^+) = \frac{v(t_i^+) \wedge \ddot v(t_i^+)}{\|v(t_i^+)\|^2}, \qquad (A.12)$$

for $v(t) := (e^{-tY})_* G(e^{tY}(0)) = v^+(G(\gamma^+(t), t; 0)$, so that

$$\dot v(t) = (e^{-tY})_* [F, G](e^{tY}(0)), \qquad \ddot v(t) = (e^{-tY})_* [Y, [F, G]](e^{tY}(0)). \quad (A.13)$$

Indeed

$$(v \wedge \dot v)(t_i^+) = \Delta_B(\gamma^+(t_i^+)) = 0. \qquad (A.14)$$

At the point $p_i \doteq e^{t_i^+ Y}(0) = \gamma^+(t_i^+)$ we now have

$$\begin{aligned} \nabla\Delta_B \cdot Y &= (\nabla G) Y \wedge [F, G] + G \wedge (\nabla [F, G]) Y \\ &= [Y, G] \wedge [F, G] - (\nabla Y) G \wedge [F, G] + G \wedge [Y, [F, G]] - G \wedge (\nabla Y)[F, G] \\ &= G \wedge [Y, [F, G]]. \qquad (A.15) \end{aligned}$$

Indeed, (A.14) implies $G = k[F, G]$ for some k, hence

$$(\nabla Y) G \wedge [F, G] + G \wedge (\nabla Y)[F, G] = (\nabla Y) G \wedge kG + G \wedge (\nabla Y) kG = 0. \quad (A.16)$$

Together, (A.12), (A.13), (A.14), (A.15), (A.16) yield

$$\nabla \Delta_B \cdot Y(p_i) < 0. \tag{A.17}$$

In particular, this implies $\nabla \Delta_B(p_i) \neq 0$. By the implicit function theorem, the equation $\Delta_B = 0$ locally defines a smooth curve S, passing through the point $p_i = e^{t_i^+ Y}(0) = \gamma^+(t_i^+)$. From **(GA5)** we have $\nabla \Delta_B \cdot X(p_i) \neq 0$. If now $\nabla \Delta_B \cdot X(p_i) > 0$, then we must have $|\nabla \Delta_B \cdot F(p_i)| < |\nabla \Delta_B \cdot G(p_i)| \neq 0$, hence the function φ_S in (2.13) is well defined and satisfies $|\varphi_S(x)| < 1$ in a neighborhood of p_i. For $\varepsilon > 0$ suitably small, the solution of the Cauchy problem

$$x(0) = 0, \qquad \dot{x}(t) = \begin{cases} Y(x) & if \quad t \in [0,\, t_i^+], \\ F(x) + \varphi(x)G(x) & if \quad t \in [t_i^+,\, t_i^+ + \varepsilon], \end{cases}$$

is thus an admissible, extremal trajectory of the control system. To show that S is a turnpike, it remains to check the sign of the function f in Definition 20, p. 44. Here $\Delta_A = F \wedge G > 0$ because of (A.9). Hence, if U is a small open ball centered at p_i, divided by S into the connected components U_X, U_Y, recalling (2.12) we have

$$sgn(f(x)) = -sgn(\nabla \Delta_B \cdot Y(p_i)) > 0 \qquad \forall x \in U_Y,$$

$$sgn(f(x)) = -sgn(\nabla \Delta_B \cdot X(p_i)) < 0 \qquad \forall y \in U_X.$$

Now consider the case where $\nabla \Delta_B \cdot X$ and $\nabla \Delta_B \cdot Y$ are both negative at the point $p_i \doteq e^{t_i^+ Y}(0) = \gamma^+(t_i^+)$. For $\varepsilon_1, \varepsilon_2$ in a neighborhood of the origin, define the function

$$\alpha(\varepsilon_1, \varepsilon_2) \doteq \arg\Big(G(0),\ \ (e^{(\varepsilon_1 - t_i^+)Y})_*(e^{-\varepsilon_2 X})_* G\big(e^{\varepsilon_2 X} e^{(t_i^+ - \varepsilon_1)Y}(0)\big)\Big). \tag{A.18}$$

Thus $\alpha(\varepsilon_1, \varepsilon_2) = \theta^\gamma(t - \varepsilon_1 + \varepsilon_2)$ where γ is the concatenation of the Y–trajectory $\gamma^+|[0, t_i^+ - \varepsilon_1]$ with an X–trajectory.

Since α is twice continuously differentiable, we can define the C^1 function β by setting

$$\beta(\varepsilon_1, \varepsilon_2) \doteq \begin{cases} \frac{\alpha(\varepsilon_1,\varepsilon_2) - \alpha(\varepsilon_1, 0)}{\varepsilon_2} & if \quad \varepsilon_2 \neq 0, \\ \frac{\partial \alpha(\varepsilon_1, \varepsilon_2)}{\partial \varepsilon_2} & if \quad \varepsilon_2 = 0. \end{cases}$$

By (A.11), at $(\varepsilon_1, \varepsilon_2) = (0, 0)$, we have

$$\beta = \frac{\partial \alpha}{\partial \varepsilon_2} = \frac{(e^{-t_i^+ Y})_* G(p_i) \wedge (e^{-t_i^+ Y})_* [F, G](p_i)}{\|(e^{-t_i^+ Y})_* G(p_i)\|^2},$$

$$\frac{\partial \beta}{\partial \varepsilon_1} = \frac{\partial^2 \alpha}{\partial \varepsilon_1 \partial \varepsilon_2} = -\frac{\left(e^{-t_i^+ Y}\right)_* G(p_i) \wedge \left(e^{-t_i^+ Y}\right)_* [Y,[F,G]](p_i)}{\|\left(e^{-t_i^+ Y}\right)_* G(p_i)\|^2},$$

$$\frac{\partial \beta}{\partial \varepsilon_2} = \frac{\partial^2 \alpha}{\partial \varepsilon_2^2} = \frac{\left(e^{-t_i^+ Y}\right)_* G(p_i) \wedge \left(e^{-t_i^+ Y}\right)_* [X,[F,G]](p_i)}{\|\left(e^{-t_i^+ Y}\right)_* G(p_i)\|^2}.$$

From (A.16), we have $\nabla \Delta_B \cdot Y = G \wedge [Y,[F,G]]$ and similarly we obtain $\nabla \Delta_B \cdot X = G \wedge [X,[F,G]]$. Since the matrix $(e^{-t_i^+ Y})_*$ preserves the orientation it follows $\beta = 0$ and:

$$sgn\left(\frac{\partial \beta}{\partial \varepsilon_1}\right) = sgn((-Y)\cdot \nabla \Delta_B(p_i)), \qquad sgn\left(\frac{\partial \beta}{\partial \varepsilon_2}\right) = sgn(X \cdot \nabla \Delta_B(p_i)).$$

By the implicit function theorem, we can now locally solve the equation $\beta(\varepsilon_1, \varepsilon_2) = 0$ and determine a function $\varepsilon_2 = \psi(\varepsilon_1)$, with

$$sgn\left(\frac{\partial \psi}{\partial \varepsilon_1}(0)\right) = sgn\left(\frac{\nabla \Delta_B \cdot X(p_i)}{\nabla \Delta_B \cdot Y(p_i)}\right) = 1.$$

From the previous analysis, it follows that for $\varepsilon \geq 0$ suitably small, there exists $t^\dagger >]t_i^+ - \varepsilon + \psi(\varepsilon)$ such that the trajectory corresponding to the control

$$u(t) = \begin{cases} 1 & if \quad t \in [0,\ t_i^+ - \varepsilon] \cup (t_i^+ - \varepsilon + \psi(\varepsilon),\ t^\dagger], \\ -1 & if \quad t \in (t - \varepsilon,\ t - \varepsilon + \psi(\varepsilon)], \end{cases}$$

is extremal. The parametrized curve

$$\varepsilon \mapsto e^{\psi(\varepsilon) X} e^{(t_i^+ - \varepsilon) Y}(0) \tag{A.19}$$

is the switching curve of conjugate points, originating to the right of γ^+.

To prove (IV), assume $\dot\theta^+(t_f^+) > 0$, the other case being entirely similar. By Lemma 8, p. 43, this implies $\Delta_B(x) > 0$ for all x in a neighborhood of $\gamma^+(t_f^+)$. Let j be the largest index for which t'_j is defined. By the definitions of t_f^+ and the points $t_i'^+$, we thus have $\theta^+(t_f^+) - \theta^+(t'_j) = \pi$, hence t_f^+ and t'_j are negatively conjugate. If $j > 1$ then by (III), from $p'_j = e^{t'_j Y}(0)$ it originates either a turnpike or a curve of conjugate points. To set the ideas, consider the turnpike case. Then a left neighborhood of the arc $\gamma^+|_{[t'_j, t_f^+]}$ is covered by extremal trajectories of the form

$$t \mapsto e^{tY} e^{\varepsilon(F + \varphi G)}(p'_j).$$

Since t_f^+ is conjugate to t'_j and $\Delta_B(x) > 0$ near $\gamma^+(t_f^+)$, by the implicit function theorem for each $\varepsilon > 0$ sufficiently small there exists a unique $t(\varepsilon)$ close to $t_f^+ - t'_j$ such that the points

$$\Lambda'(\varepsilon) \doteq e^{\varepsilon(F+\varphi G)}(p'_j), \qquad \Lambda''(\varepsilon) \doteq e^{t(\varepsilon)Y} e^{\varepsilon(F+\varphi G)}(p'_j)$$

are conjugate along an integral curve of Y. The map Λ'' now parametrizes the desired curve of conjugate point. If $j = 1$ we can repeat the same argument using the extremal trajectories:

$$t \mapsto e^{tY} e^{\varepsilon X}(0).$$

Since the system is locally controllable at the origin, it is well known that for small ε, t these trajectories are time optimal. This completes the proof of (IV). ■

A.4 Proof of Proposition 4, p. 94

Fix some time s_i^+ with $i \geq 2$, the proof for $s_i'^+$ being similar. By (III), at the point $p_{i-1} := e^{t_{i-1}Y}(0) = \gamma^+(t_{i-1})$ initiates either a turnpike or a curve of conjugate points.

We study the turnpike case first. Consider the equation

$$\Psi(\sigma_1, \sigma_2, \sigma_3) \doteq e^{(\sigma_3 - \sigma_1)Y} e^{\sigma_1(F + \varphi G)}(p_{i-1}) - e^{\sigma_2 X} e^{(\sigma_3 - \sigma_2)Y}(p_{i-1}) = 0. \quad \text{(A.20)}$$

A trivial branch of solutions is $\sigma_1 = \sigma_2 = 0$. Observing that

$$\frac{\partial \Psi}{\partial \sigma_1}(0, 0, \sigma_3) = \left(e^{\sigma_3 Y}\right)_* (\varphi G - G)(p_{i-1}), \qquad \frac{\partial \Psi}{\partial \sigma_2}(0, 0, \sigma_3) = 2G\left(e^{\sigma_3 Y}(p_{i-1})\right), \quad \text{(A.21)}$$

since $|\varphi(p_{i-1})| < 1$, it is clear that a nontrivial branch of solutions of (A.21) can bifurcate only when $t_{i-1} + \bar{\sigma}_3$ is conjugate to t_{i-1} along γ^+. At $\sigma_3 = s_i'^+ - t_{i-1}$ we have

$$\frac{d}{d\sigma_3}\Big(G(p_{i-1}) \wedge \left(e^{-\sigma_3 Y}\right)_* G\left(e^{\sigma_3 Y}(p_{i-1})\right)\Big) > 0$$

because of the stability assumptions **(GA3)**, **(GA4)**. Therefore,

$$\frac{\partial}{\partial \sigma_3}\left(\frac{\partial \Psi}{\partial \sigma_1} \wedge \frac{\partial \Psi}{\partial \sigma_2}\right) \neq 0.$$

A standard result in bifurcation theory [52] now implies the existence of a $\mathcal{C}^1$ function $\varepsilon \mapsto \big(\sigma_1(\varepsilon), \sigma_2(\varepsilon), \sigma_3(\varepsilon)\big)$ such that

$$(\sigma_1, \sigma_2, \sigma_3)(0) = (0, 0, s_i^+ - t_{i-1}^+), \qquad \Psi\big(\sigma_1(\varepsilon), \sigma_2(\varepsilon), \sigma_3(\varepsilon)\big) = 0 \qquad \forall \varepsilon,$$

and such that the nontrivial vector

$$\left(\frac{\partial \sigma_1}{\partial \varepsilon}, \frac{\partial \sigma_2}{\partial \varepsilon}\right)$$

is in the kernel of the 2×2 matrix

$$A = \left(\frac{\partial \Psi}{\partial \sigma_1}, \frac{\partial \Psi}{\partial \sigma_2} \right).$$

Because of (A.21), the vectors $\partial\Psi/\partial\sigma_1$, $\partial\Psi/\partial\sigma_2$ have opposite orientations at $(0, 0, s_i^+ - t_{i-1}^+)$. We can thus assume that the nontrivial branch of solutions is parametrized so that the maps $\varepsilon \mapsto \sigma_1(\varepsilon)$, $\varepsilon \mapsto \sigma_2(\varepsilon)$ are both increasing. The assignment

$$\varepsilon \mapsto e^{\sigma_2(\varepsilon)X} e^{\left(\sigma_3(\varepsilon)-\sigma_2(\varepsilon)\right)Y}(p_{i-1}) \qquad \varepsilon \in [0, \varepsilon_0] \tag{A.22}$$

locally parametrizes the overlap curve, for ε_0 small enough.

At this stage, two lines through q_i have been constructed: the curve Γ_i of conjugate points at (2.63) and the overlap curve Λ_i in (A.22). Of these two, only one actually occurs in the time optimal synthesis. To decide which one, observe that by **(GA7)** the vector field X is not tangent to Γ_i. If X, Y point to the same side of Γ_i, then there is a neighborhood $\mathcal{N}$ of q_i such that all points in $\mathcal{N}$ to the right of γ^+ can be covered by extremal trajectories which either make a switching on the curve Γ_i, or else follow γ^+ up to some time $t \geq s_i^+$ and then make a switching. By a sufficiency argument, see [111], these trajectories are optimal. On the other hand, if X, Y point to opposite sides of Γ_i, then the trajectories of the form

$$t \mapsto e^{tY} e^{\varepsilon(F+\varphi G)} e^{t_{i-1}Y}(0) \tag{A.23}$$

cross the curve

$$t \mapsto e^{tX} e^{s_i^+ Y}(0) \tag{A.24}$$

before hitting the curve Γ_i. In this case, the curves (A.23) remain optimal for a short time beyond the crossing of the trajectory (A.24). This implies that in (A.22) one has

$$\frac{d}{d\varepsilon}\left(\sigma_3(\varepsilon) - \sigma_2(\varepsilon)\right) > 0,$$

hence the trajectories that reach the overlap curve make their switching after time s_i^+, and are thus extremal. Again a sufficiency theorem ensures that, such a local feedback is optimal.

In the case where at p_{i-1} starts a curve of conjugate points, let

$$\varepsilon \mapsto e^{\psi(\varepsilon)X} e^{-\varepsilon Y}(p_{i-1}) \tag{A.25}$$

be a parametrization of such curve, with ψ as in (A.19). Consider the equation

$$\Psi(\sigma_1, \sigma_2, \sigma_3) \doteq e^{(\sigma_3 - \psi(\sigma_1) + \sigma_1)Y} e^{\psi(\sigma_1)X} e^{-\sigma_1 Y}(p_{i-1}) - e^{\sigma_2 X} e^{(\sigma_3 - \sigma_2)Y}(p_{i-1}) = 0. \tag{A.26}$$

Again, $\sigma_1 = \sigma_2 = 0$ is a trivial branch of solutions.

We now have

$$\frac{\partial \Psi}{\partial \sigma_1}(0,0,\sigma_3) = \left(e^{\sigma_3 Y}\right)_* (-2\psi'(0)) G(p_{i-1}), \qquad \frac{\partial \Psi}{\partial \sigma_2}(0,0,\sigma_3) = 2G\left(e^{\sigma_3 Y}(p_{i-1})\right)$$

Therefore, when $\sigma_3 = s_i^+ - t_{i-1}^+$, the assumptions **(GA3)**, **(GA4)** imply

$$\frac{\partial \Psi}{\partial \sigma_1} \wedge \frac{\partial \Psi}{\partial \sigma_2} = 0, \qquad \frac{\partial}{\partial \sigma_1}\left(\frac{\partial \Psi}{\partial \sigma_1} \wedge \frac{\partial \Psi}{\partial \sigma_2}\right) \neq 0.$$

As in the previous case, standard bifurcation theory now yields the existence of a nontrivial branch of solutions $\varepsilon \to (\sigma_1, \sigma_2, \sigma_3)(\varepsilon)$ of (A.26). The assignment

$$\varepsilon \mapsto e^{\sigma_2(\varepsilon)X} e^{\left(\sigma_3(\varepsilon)-\sigma_2(\varepsilon)\right)Y}(p_{i-1}) \qquad \varepsilon \in [0, \varepsilon_0]$$

locally parametrizes the overlap curve.

As in the turnpike case, this overlap curve is actually present in the optimal feedback synthesis if the trajectories

$$t \mapsto e^{tY} e^{\psi(\varepsilon)X} e^{-\varepsilon Y}(p_{i-1})$$

cross the curve (A.24) before reaching Γ_i. This completes the proof. ■

A.5 Proof of Theorem 20, p. 113

We construct $\Sigma = (F, G)$ defining it on a finite collection of open sets that cover $\mathcal{G}$ and then gluing together along the intersections. We proceed defining Σ and a synthesis Γ for Σ at the same time. Moreover, every trajectory $\gamma \in \Gamma$ is endowed with an adjoint covector. At the end of the construction, we have $\Gamma \equiv \Gamma_{\mathcal{A}}(\Sigma)$. It can happen that Σ is determined defining two of the fields $F, G, X = F - G, Y = F + G$.

Let $\tilde{F}$ be the union of the elements of the equivalence class of F-edges described in ($\mathcal{G}4$). Consider the connected components of the complement, in $\mathbb{R}^2$, of the union of F-edges of $\mathcal{G}$. There is only one such component $\mathcal{R}$ that is contained in the region enclosed by $\tilde{F}$ and such that $\tilde{F} \subset Cl(\mathcal{R})$. We have to construct Σ only on $\mathcal{R}$.

From ($\mathcal{G}2$), we have that there is one origin O and O is also the origin for Σ. It is clear that, possibly translating $\mathcal{G}$, we can assume that O is the origin of $\mathbb{R}^2$. Consider a differentiable change of coordinate such that η^+ corresponds, in the new coordinates, to the line $\{(x_1, x_2) : x_2 = 0, 0 \leq x_1 \leq a\}$ for some $a > 0$. We define the field $Y = F + G$ to be the constant field $(1, 0)$ on a neighborhood N^+ of η^+ that contains only the points of $\mathcal{G}$ that are in η^+. Since η^+ is admissible there exists a function $\tilde{\theta}^+$ such that the points

$\{t_i^+, t_i'^+, s_i^+, s_i'^+\}$ of Definition 38, p. 90 determine the same sequence of frame points of η^+. We can define a vector field $G(x_1)$ on N^+ such that the function θ^+ verifies $\theta^+(t) = \tilde{\theta}^+(t)$. This is easy because from the definition of Y we have that

$$\theta^+(t) = \theta^+(x_1) = arg(G(0), G(x_1)).$$

Since the synthesis is determined by the sequence of maxima and minima of θ^+ and not by the values at these points, we can assume that $|\theta^+| < \pi/2$ at every point $t_i^+, t_i'^+, s_i^+, s_i'^+$. Therefore, if $G(x_1) = (\alpha(x_1), \beta(x_1))$ then

$$\nabla \Delta_B \cdot X = (1 - 2\alpha)\, \nabla \Delta_B \cdot Y.$$

Indeed, we have:

$$\nabla \Delta_B = \begin{pmatrix} \alpha \frac{\partial^2 \beta}{\partial x_1^2} - \beta \frac{\partial^2 \alpha}{\partial x_1^2} \\ 0 \end{pmatrix}, \qquad X = \begin{pmatrix} 1 - 2\alpha \\ -2\beta \end{pmatrix}.$$

The choice of θ^+ uniquely determines the direction of the vector G, but not its norm. Hence, we can choose α in such a way that $\nabla \Delta_B \cdot Y$, $\nabla \Delta_B \cdot X$ have the same (resp. the opposite) sign at the points $p_i = \gamma^+(t_i^+), p_i' = \gamma^+(t_i'^+)$ if at the corresponding points of η^+ there is a C-edge (resp. a S-edge). From (III) of Proposition 3, p. 92, it follows that there is a canonical correspondence at the points p_i, p_i'.
We can again modify θ^+, G in such a way that $\dot{\Gamma}_i(0)$, $i > 1$, see (2.63), (2.64), **(GA7)** 94, lies in the cone determined by $Y(q_i), G(q_i)$, $q_i = \gamma^+(s_i^+)$. We proceed in the following way. For every y sufficiently small there exists an extremal trajectory γ_y, with second coordinate constantly equal to y after the last switching, that switches along Γ_i. Let λ_y be its associated covector. Now, Y is constant then λ_y is also constant, after the last switching time of γ_y, and there exists $\zeta_1(y)$ such that $\lambda_y \cdot G(\zeta_1(y)) = 0$. Since θ^+ is increasing near s_1^+, we have that $\lambda_y \cdot G(x_1)$ is a monotone function of x_1 in $[s_1^+ - \varepsilon, s_1^+ + \varepsilon]$ for some $\varepsilon > 0$ and then $\zeta_1(y)$ exists unique for y small. Assume we want to modify G in such a way that Γ_i is described by the points $(\zeta_2(y), y)$. Let $\xi(y, x_1)$ be a smooth function, monotone in x_1 for every y, verifying

$$\xi(y, s_i^+ \pm \varepsilon) = s_i^+ \pm \varepsilon, \qquad \xi(y, \zeta_2(y)) = \zeta_1(y).$$

We redefine G in such a way that if $\tilde{\theta}^+(x_1, x_2) = arg(G(0), G(x_1, x_2))$ then $\tilde{\theta}^+(x_1, x_2) = \theta^+(\xi(x_2, x_1))$. From the definition of ξ and its monotonicity we have that γ_y switches at $(\zeta_2(y), y)$.
Now, choosing the module of $G(q_i)$ in a suitable way, we can assume that X, Y point to the same side, resp. to opposite side, of Γ_i if at the corresponding points of η^+ there is a C-edge, resp. a K-edge. We can repeat the same construction for $q_i' = \gamma^+(s_i'^+)$.
Finally, possibly changing θ^+, G, we can assume that $\delta > 0$, resp. < 0, see **(GA8)**, p. 95, for the definition of δ, if at the point of η^+ corresponding to q_1 there is a C-edge, resp. a K-edge. We repeat the same arguments for q_1'.

Therefore from Propositions 3, p. 92, Proposition 4, p. 94, Proposition 5, p. 95, we have that η^+ corresponds to γ^+ in the canonical way. Since we have defined Y and G the system Σ is determined.
Now consider η^- and a change of coordinate as for η^+. Possibly restricting N^+, we can define X and G on a neighborhood N^- of η^-, in such a way that they coincide on N^+ with the previous definitions and such that γ^- correspond to η^- in the canonical way. In this way, we have defined Σ on $N^+ \cup N^-$, that is a neighborhood of $\eta^+ \cup \eta^-$. We define $\Gamma = \Gamma_{\mathcal{A}}(\Sigma)$ on $N^+ \cup N^-$ and to every $\gamma \in \Gamma$ we associate the covector field constructed by $\mathcal{A}$.

Now, let x' be a point of $\mathcal{G}$ that is not in $N^+ \cup N^-$. From ($\mathcal{G}1$), there exists a frame point x, corresponding to x', that is of one of the types classified in Section 2.6.2, p. 60. We have shown, in Section 2.6.4, p. 62, an example for every classified point, hence there exists a system $\Sigma(x')$, a synthesis $\Gamma(x')$ both defined on an open set $U(x')$ and a frame point $x \in \Gamma(x')$ that corresponds to x' in the canonical way. Consider an open neighborhood U' of x' that does not contain any other frame point and define a diffeomorphism $\Psi : U(x') \to U'$ in such a way that Ψ maps frame points and curves to corresponding points and edges. Moreover, Ψ maps some constructed trajectories to the corresponding lines. Using Ψ, we define Σ and Γ on U' and we associate a covector field to every $\gamma \in \Gamma$.
From ($\mathcal{G}3$) it follows that every C-edge E is admissible. However, it may happen that, if x', y' are the points belonging to E, the functions Δ_A, Δ_B do not have the required signs on $U'(x'), U'(y')$. If $\Sigma(x') = (F, G)$ is one of the system of the examples of 2.6.4, p. 62, we can consider the systems

$$\Sigma_1 = (F, -G), \qquad \Sigma_2 = (-F, G), \qquad \Sigma_3 = (-F, -G).$$

Let Δ_A^i, Δ_B^i be the functions Δ_A, Δ_B for Σ_i. We have that

$$\Delta_A^1, \Delta_A^2 = -\Delta_A; \qquad \Delta_A^3 = \Delta_A; \qquad \Delta_B^1 = \Delta_B; \qquad \Delta_B^2, \Delta_B^3 = -\Delta_B.$$

The systems Σ_i have the same type of synthesis of Σ (choosing the dual vectors in a suitable way). Therefore we can define $\Sigma(x'), \Sigma(y')$ in such a way that the functions Δ_A, Δ_B have the correct signs.

Next, we define Σ on neighborhoods of frame curves. Let E be a frame curve, not of X or Y type, connecting the points x', y'. We choose a differentiable change of coordinates Ψ in such a way that E corresponds to the line $\{(x_1, x_2) : x_2 = 0, 0 \le x_1 \le a\}$ for some $a > 0$. If E is of C, S or K type then we define Ψ in such a way that the vector field Y (defined on $U'(x') \cup U'(y')$) corresponds to the vector field $(0, 1)$. If E is of F type and the region on one side of F is positive then again we let Y corresponds to $(0, 1)$, otherwise we let X correspond to $(0, 1)$. For each type of curve we have shown an example in Section 2.6.4, p. 62. We choose the system $\Sigma(E)$ that gives an example of frame curve D of the same type of E and is defined on an open set $U(E)$. If E is of C type, we can choose $\Sigma(E)$ in such a way that Δ_A, Δ_B have the right sign, i.e. compatible with the systems $\Sigma(x'), \Sigma(y')$. We define a diffeomorphism $\Psi' : U(E) \to U'(E)$, where $U'(E)$ is a neighborhood of E, in such a way that Ψ' establishes a canonical correspondence between D and E and its differential $d\Psi'$ sends either the vector field Y or X onto the vector field $(0,1)$, following the same rules used for Ψ.

We now glue together the systems defined near points and edges. Let V_1, V_2 be two open neighborhoods of x' verifying

$$Cl(V_1) \subset V_2 \subset Cl(V_2) \subset U'(x')$$

and consider a smooth function $h_{x'}$ defined on

$$U = U'(x') \cup U'(y') \cup U'(E)$$

such that

$$h_{x'}|_{V_1} \equiv 1, \qquad h_{x'}|_{U \setminus V_2} \equiv 0.$$

We define $h_{y'}$ in the same way for y'. Let $(F', G'), (F'', G'')$ be the vector fields already defined on $U'(x') \cup U'(y'), U'(E)$ respectively, and define them to be zero elsewhere in U. We set:

$$\tilde{F} \doteq (h_{x'} + h_{y'})F' + (1 - h_{x'} - h_{y'})F'', \qquad \tilde{G} \doteq (h_{x'} + h_{y'})G' + (1 - h_{x'} - h_{y'})G''.$$

In this way we have defined a system $\tilde{\Sigma} = (\tilde{F}, \tilde{G})$ on U. Since the syntheses corresponding to $\Sigma(x'), \Sigma(y')$ and $\Sigma(E)$ coincide on the set of intersections, Γ is well defined on U. However, if E is of C or of S type, it may happen that in the set where $h_{x'}, h_{y'} \neq 0, 1$, the functions $\tilde{\Delta}_A, \tilde{\Delta}_B$, have not the required properties.
Consider first the case in which E is an S-edge. From E there originate Y-trajectories that enter the half plane $\{(x_1, x_2) : x_2 > 0\}$. In this case:

$$\tilde{X}_1 > 0, \qquad \tilde{X}_2 < 0, \qquad \tilde{G}_1 < 0, \qquad \tilde{\Delta}_A > 0.$$

We define a new system Σ by setting

$$Y := \tilde{Y} + (0, \alpha), \qquad X := \tilde{X}$$

where $|\alpha| < 1$. We have $\Delta_A = (1/2)(1+\alpha)\tilde{X}_1 > 0$. If $\alpha(x_1, 0) \equiv 0$ then, after straightforward calculations, we obtain:

$$\Delta_B(x_1, 0) = \frac{1}{2}\left(2\tilde{\Delta}_B + \frac{\partial \alpha}{\partial x_2}\tilde{G}_1 \tilde{X}_2\right) \tag{A.27}$$

and then we can choose $(\partial\alpha/\partial x_2)(x_1, 0)$ in such a way that $\Delta_B(x_1, 0) \equiv 0$. Moreover:

$$\nabla\Delta_B(x_1, 0) = \nabla\tilde{\Delta}_B(x_1, 0) + \frac{1}{2}\Theta_1 + \frac{1}{2}\Theta_2 \qquad \Theta_1 = \begin{pmatrix} 0 \\ \frac{\partial^2 \alpha}{\partial x_2^2}\tilde{G}_1\tilde{X}_2 \end{pmatrix}$$

$$\Theta_2 = \frac{\partial \alpha}{\partial x_2}\left[\nabla(\tilde{G}_1\tilde{X}_2) + \begin{pmatrix} 0 \\ \frac{1}{2}(2 - \tilde{X}_2)\frac{\partial \tilde{X}_1}{\partial x_2} - \tilde{G}_1\frac{\partial \tilde{X}_2}{\partial x_2} \end{pmatrix}\right] + \begin{pmatrix} \frac{\partial^2 \alpha}{\partial x_1 \partial x_2}\tilde{G}_1\tilde{X}_2 \\ \frac{\partial^2 \alpha}{\partial x_2 \partial x_1}\tilde{G}_1\tilde{X}_1 \end{pmatrix}$$

hence Θ_2 is determined by the previous choices but we can define α choosing:

$$\frac{\partial^2 \alpha}{\partial x_2^2}(x_1, 0)$$

in such a way that $\nabla \Delta_B(x_1, 0) \neq 0$. From the compactness of E, it follows that there exists a neighborhood U' of E such that $\{x \in U' : \Delta_B(x) = 0\} = \{(x_1, x_2) : x_2 = 0\}$. Then we consider Σ restricted to U'.
Consider now the case in which E is a C-edge. Assume that from E start Y-trajectories that enter the half plane $\{(x_1, x_2) : x_2 > 0\}$ and that $\tilde{X}_1 > 0$ ($\tilde{X}_2 > 0$ follows from $\tilde{Y}_2 > 0$). Again we define $Y = \tilde{Y} + (0, \alpha), X = \tilde{X}$. If we set $\alpha(x_1, 0) = 0$ then (A.27) holds and we can choose $(\partial\alpha/\partial x_2)$ in such a way that $\Delta_B(x_1, 0) \neq 0$. Again by the compactness of E, there exists a neighborhood U' of E in which Δ_B does not vanish. We consider Σ restricted to U'.
Finally, we want to associate to every trajectory γ of Γ a covector field. If γ is contained in $V_1(x')$ or $V_1(y')$ or in $U'(E) \setminus (V_2(x') \cup V_2(y'))$ (see the definitions above), we can associate a dual variable to γ using Ψ or Ψ', because γ corresponds to a trajectory of the synthesis of $\Sigma(x')$ or $\Sigma(y')$ or $\Sigma(E)$. Otherwise assume that γ verifies $\gamma(t_x) = x \in E \setminus \partial E$. If E is either an F- or K-edge and γ is a Y-trajectory, resp. X-trajectory, then we choose λ_γ such that $\lambda_\gamma \cdot G(x) > 0$, resp. < 0. If E is either an S- or a C-edge and γ is a Y-trajectory, resp. X-trajectory, after t_x then we choose λ_γ in such a way that $\lambda_\gamma \cdot G(x) = 0$ and, if E is a C edge, $\lambda_\gamma \cdot [F, G](x) > 0$, resp. < 0. We associate to γ the adjoint variable that verifies $\lambda(t_x) = \lambda_\gamma$. It is clear that if $\gamma(I)$ is not a turnpike for every $I \subset Dom(\gamma)$ then (γ, λ) satisfies the PMP on some neighborhood of t_x. Assume now that $\gamma(I)$ is a turnpike, $I = [a, b]$. Let φ be the control defined in (2.13) and consider the system:

$$\begin{cases} \dot{x} = F(x) + \varphi(x)G(x) \\ \dot{\lambda} = -\lambda \cdot (\nabla F(x) + \varphi(x)\nabla G(x)) \end{cases} \tag{A.28}$$

and the following submanifold of $\mathbb{R}^4$:

$$Z = \{(x, \lambda) : \lambda \cdot G(x) = 0\}.$$

From the definition of λ, we have $\lambda(b) \cdot G(\gamma(b)) = 0$. Since $\Delta_B(\gamma(t)) = 0$ for $t \in [a, b]$, from

$$\frac{d}{dt}(\lambda \cdot G) = \lambda \cdot [F, G],$$

we have:

$$\lambda(t) \cdot G(\gamma(t)) = 0 \quad \Rightarrow \quad \frac{d}{ds}\Big(\lambda(s) \cdot G(\gamma(s))\Big)\Big|_{s=t} = 0.$$

By the standard theory of O.D.E. on closed set, we obtain the existence of a solution (x, μ) that verifies $x(b) = \gamma(b), \mu(b) = \lambda(b)$ and $(x(t), \mu(t)) \in Z$ for

every $t \in [a,b]$. Since the righthandside of (A.28) are Lipschitz continuous, there is a unique solution for every initial data. Hence $\lambda(t) \cdot G(\gamma(t)) = 0$ for every $t \in [a,b]$. We conclude that (γ, λ) satisfies the PMP.
From the compactness of E there exists a neighborhood U'' of E such that every $\gamma \in \Gamma$ restricted to U'' is extremal. We consider Σ restricted to U''.

In this way we have defined Σ, Γ on an open set that contains all frame points and curves. Now we complete the definition of Σ, Γ considering the regions enclosed by edges.

For every region $A \subset \mathcal{R}$ let $B_i(A)$, $i = 1, \ldots, n(A)$, be the connected components of $A \setminus L(A)$, where $L(A)$ is the union of lines in A. Let $\mathcal{B}$ be the set of all $B_i(A)$, $i = 1, \ldots, n(A)$, as A ranges over the set of regions contained in $\mathcal{R}$. We define Σ on every B by induction. From ($\mathcal{G}5$) we have that every $Cl(B)$, $B \in \mathcal{B}$, contains exactly one entrance $E(B)$. The induction hypotheses is that for every $x \in E(B)$ there exists $\gamma_x : [0, t_x] \to \mathbb{R}^2$, $\gamma_x \in \Gamma$, such that $\gamma_x(t_x) = x$, i.e. the system Σ is constructed along γ_x backward in time. We start defining Σ on the regions B for which $E(B)$ is of X or Y type. Then we consider the regions B such that on the region B', that lies on the other side of $E(B)$, the system Σ is already defined. If $E(B)$ is of S type and x is the initial point of $E(B)$, then we consider B if there is a trajectory γ_x that verifies the induction hypothesis. In a finite number of steps we define Σ on every $B \in \mathcal{B}$.
Fix, now, a region $B \in \mathcal{B}$ and assume that the induction hypothesis holds. From ($\mathcal{G}5$) we have that $Cl(B)$ contains exactly one entrance E_1 and one exit E_2. If $E_1 \sim E_2$ then B is enclosed by E_1, E_2 and either a line l or a side E_3. Otherwise, B is enclosed by E_1, E_2, a line l_1 and either another line l_2 or a side E_3. We define Σ on B defining Y or X, and G. Indeed, we define Σ also on a neighborhood of the lines in B if the system is not already defined near these lines. Consider the case $E_1 \sim E_2$ and assume that B is positive, being similar the other case. Possibly using a change of coordinates, we can assume that

$$E_1 = \{(x_1, x_2) : x_1 = 0, 0 \le x_2 \le a\}, \qquad E' = \{(x_1, x_2) : x_2 = 0, 0 \le x_1 \le b\},$$

where either $E' = l$ or $E' = E_3$, and that Y is the constant vector field $(1,0)$. We could define $Y \equiv (1,0)$ on B and let Γ be formed by Y-trajectories, but we have to make some modifications to ensure that every $\gamma \in \Gamma$ is extremal. Consider $\gamma_y \in \Gamma$ that verifies $\gamma_y(t_1) = (0,y)$, $\gamma_y(t_2) \in E_2$. By the induction hypothesis such a trajectory γ_y exists defined on $[0, t_2]$ for every $y \in [0,a]$. Since we have already defined Σ on a neighborhood of $E_1 \cup E_2$, there is a covector field λ_y associated to γ_y that is defined on

$$I = [t_1, t_1 + \mu_1] \cup [t_2 - \mu_2, t_2]$$

for some positive μ_1, μ_2. It can happen that $t_1 + \mu_1 = t_2 - \mu_2$, e.g. if we are near the point $E_1 \cap E_2$. We want to define Y in such a way that we can associate to γ_y a covector field, defined on $Dom(\gamma_y)$, that coincides with λ_y on I. This ensures, choosing G in a suitable way, that every γ_y is extremal.

Consider a region

$$\Omega = [\delta_1, \delta_2] \times [\varepsilon, a - \delta_3], \qquad \delta_3 > 0, \qquad 0 < \delta_1 < \delta_2$$

such that the following holds: $\Omega \subset A$, where A is the region containing B, and $\Omega \cap (E_1 \cup E_2) = \emptyset$. See Figure A.1 where Ω is the darkened region.

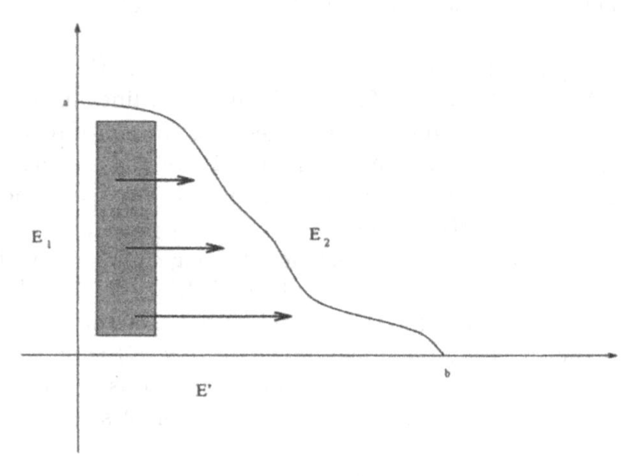

Fig. A.1. Proof of Theorem 20, p. 113

Let B' be the region on the other side of E'. If Σ is already defined on B' then $\varepsilon > 0$ otherwise $\varepsilon < 0$. Notice that if E' is a side (see definition in section 2.8.4, p. 104) then it is of X or of Y type and the former case holds. We choose ε, δ_3 in such a way that Σ is already defined on $B \cap \{(x_1, x_2) : a - 2\delta_3 \leq x_2 \leq a\}$ and if $\varepsilon > 0$ then Σ is already defined on $B \cap \{(x_1, x_2) : 0 \leq x_2 \leq 2\varepsilon\}$. For every $y \in [\varepsilon, a - \delta_3]$, let $\gamma_y^1 \in \Gamma$ be the trajectory that verifies $\gamma_y^1(t_1(y)) = (0, y)$ and let l_y^1 be the covector field associated to γ_y^1. We have that γ_y^1, l_y^1 are defined on a neighborhood of $t_1(y)$. Consider the Mayer problem with final target E_2 and the cost function:

$$\psi(T, x(T)) = -T + \psi_0(x(T)) \tag{A.29}$$

depending on terminal point and time, where we want to maximize ψ. For every $y \in [\varepsilon, a - \delta_3]$ let $\bar{x}(y)$ be such that $(\bar{x}(y), y) \in E_2$. There exists trajectories $\gamma_y^2 \in \Gamma$ that reach $(\bar{x}(y), y)$ with an associated covector field l_y^2. Let $t_2(y)$ be such that $\gamma_y^2(t_2(y)) = (\bar{x}(y), y)$. Observe that γ_y^2, l_y^2 are defined on a neighborhood of $t_2(y)$. We can define ψ in such a way that (γ_y^2, l_y^2) satisfies the PMP and the final transversality condition for the Mayer problem, see [92]. Indeed the PMP is satisfied because γ_y^2 is extremal for the time optimal problem.

To satisfy the transversality condition, in view of (A.29) we need to find λ_0, λ_1 solution to:

$$\lambda_0 = \max_{|\omega|\leq 1} \lambda_y^2(t_2(y)) \cdot (F + \omega G)(\bar{x}(y), y) \tag{A.30}$$

$$\lambda_y^2(t_2(y)) = \lambda_0 \nabla \psi_0(\bar{x}(y), y) + \lambda_1 n_2(\bar{x}(y), y) \tag{A.31}$$

where n_2 is a unit normal vector to E_2. Hence, (A.30) determines λ_0 and (A.31) gives a condition for λ_1, ψ_0.
Choose ν_1, ν_2, T_1, T_2 such that $\delta_1 < \nu_1 < \nu_2 < \delta_2$ and:

$$T_1 > \sup\{t_1(y) : y \in [\varepsilon, a - \delta_3]\} \qquad T_2 < \inf\{\psi_0((\bar{x}(y), y)) : y \in [\varepsilon, a - \delta_3]\}.$$

We define $Y = (\alpha, 0)$ on Ω, α continuous and positive, $\alpha \equiv 1$ on $\partial\Omega \cup [\nu_1, \nu_2] \times [\varepsilon, a - \delta_3]$, and we let $Y = (1, 0)$ outside Ω. We choose α in such a way that the following holds. For every y we have $\gamma_y^1(T_1) = (\nu_1, y)$. If $T_2(y) < t_2(y)$ is the time at which γ_y^2 reaches, backward in time, the point (ν_2, y) then

$$\psi(t_2(y) - T_2(y), (\bar{x}(y), y)) = T_2.$$

With this definition of Y we prolong $\gamma_y^{1,2}, l_y^{1,2}$ defining them on the whole set B. Consider the reachable set $\mathcal{R}(T_1)$, we have that

$$\{(\nu_1, y) : \varepsilon \leq y \leq a - \delta_3\} \subset \partial\mathcal{R}(T_1).$$

Since $l_y^1(T_1)$ has to be perpendicular to $\partial R(T_1)$, it follows that $l_y^1(T_1)$ has the second component equal to zero. From Theorem 8.2 of Chapter IV of [62], we have that l_y^2 has to be perpendicular to the level set of the function:

$$\psi'(x, y) = \psi(t_2(y) - t(x, y), (\bar{x}(y), y)),$$

where $t(x, y)$ is defined by $\gamma_y^2(t(x, y)) = (x, y)$. Hence also the second component of $l_y^2(T_2(y))$ has to be zero. By the PMP, since the Hamiltonian is positive (see ii) of PMP), the first components of $l_y^1(T_1), l_y^2(T_2(y))$ have the same sign. Since $\alpha = 1$ on $[\nu_1, \nu_2] \times [\varepsilon, a - \delta_3]$, we obtain that l_y^1, l_y^2 coincide up to a scalar multiple. We can now associate to every γ_y^1 the covector field l_y^1 and define G in such a way that G is of class C^3 and every γ_y^1 is extremal. It may happen, however, that α is not smooth and hence Σ is not smooth. Since α is continuous there exists a sequence α_n of smooth functions converging uniformly to α. Let Σ_n, Γ_n be the system and synthesis associated to α_n. If E_2 is of K or F type then for n large every $\gamma \in \Gamma_n$ is extremal and we are done. Indeed in this case no trajectory of Γ switches on E_2 and by compactness the same holds, if n is sufficiently large, for Γ_n. If E_2 is of C type then Σ_n has a switching curve C_n near to E_2. Since Σ has not already been defined on the region B' that lies on the other side of E_2, we can define $\Sigma = \Sigma_n$ for n sufficiently large. The only change is that we construct the system on a graph equivalent to $\mathcal{G}$, not exactly on $\mathcal{G}$.

The other case, that is when B is enclosed by E_1, E_2 and either two lines or one line and one side, can be treated in an entirely similar manner. This concludes the construction on the regions $B \in \mathcal{B}$ and then we have defined Σ and Γ on the whole $\mathcal{R}$.

We can again modify Σ on the regions $B \in \mathcal{B}$, using the same techniques described above, in such a way the following holds. If $\gamma \in \Gamma$ reaches $Fr(\mathcal{R})$ then it reaches $Fr(\mathcal{R})$ at time τ. If x belongs to an overlap curve, $\gamma_1, \gamma_2 \in \Gamma$, $\gamma_1(t_1) = x = \gamma_2(t_2)$ then $t_1 = t_2 \leq \tau$, with equality holding only if $x \in Fr(\mathcal{R})$.

Using a sufficiency argument as in [111], we can conclude that every $\gamma \in \Gamma$ is optimal and then $\mathcal{R} = \mathcal{R}(\tau)$, the reachable set in time τ for Σ. It is also possible to use a dynamic programming argument. Indeed the time along the set of trajectories Γ satisfies the Hamilton–Jacobi–Bellman equation for the value function inside $\mathcal{R}$ and is constant on its frontier, see [22, 61]. It can happen that some points are reached by more than one trajectory of Γ. However, we can construct a synthesis from Γ, that we call again Γ, following the procedure described in Section 2.5, p. 56. We obtain $\Gamma = \Gamma_{\mathcal{A}}(\Sigma)$. From the construction it is clear that $\mathcal{G}$ corresponds to Σ in the canonical way. ■

B

Bidimensional Sources

The aim of this Appendix is to show how to extend the theory, developed in the various chapters for the two points model problem, to the case of two dimensional initial source or final target. Being the two cases equivalent (it is enough to reverse time), we treat only the case of a two dimensional source.

From now on we fix a source $\mathcal{S}$ that is assumed to be a smooth two dimensional manifold with smooth boundary, and consider again the minimum time problem from $\mathcal{S}$ for a control system:

$$\dot{x} = F(x) + uG(x), \qquad |u| \leq 1, \tag{B.1}$$

where $x \in \mathbb{R}^2$. The conclusions are valid, mutatis mutandis, in the general case of a two dimensional manifold.

Notice that the condition $F(0) = 0$ is no more meaningful. The key condition here is the one of local controllability from the source.

Definition 77 *We say that the system (B.1) is locally controllable from the source at a point $x \in \partial\mathcal{S}$ if, for every y in a neighborhood of x in $\partial\mathcal{S}$, there exists u_y, $|u_y| \leq 1$, such that $(F(y) + u_y\, G(y)) \cdot \mathbf{n}_y > 0$, where $\mathbf{n}_y$ is the outer normal to $\mathcal{S}$ at y. We say that the system (B.1) is locally controllable from the source if it is locally controllable at every point of $\partial\mathcal{S}$.*

Remark **60** In the case of a two dimensional source the transversality condition *(PMP3)* of Definition 8, p. 22 becomes essential (see also Remark 12, p. 23):

$$\text{for every } v \in T_{\gamma(0)}\mathcal{S}, \quad \lambda(0) \cdot v = 0. \tag{B.2}$$

This means that if $\mathbf{n}_{\gamma(0)}$ is the outer normal to $\mathcal{S}$ at $\gamma(0)$, we have $\lambda(0) = \alpha\mathbf{n}_{\gamma(0)}$, for some $\alpha \in \mathbb{R} \setminus \{0\}$. From the condition of positivity of the Hamiltonian (see condition *ii)* of Theorem 10, p. 35) we get that $\alpha > 0$ (in the case of a two dimensional target, $\alpha < 0$, see Figure B.1).

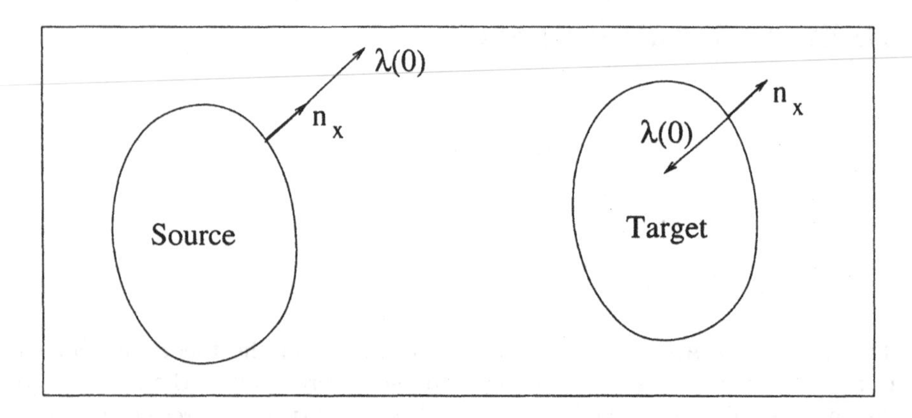

Fig. B.1. Transversality conditions for a two dimensional source and a two dimensional target. The covector points outside the source and inside the target.

B.1 Local Optimal Synthesis at the Source

We start analyzing the local optimal synthesis at a generic point of $\partial\mathcal{S}$. Fix a point $x \in \partial\mathcal{S}$ and let $\mathbf{n}_x$ be the outer normal. We distinguish three cases that happen at a generic point:

a) $\mathbf{n}_x \cdot Y(x) > 0$ and $\mathbf{n}_x \cdot X(x) > 0$;
b) $\mathbf{n}_x \cdot Y(x) > 0$ and $\mathbf{n}_x \cdot X(x) < 0$ (or viceversa);
c) $\mathbf{n}_x \cdot Y(x) < 0$ and $\mathbf{n}_x \cdot X(x) < 0$;

From Remark 60 we get the following. For case **a)** the synthesis is determined by the sign of $\phi(0) = \lambda(0) \cdot G(\gamma(0))$. More precisely if $\phi(0) > 0$ then the local synthesis is formed of Y trajectories, while if $\phi(0) < 0$ it is formed by X trajectories.
Assume now that $\phi(0) = 0$. Under generic assumptions on the system, we have $\Delta_B(x) \neq 0$, indeed $sgn(\phi(0)) = -sgn(\mathbf{n}_x \cdot G(x))$ and generically the two set of zeroes do not intersect. Hence there is no turnpike starting at x. The synthesis can be studied by the same methods of Chapter 2: two possible cases appear with a C curve or a K curve, shown in Figure B.2, Cases **a.1** and **a.2**. Notice that, in this case, there are not abnormal extremals. Indeed the condition $\Delta_A = 0$ is generically verified at isolated points of $\partial\mathcal{S}$ not coinciding with points where $\mathbf{n}_x \cdot G(x) = 0$.
In case **b)** the synthesis is formed by Y trajectories (X if the viceversa happens). See Figure B.2, case **b**. This analysis covers the points at which the system is locally controllable.
Finally the synthesis is empty in case **c)**, because no trajectory is exiting $\mathcal{S}$. This ends the treatment of synthesis at a generic point.

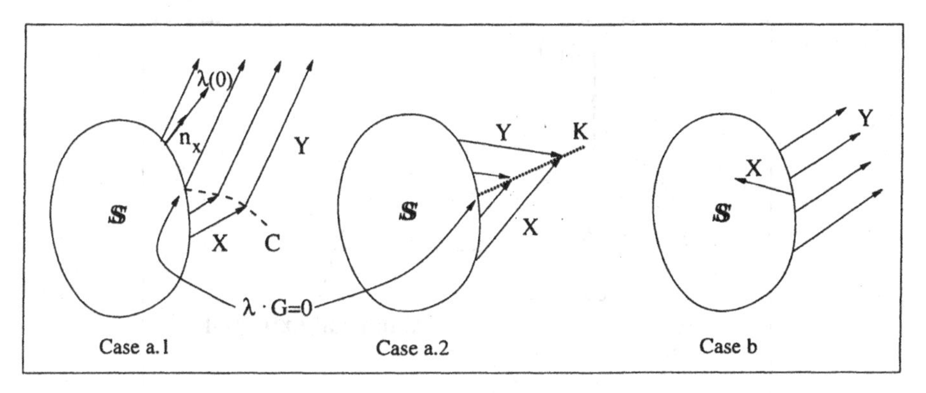

Fig. B.2. Optimal syntheses in a neighborhood of a point in which the system is locally controllable.

Let us pass to the case of not generic points that can happen for a system that is not locally controllable. Assume, for example, that:

$$\mathbf{n}_x \cdot X(x) < 0, \qquad \mathbf{n}_x \cdot Y(x) = 0,$$

and by genericity that, denoting by $s \to \zeta(s)$, $\zeta(0) = x$, a local parametrization of $\partial\mathcal{S}$:

$$\left. \frac{d}{ds} \mathbf{n}_{\zeta(s)} \cdot Y(\zeta(s)) \right|_{s=0} \neq 0.$$

Clearly the local optimal synthesis covers only part of a neighborhood of x with Y trajectories. See Figure B.3. Notice that the Y trajectory exiting x is an abnormal extremal. Indeed the condition $\mathbf{n}_x \cdot Y(x) = 0$ implies that the Hamiltonian is vanishing.

B.2 The Locally Controllable Case and Semiconcavity of the Minimum Time Function

In the locally controllable case, the optimal synthesis can be constructed in similar way to the point source case. Now there are no more curves $\gamma^{\pm}$ that play a special role. Thus we obtain singularities entirely similar to the case of a zero dimensional source, except for the singular points along $\gamma^{\pm}$.
Also the classification program can be carried out with the obvious changes (presence of the source $\mathcal{S}$ and absence of $\gamma^{\pm}$). Similarly the extremal synthesis shares exactly the same properties.

Finally, projection singularities are simpler. Indeed, there are no more vertical and ribbon singularities for Π_2 and no more ribbon for Π_4.

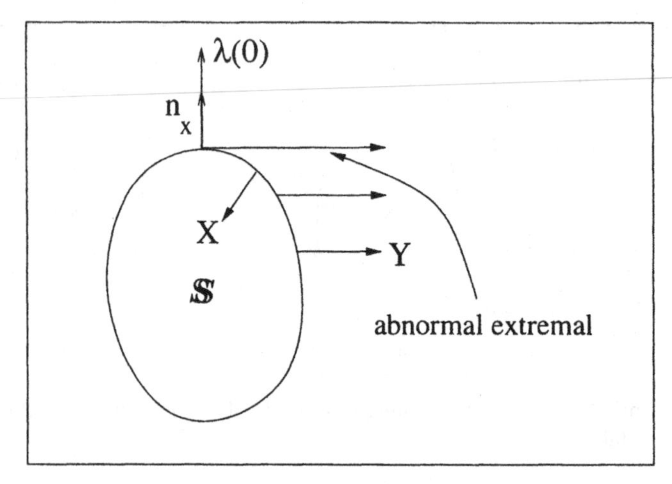

Fig. B.3. Optimal syntheses in a neighborhood of a point in which the system is not locally controllable.

The minimum time function shares the same regularity: it is piecewise smooth and topologically a Morse function. However, we have some more regularity because the non differentiability occurs only along K Frame Curves (not along $\gamma^{\pm}$). This permits to prove the semiconcavity of the minimum time function in the case in which $M = \mathbb{R}^2$.

Definition 78 *A function $f : A \to \mathbb{R}$, $A \subset \mathbb{R}^2$ open, is said to be semiconcave if for every compact convex $\mathcal{K} \subset A$, there exists $c_{\mathcal{K}} > 0$ such that $f(x) - c_{\mathcal{K}}\,|x|^2$ is concave on $\mathcal{K}$.*

Theorem 44 *Assume that the system (B.1) is locally controllable from a source $\mathcal{S}$, then the minimum time function $\mathbf{T}(\cdot)$, with initial set $\mathcal{S}$, is semiconcave.*

Proof. The assumption of local controllability ensures that the minimum time function $\mathbf{T}(\cdot)$ is Lipschitz continuous.

Obviously a smooth function is semiconcave, hence we have only to consider Frame Curves and Frame Points. It is easy to check that points, at which $\mathbf{T}(\cdot)$ is topologically equivalent to a linear function, satisfy the assumptions of semiconcavity. The same happens for maxima.

Thus we have only to check minima and saddles. In the case of a zero dimensional source, the only minimum of $\mathbf{T}(x)$ is the origin, while now all the points of $\partial\mathcal{S}$ are minima and there are no other minima.

Moreover, from the previous analysis we know that the possible saddle points are internal points of overlap curves K or Frame Points of kind $(Y, K)_{2,3}$. In the first case the function $\mathbf{T}(\cdot)$ is smooth along K and again we retrieve semiconcavity, while the second case does not happen (there are no $\gamma^{\pm}$ curves now). ■

Remark **61** In the non-locally controllable case, the minimum time function is discontinuous. This case is more delicate, because new singularities are appearing, both due to discontinuities of the minimum time and due to the presence of abnormal extremals showed above.

References

1. A.A. Agrachev, B. Bonnard, M. Chyba, I. Kupka, "Sub-Riemannian sphere in Martinet flat case", *ESAIM Control Optim. Calc. Var.* 2 (1997), pp. 377–448.
2. A.A. Agrachev, Yu.L. Sachkov, "Lectures on Geometric Control Theory", Preprint SISSA 38/2001/M, May 2001, SISSA, Trieste. .
3. A.A. Agrachev and R.V. Gamkrelidze, "Symplectic Geometry for Optimal Control", in "Non-Linear Controllability and Optimal Control", Monogr. Textbooks Pure Appl. Math., 133, Dekker, New York, pp. 263-277, (1990).
4. A.A. Agrachev and R.V. Gamkrelidze, "Symplectic methods for optimization and control", in "Geometry of feedback and optimal control", Monogr. Textbooks Pure Appl. Math., 207, Dekker, New York, pp. 19–77, (1998).
5. A.A. Agrachev, G. Charlot, J.P. Gauthier, V.M. Zakalyukin, "On sub-Riemannian caustics and wave fronts for contact distributions in the three-space, *J. Dynam. Control Systems* 6 (2000), no. 3, 365–395.
6. A.A. Agrachev, J.P. Gauthier, "On the subanalyticity of Carnot-Caratheodory distances." *Ann. Inst. H. Poincar Anal. Non Linaire* 18 (2001), no. 3.
7. A.A. Agrachev, J.P. Gauthier, "On the Dido problem and plane isoperimetric problems." *Acta Appl. Math.* 57 (1999), no. 3, 287–338.
8. A. A. Agrachev, D. Pallaschke, S. Scholtes, "On Morse theory for piecewise smooth functions", *J. Dynam. Control Systems*, Vol 3 no. 4 (1997), pp. 449–469.
9. A.A. Agrachev, A. Sarychev, "Abnormal Sub-Riemannian Geodesics: Morse Index and Rigidity," *Ann. Inst. H. Poincaré Anal. Non Linéaire* Vol. 13, No. 6 (1996), pp. 635-690.
10. A.A. Agrachev, A. Sarychev, " On Abnormal Extremals for Lagrange Variational Problems", *J. Math. Systems Estim. Control,* Vol. 8, No. 1 (1998), pp. 87-118.
11. A.A. Agrachev, M. Sigalotti, "On the Local Structure of Optimal Trajectories in $\mathbb{R}^3$", *SIAM J. Control Optim*, Vol. 42, No.2 (2003), pp. 513-531.
12. A.A. Agrachev, G. Stefani, P. Zezza, "A Hamiltonian approach to strong minima in optimal control", in *Differential geometry and control* (Boulder, CO, 1997), 11–22, Proc. Sympos. Pure Math., 64, Amer. Math. Soc., Providence, 1999.
13. A.A. Agrachev, G. Stefani, P. Zezza, "Strong optimality for a bang-bang trajectory." *SIAM J. on Control and Optimization,* Vol. 41, No.4 (2002), pp. 991-1014.

14. R. El Assoudi, J.P. Gauthier, I.A.K. Kupka, "Controllability of right invariant systems on semi-simple Lie groups." *Geometry in nonlinear control and differential inclusions* (Warsaw, 1993), 199–208, Banach Center Publ., 32, Polish Acad. Sci., Warsaw, 1995.
15. R. El Assoudi, J.P. Gauthier, I.A.K. Kupka, "On subsemigroups of semisimple Lie groups." *Ann. Inst. H. Poincar Anal. Non Linaire* 13 (1996), no. 1, 117–133.
16. V.I. Arnold, "Catastrophe theory", Springer-Verlag, 1992.
17. V.I. Arnold, S.M. Gusein-Zade, A.N. Varchenko, "Singularities of differentiable maps", Birkhauser (1985), 2v.
18. V.I. Arnold, "Geometric Method in the Theory of ODE", Springer-Verlag, New York, 1983.
19. M. M. Baitmann, "Controllability Regions on the Plane", *Differential Equations*, Vol.14 (1978), pp. 407-417.
20. M. M. Baitmann, "Switching Lines in the Plane", *Differential Equations*, Vol.14 (1978), pp. 1093-1101.
21. A. Ballucchi, A. Bicchi, B. Piccoli, P. Soueres, "Stability and Robustness of Optimal Synthesis for Route Tracking by Dubins' Vehicles", in *Proceedings of the 39th IEEE Conference on Decision and Control, Sidney* (2000).
22. M.Bardi, "A boundary value problem for the minimum-time function, *SIAM J. Control Optim*, 27 (1989), pp. 776-785.
23. M. Bardi and I. Capuzzo-Dolcetta, "Optimal control and viscosity solutions of Hamilton-Jacobi-Bellman equations", *Systems & Control: Foundations & Applications*, Birkhäuser, Boston, 1997.
24. A. Bellaiche, "The tangent space in sub-Riemannian geometry", Sub-Riemannian geometry, 1–78, Progr. Math., 144, Birkhäuser, Basel, 1996.
25. R. M. Bianchini and G. Stefani, "Graded Approximations and Controllability along a Trajectory", *SIAM J. Control and Optimization*, Vol.28, 1990, pp. 903–924.
26. A.M. Bloch, P.E. Crouch, "Optimal control, optimization, and analytical mechanics", in *Mathematical control theory*, 268–321, Springer, New York, 1999.
27. V. Boltyanskii, "Sufficient Condition for Optimality and the Justification of the Dynamic Programming Principle", *SIAM J. Control and Optimization*, Vol.4, (1966), pp. 326-361.
28. B. Bonnard, "Controllabilité de systemes mécaniques sur les groupes de Lie." (French) *SIAM J. Control Optim.* 22 (1984), no. 5, 711–722.
29. B. Bonnard, M. Chyba, "The role of singular trajectories in control theory", Springer SMAI, Vol.40.
30. B. Bonnard, M. Chyba, E. Trelat, "Sub-Riemannian geometry, one-parameter deformation of the Martinet flat case." *J. Dynam. Control Systems* 4 (1998), no. 1, 59–76.
31. B. Bonnard, V. Jurdjevic, I. Kupka, G. Sallet, "Transitivity of families of invariant vector fields on the semidirect products of Lie groups", *Trans. Amer. Math. Soc.* 271 (1982), no. 2, 525–535.
32. B. Bonnard, I. Kupka, "Théorie des singularites de l'application entree/sortie et optimalite des singulieres dans le probleme du temps minimal", *Forum Mathematicum* Vol.5 (1993), pp. 111-159.
33. U. Boscain, Y. Chitour, "On the minimum time problem for driftless left-invariant control systems on $SO(3)$" *Communications on Pure and Applied Analysis*, Vol 1, no 3, pp. 285–312, (2002).

34. U. Boscain, Y. Chitour, "Time Optimal Synthesis for Left–Invariant Control Systems on $SO(3)$", submitted.
35. U. Boscain, I. Nikolaev, B. Piccoli, "Classification of Stable Time–Optimal Controls on 2-manifolds", submitted.
36. U. Boscain, B. Piccoli, "Geometric Control Approach To Synthesis Theory" , *Rendiconti del Seminario Matematico dell'Università e del Politecnico di Torino,* Torino, 1998, V.56, 4 (1998), pp. 53-67.
37. U. Boscain, B. Piccoli, "Projection singularities of extremals for planar systems", *Proceeding of the 38th IEEE Conference on Decision and Control,* Phoenix, Arizona, USA, December 7–10, 1999, pp. 2936-2941.
38. U. Boscain and B. Piccoli, "Extremal syntheses for generic planar systems", *Journal of Dynamical and Control Systems* Vol.7 No.2 (2001), pp. 209-258.
39. U. Boscain and B. Piccoli, "On automaton recognizability of abnormal extremals", *SIAM J. Control and Opt.*, Vol. 40 No.5 pp. 1333-1357. (2002).
40. U. Boscain and B. Piccoli, "Morse Properties for the Minimum Time Function on 2-D Manifolds", *Journal of Dynamical and Control Systems* Vol. 7, No. 3, pp. 385-423, (2001).
41. A. Bressan, "A high order test for optimality of bang-bang controls", *SIAM J. Control Optim.* 23 (1985), no. 1, pp. 38–48.
42. A. Bressan, "The Generic Local Time-Optimal Stabilizing controls in Dimension 3, *SIAM J. Control Optim.* 24 (1986), pp. 177–190.
43. A. Bressan and B. Piccoli, "Structural Stability for Time–Optimal Planar Syntheses", *Dynamics of Continuous, Discrete and Impulsive Systems,* No.3 (1997), pp. 335-371.
44. A. Bressan and B. Piccoli, "A Generic Classification of Time Optimal Planar Stabilizing Feedbacks", *SIAM J. Control and Optimization,* Vol.36 No.1 (1998), pp. 12-32.
45. P. Brunovsky, "Existence of Regular Syntheses for General Problems," *J. Diff. Equations 38 (1980), pp. 317-343.*
46. P. Brunovsky, "Every Normal Linear System Has a Regular Time–Optimal Synthesis", Math. Slovaca 28, 1978, pp. 81-100.
47. N. N. Butenina, "The structure of the boundary curve for planar controllability domains", *Methods of qualitative theory of differential equations and related topics,* 73–86, Amer. Math. Soc. Transl. Ser. 2, 200, Amer. Math. Soc., Providence, RI, 2000.
48. P. Cannarsa, A. Mennucci, C. Sinestrari, "Regularity Results for Solutions of a Class of Hamilton Jacobi equations," *Archive for Rational Mechanics and Analysis,* 140 (1997) pp. 197-223.
49. N. Caroff, H. Frankowska, "Conjugate points and shocks in nonlinear optimal control", *Trans. Amer. Math. Soc.* 348 (1996), no. 8, pp. 3133–3153.
50. L. Cesari, "Optimization-theory and applications: problems with ordinary differential equations". New York : Springer-Verlag, c1983.
51. G. Charlot, "Quasi-contact S-R metrics: normal form in $\mathbb{R}^{2n}$, wave front and caustic in $\mathbb{R}^4$." Acta Appl. Math. 74 (2002), no. 3, 217–263.
52. S.N. Chow and J.K. Hale, "Methods of Bifurcation Theory", Springer-Verlag, New York, 1982.
53. F.H. Clarke, "The maximum principle under minimal hypotheses." *SIAM J. Control Optimization* 14 (1976), no. 6, pp. 1078–109.
54. A.A. Davydov, "Qualitative Theory of Control Systems", *Translations of Mathematical Monographs,* American Mathematical Society (1994).

55. A.A. Davydov, "Controllability of generic control systems on surfaces." *Geometry of feedback and optimal control*, 111–163, Monogr. Textbooks Pure Appl. Math., 207, Dekker, New York, 1998.
56. A.A. Davydov, "Singularities of the boundary of accessibility in two-dimensional control systems." (Russian) *Uspekhi Mat. Nauk* 37 (1982), no. 3(225), 183–184.
57. A.A. Davydov, "Singularities of limit direction fields of two-dimensional control systems." (Russian) Mat. Sb. (N.S.) 136(178) (1988), no. 4, 478–499, 590; translation in Math. USSR-Sb. 64 (1989), no. 2, 471–493.
58. A.A. Davydov, "Structural stability of control systems on orientable surfaces." (Russian) Mat. Sb. 182 (1991), no. 1, 3–35; translation in Math. USSR-Sb. 72 (1992), no. 1, 1–28.
59. Dubins L.E. "On curves of minimal length with a constraint on average curvature and with prescribed initial and terminal position and tangents", *Amer. J. Math*, V.79 (1957), pp.497-516.
60. S. Eilenberg, "Automata, Languages and Machines", Vol.A. and Vol. B, *Academic Press* (1974).
61. L.C. Evans, M.R. James, "The Hamilton-Jacoby-Bellman equation for the time-optimal control." *SIAM J. Control Optim.* 27 (1989), pp.1477-1489.
62. W.H. Fleming and R.W. Rishel, "Deterministic and Stochastic Optimal Control", Springer-Verlag, New York, 1975.
63. W.H. Fleming and M. Soner, "Controlled Markov Processes and Viscosity Solutions," Springer–Verlag, New York, 1993.
64. A.T. Fuller, "Relay Control Systems Optimized for Various Performance Criteria", *Automatic and remote control, Proc. first world congress*, IFAC Moscow, vol. 1, Butterworths, 1961, pp. 510-519.
65. M. Garavello, "Verification Theorems for Hamilton-Jacobi-Bellman equations", submitted.
66. J.P. Gauthier and G. Bornard. "Controlabilite des sytemes bilineaires", SIAM J. Control and Optimization, 20:377-384, 1982.
67. M. Gromov, "Carnot-Carathodory spaces seen from within," Sub-Riemannian geometry, 79–323, Progr. Math., 144, Birkhäuser, Basel, 1996.
68. J. L. Gross, Thomas W. Tucker, "Topological Graph Theory", *John Wiley and Sons, Inc* (1987).
69. H. Hermes and J.P. LaSalle, "Functional analysis and time optimal control", *Mathematics in Science and Engineering*, Vol. 56. Academic Press, New York-London, 1969.
70. L. Heffter, "Ueber das Problem der Nachbargebeite," *Mat. Annalen* 38 (1891), pp. 477–508.
71. M.W. Hirsch, "Differential Topology", Springer-Verlag, New-York, 1976.
72. A. Isidori. "Nonlinear control systems: an introduction", Springer-Verlag, (1985).
73. B. Jakubczyk, W. Respondek, "Feedback equivalence of planar systems and stabilizability". Robust control of linear systems and nonlinear control (Amsterdam, 1989), 447–456, Progr. Systems Control Theory, 4, Birkhuser Boston, Boston, MA, 1990.
74. B. Jakubczyk, W. Respondek, "Feedback classification of analytic control systems in the plane". Analysis of controlled dynamical systems (Lyon, 1990), 263–273, Progr. Systems Control Theory, 8, Birkhuser Boston, Boston, MA, 1991.

75. B. Jakubczyk, W. Respondek, "Bifurcations and phase portraits of control-affine systems in the plane", preprint.
76. F. Jean, "Uniform estimation of sub-Riemannian balls". J. Dynam. Control Systems 7 (2001), no. 4, 473–500.
77. V. Jurdjevic, "The geometry of the plate-ball problem", *Arch. Rational Mech. Anal.* 124 (1993), no. 4, pp. 305–328.
78. V. Jurdjevic, "Geometric Control Theory", *Cambridge University Press*, (1997).
79. V. Jurdjevic, "Optimal control problems on Lie groups: crossroads between geometry and mechanics", in *Geometry of feedback and optimal control*, 257–303, Monogr. Textbooks Pure Appl. Math., 207, Dekker, New York, 1998.
80. V. Jurdjevic and I.K. Kupka, "Control Systems subordinated to a group action: accessibility", *J. Diff. Eq.* Vol.39, pp. 186–211.
81. V. Jurdjevic and I.K. Kupka, "Control Systems on Semisimple Lie Groups and Their Homogeneous Spaces", *Ann. Inst. Fourier*, Vol.31, pp. 151–179.
82. V. Jurdjevic and H.J. Sussmann, "Controllability of Non-Linear systems", *J. Diff. Eq.* Vol.12, pp. 95–116.
83. M. Kiefer and H. Schattler, "Cut-Loci, and Cusp Singularities in Parameterized Families of Extremals," *Optimal Control (Gainesville, FL, 1997), pp. 250-277, Appl. Optim., 15, Kluwer Acad. Publ., Dordrecht, (1998).*
84. M. Kiefer and H. Schattler, "Parametrized Families of Extremals and Singularities in Solution to the Hamilton Jacobi Bellman Equation" *SIAM J. Control and Opt.*, 37 (1999), pp. 1346-1371.
85. W.-S. Koon, J.E. Marsden, "Optimal control for holonomic and nonholonomic mechanical systems with symmetry and Lagrangian reduction", *SIAM J. Control Optim.* 35 (1997), no. 3, 901–929.
86. V. Kostov and E. Degtiariova-Kostova "Suboptimal paths in the problem of a planar motion with bounded derivative of the curvature." *Comptes rendus de l'acadmie des sciences* 321, 1995, pp. 1441–1447.
87. V. Kostov and E. Degtiariova-Kostova, "Irregularity of Optimal Trajectories in a Control Problem for a Car-like Robot", Research Report 3411, INRIA, 1998.
88. A.J. Krener, "A generalization of Chow's and the bang bang theorem to non-linear control problems". *SIAM J. Control Optimization* 12 (1974), pp. 43-51.
89. A.J. Krener,"The high order maximal principle and its application to singular extremals", *SIAM J. Control Optimization* 15 (1977), no. 2, pp. 256–293.
90. A.J. Krener, H. Schättler, "The structure of small-time reachable sets in low dimensions", *SIAM J. Control Optim.* 27 (1989), no. 1, pp. 120–147
91. I.A.K. Kupka, "The ubiquity of Fuller's phenomenon", in *Nonlinear controllability and optimal control*, 313–350, Monogr. Textbooks Pure Appl. Math., 133, Dekker, New York, 1990.
92. M.M. Lee, L. Markus, "Foundations of Optimal Control Theory", Whiley, New York, 1967.
93. C. Lobry, "Controllability of non-linear systems on compact manifolds", *SIAM J. Control Optim.*, Vol.1, (1974) pp.1–4.
94. S. Lojasiewicz, H.J. Sussmann, "Some examples of reachable sets and optimal cost functions that fail to be subanalytic". *SIAM J. Control Optim.* Vol. 23, no. 4, pp. 584–598 (1985).
95. A.A. Markov, "Some examples of the solution of a special kind of problem in greatest and least quantities", (in Russian) *Soobshch. Karkovsk. Mat. Obshch.* 1, 1887, pp. 250–276.

96. C. Marchal, "Chattering arcs and chattering controls", *J. Optimization Theory Appl.* 11 (1973), 441–468.
97. A. Marigo, B. Piccoli "Regular syntheses and solutions to discontinuous ODEs" *ESAIM: Control, Optimisation and Calculus of Variations* Vol. 7 (2002) pp. 291-308
98. V.I. Matov "Topological Classification of the Germ of Functions of the Maximum and Minimax Functions of a Family of Functions in General Position". *Uspekhi Mat. Nauk* (Russian), Vol. 37, no.4 (1982), pp.167-168.
99. W.S. Massey, "Algebraic Topology: An Introduction," Springer-Verlag, New-York, 1973.
100. D. McCaffrey, S.P. Banks, "Riemannian comparison and conjugate locus in optimal control", *IMA J. Math. Control Inform.* 17 (2000), no. 2, pp. 123–145.
101. J. Milnor, "Morse theory", based on lecture notes by M. Spivak and R. Wells. Annals of Mathematics Studies, No. 51 Princeton University Press, Princeton, 1963.
102. R. Montgomery, "A survey of singular curves in sub-Riemannian geometry",*J. Dynam. Control Systems* 1 (1995), no. 1, pp. 49–90.
103. R. Montgomery, "A Tour of Subriemannian Geometries, Their Geodesics and Applications", American Mathematical Society, (2001).
104. I. Nikolaev, "Foliations on surfaces", Ergebnisse der Mathematik und ihrer Grenzgebiete 41, Springer-Verlag, Berlin, 2001.
105. I. Nikolaev, E. Zhuzhoma, "Flows on 2-dimensional manifolds: an overview", Lecture notes in mathematics. v.1705, Berlin, Springer-Verlag, 1999.
106. M.M. Peixoto, "Structural stability on two-dimensional manifolds", *Topology* 1 (1962), pp. 101–120.
107. M.M. Peixoto, "On the Classification of Flows on 2-Manifolds", in *Dynamical Systems*, M.M. Peixoto, ed., Academic Press, New York, 1973, pp. 389-419.
108. B. Piccoli, "A Generic Classification of Time Optimal Planar Stabilizing Feedback", *Ph.D. Thesis, SISSA*, Trieste (1994).
109. B. Piccoli, "Regular Time–Optimal Syntheses for Smooth Planar Systems," *Rend. Sem Mat. Univ. Padova*, Vol.95 (1996), pp. 59-79.
110. B. Piccoli, "Classifications of Generic Singularities for the Planar Time-Optimal Synthesis", *SIAM J. Control and Optimization*, Vol.34 No.6 (December 1996), pp. 1914-1946.
111. B. Piccoli and H.J. Sussmann, "Regular Synthesis and Sufficiency Conditions for Optimality", *SIAM J. Control and Optimization*, Vol. 39 No. 2 pp. 359-410, (2000).
112. L.S. Pontryagin, V. Boltianski, R. Gamkrelidze and E. Mitchtchenko, "The Mathematical Theory of Optimal Processes", *John Wiley and Sons, Inc, 1961.*
113. J.A. Reeds and L.A. Shepp, "Optimal Path for a car that goes both forwards and backwards", *Pacific J. Math*, Vol. 145, pp. 367-393, (1990).
114. Y. Sachkov, "Controllability of Invariant Systems on Lie Groups and Homogeneous Spaces," J. Math. Sci., Vol.100, n.4, pp.2355-2427, (2000).
115. H. Schattler, "On the Local Structure of Time Optimal Bang-Bang Trajectories in $\mathbb{R}^3$," *SIAM J. Control Optim.* 26, No.1 (1988), pp. 186-204.
116. H. Schattler, "The Local Structure of Time Optimal Trajectories in Dimension 3, Under Generic Conditions," *SIAM J. Control Optim.* 26, No.4 (1988), pp. 899-918.
117. E.D. Sontag, "Mathematical control theory: Deterministic finite dimensional systems", *Springer–Verlag*, New York, 1990.

118. P. Soueres and J.P. Laumond, Jean-Paul, "Shortest paths synthesis for a car-like robot", *IEEE Trans. Automat. Control* 41 (1996), no. 5, pp. 672–688.
119. H.J. Sussmann, "Analytic stratifications and control theory", in *Proceedings of the International Congress of Mathematicians* (Helsinki, 1978), pp. 865–871, Acad. Sci. Fennica, Helsinki, 1980.
120. H.J. Sussmann, "Geometry and Optimal Control" in *Mathematical Control Theory*, Springer–Verlag, New York, 1998, pp. 140–198.
121. H.J. Sussmann, "Subanalytic sets and feedback control." *J. Differential Equations* Vol. 31 (1979), no. 1, pp. 31–52.
122. H.J. Sussmann, "Envelopes, conjugate points, and optimal bang-bang extremals", in *Algebraic and geometric methods in nonlinear control theory*, 325–346, Math. Appl., 29, Reidel, Dordrecht, (1986).
123. H.J. Sussmann, "The Markov-Dubins Problem with Angular Acceleration Control," Proceedings of the 36th IEEE, Conference on Decision and Control, San Diego, CA, Dec. 1997. IEEE Publications, New York, 1997, pp. 2639-2643.
124. H.J. Sussmann, "Envelopes, higher–order optimality conditions and Lie Brackets", in *Proc. 1989 I.E.E.E, Conf. Decision and Control.*
125. H.J. Sussmann, "The structure of time-optimal trajectories for single-input systems in the plane: the general real analytic case", *SIAM J. Control Optim.* 25 No. 4 (1987), pp. 868–904
126. H.J. Sussmann, "The Structure of Time-Optimal Trajectories for Single-Input Systems in the Plane: the C^∞ Nonsingular Case," *SIAM J. Control Optim.* 25, No.2 (1987), pp. 433-465.
127. H.J. Sussmann, "Regular synthesis for time optimal control of single–input real–analytic systems in the plane," *SIAM J. Control and Opt.*, 25 (1987), pp.1145-1162.
128. H.J. Sussmann, G. Q. Tang, "Shortest Paths for the Reeds-Shepp Car: A Worked Out Example of the Use of Geometric Techniques in Nonlinear Optimal Control", *Rutgers Center for Systems and Control Technical Report* 91-10, September 1991, to appear in SIAM J. Control.
129. E. Trelat "Some properties of the value function and its level sets for affine control systems with quadratic cost." *Journal of Dynamical and Control Systems.* Vol.6 (2000), N.4, pp. 511–541.
130. H. Whitney, "On Singularities of Mappings of Euclidean Spaces. I, Mappings of the Plane into the Plane", *Ann. Math.* 62 (1955), 374-410.
131. M.I. Zelikin, "Synthesis of optimal trajectories on spaces of representations of Lie groups" (Russian), *Mat. Sb. (N.S.)* 132(174) (1987), no. 4,541–555; translation in *Math. USSR-Sb.* 60 (1988), no. 2, pp. 533–546.
132. M.I. Zelikin and V.F. Borisov, "Theory of chattering control. With applications to astronautics, robotics, economics, and engineering", *Systems & Control: Foundations & Applications*, Birkhäuser, Boston, 1994.

Index

Déjà parus dans la même collection

---

1. T. Cazenave, A. Haraux
   Introduction aux problèmes d'évolution semi-linéaires. 1990
2. P. Joly
   Mise en oeuvre de la méthode des éléments finis. 1990

3/4. E. Godlewski, P.-A. Raviart
   Hyperbolic systems of conservation laws. 1991

5/6. Ph. Destuynder
   Modélisation mécanique des milieux continus. 1991

7. J. C. Nedelec
   Notions sur les techniques d'éléments finis. 1992
8. G. Robin
   Algorithmique et cryptographie. 1992
9. D. Lamberton, B. Lapeyre
   Introduction au calcul stochastique appliqué. 1992
10. C. Bernardi, Y. Maday
    Approximations spectrales de problèmes aux limites elliptiques. 1992
11. V. Genon-Catalot, D. Picard
    Eléments de statistique asymptotique. 1993
12. P. Dehornoy
    Complexité et décidabilité. 1993
13. O. Kavian
    Introduction à la théorie des points critiques. 1994
14. A. Bossavit
    Électromagnétisme, en vue de la modélisation. 1994
15. R. Kh. Zeytounian
    Modélisation asymptotique en mécanique des fluides newtoniens. 1994
16. D. Bouche, F. Molinet
    Méthodes asymptotiques en électromagnétisme. 1994
17. G. Barles
    Solutions de viscosité des équations de Hamilton-Jacobi. 1994
18. Q. S. Nguyen
    Stabilité des structures élastiques. 1995
19. F. Robert
    Les systèmes dynamiques discrets. 1995
20. O. Papini, J. Wolfmann
    Algèbre discrète et codes correcteurs. 1995
21. D. Collombier
    Plans d'expérience factoriels. 1996
22. G. Gagneux, M. Madaune-Tort
    Analyse mathématique de modèles non linéaires
    de l'ingénierie pétrolière. 1996
23. M. Duflo
    Algorithmes stochastiques. 1996
24. P. Destuynder, M. Salaun
    Mathematical Analysis of Thin Plate Models. 1996

Druck und Bindung: Strauss Offsetdruck GmbH